Seismic Isolation, Structural Health Monitoring, and Performance Based Seismic Design in Earthquake Engineering

Azer A. Kasimzade · Erdal Şafak
Carlos E. Ventura · Farzad Naeim
Yoichi Mukai
Editors

Seismic Isolation, Structural Health Monitoring, and Performance Based Seismic Design in Earthquake Engineering

Recent Developments

Springer

Editors
Azer A. Kasimzade
Earthquake Engineering Research Facility, Department of Civil Engineering
Ondokuz Mayıs University
Atakum, Samsun
Turkey

Erdal Şafak
Kandilli Observatory and Earthquake Research Institute
Bogazici University
Istanbul
Turkey

Carlos E. Ventura
Department of Civil Engineering
University of British Columbia
Vancouver, BC
Canada

Farzad Naeim
Farzad Naeim Incorporation
Irvine, CA
USA

Yoichi Mukai
Graduate School of Engineering
Kobe University
Kobe, Hyogo
Japan

ISBN 978-3-030-06603-1 ISBN 978-3-319-93157-9 (eBook)
https://doi.org/10.1007/978-3-319-93157-9

Printed on acid-free paper

This Springer imprint is published by the registered company Springer Nature Switzerland AG
The registered company address is: Gewerbestrasse 11, 6330 Cham, Switzerland

Preface

This book is based on and derived from chapters on the latest developments in three popular research areas of recent years in structural and earthquake engineering: Seismic Isolation Systems, Structural Health Monitoring and Earthquake Early Warning Systems, and Performance Based Seismic Design.

Part I includes three chapters on seismic isolation systems, a new passive isolation system, a semi-active system, and a simple mortar bearing to act as a base isolator. Vulnerability of high-rise and seismically isolated buildings under long period strong ground motions is discussed, and new passive and semi-active structural seismic isolation systems are proposed against such effects. These systems are validated through real-time hybrid tests on shake tables.

Part II presents seven chapters on the new developments on Structural Health Monitoring (SHM) and Earthquake Early Warning (EEW) systems. Real-life implementations of SHM systems are presented on tall buildings, historical structures, and lifelines. The lessons learned from the analysis of data from such systems are discussed. EEW and SHM systems provide rapid assessment of structural safety after an extreme event (e.g., a large earthquake), and allow taking preventive measures, such as shutting down the elevators and gas lines, before the damage occurs. Vibration data from instrumented tall buildings have shown that large distant earthquakes and surface waves, which are not accounted for in most Ground Motion Prediction Equations (GMPE), can cause long-duration shaking and damage in tall buildings.

Part III includes eight chapters on performance-based seismic design. The overview of the current performance-based design methodologies includes discussions on the design of tall buildings and the reasons why common prescriptive code provisions are not sufficient to address the needs of tall building design. Modeling and acceptance criteria associated with various performance-based design guidelines are explained and special issues such as selection and scaling of ground motion records, soil–foundation–structure interaction issues, and seismic instrumentation and peer review needs are discussed.

The book presents a concise summary of latest research findings, and will be of interest to a wide range of professionals in earthquake engineering, including instructors, researchers, graduate students, and designers.

Atakum, Turkey — Azer A. Kasimzade
Istanbul, Turkey — Erdal Şafak
Vancouver, Canada — Carlos E. Ventura
Irvine, USA — Farzad Naeim
Kobe, Japan — Yoichi Mukai

Contents

Part I
Seismic Isolation Systems

Chapter 1
New Structural Seismic Isolation System

Azer A. Kasimzade, Sertac Tuhta and Gencay Atmaca

Abstract New structural seismic isolation system named Spherical Foundation Structural Seismic Isolation (SFSSI) system which has been discovered and first prototyped on a 104 m height (26 storey) building example is presented. Creation of SFSSI system aimed to build structural seismic isolation system, of which period must be more than the predominant period ground motion of majority existing earthquakes including near-fault zones also. As known, classical period-dependent isolation systems and not isolated (fixed base-FB) buildings are vulnerable under long-period earthquakes and it is required to design resistant structural system components. For this purpose, a system has been realised as a structure with inverse pendulum system's behavior. Structure foot base and foundation contact surface are formed in spherical appearance and they are separated by known Lead Core Rubber Bearing (LCRB) or Laminate Rubber bearing (LRB) isolators which are installed through spherical contact surfaces. Dampers are installed through base (structure foot) plate counter for controlling system's response. This allows spherical foundation's turning around gyration center through rubber bearing contact and keeps the same behavior to superstructure. In the SFSSI system it is possible to keep the natural period of the structure in a large interval, which is much more than predominant period of ground motion of the majority of the existing earthquakes including near-fault zones also. In this study, behavior of the both classical base-isolated building using LCRB and not isolated-FB building has been investigated under long-period ground motion. Obtained results are compared with SFSSI system with the same stiffness and equivalent damping and number of LCRB which was installed on spherical foundation. It was shown that, SFSSI system deprived from known deficiency period-dependent isolation systems and it

A. A. Kasimzade (✉) · S. Tuhta
Department of Civil Engineering, Ondokuz Mayis University, Samsun, Turkey
e-mail: azer@omu.edu.tr

S. Tuhta
e-mail: stuhta@omu.edu.tr

G. Atmaca
Provincial Directorate of Disaster and Emergency, Samsun, Turkey
e-mail: gencayatmaca@hotmail.com

A. A. Kasimzade et al. (eds.), *Seismic Isolation, Structural Health Monitoring, and Performance Based Seismic Design in Earthquake Engineering*,
https://doi.org/10.1007/978-3-319-93157-9_1

is the progress of the new type earthquake-resistant structures. SFSSI system exhibits improved Roly-poly ('Haciyatmaz', 'Okiogary-Koboshi') behavior. We believe that as Roly-poly, structures with SFSSI system will symbolize the ability to have success overcome adversity, and recover from misfortune.

Keywords Spherical foundation structural seismic Isolation-SFSSI system Earthquake-resistant structures · Nonlinear LS-DYNA finite element modeling Seismic base isolation

1.1 Introduction

Seismic isolation separates the structure from the harmful motions of the ground by providing the flexibility and energy dissipation capability through the insertion of the isolated devices, called isolators, between the foundation and the building structure. The simplest sliding system is the pure-friction (P-F) system well known in references, which does not have any restoring force and a period. Other isolation systems with restoring force and a period include; the resilient-friction system, the Electricite de France, the friction pendulum, the sliding resilient friction, the elliptical rolling rods and other period-dependent base isolator systems well known in appropriate references.

However, seismologists have reported on the vulnerable state of buildings isolated using a period-dependent base isolator against the near-fault pulse ground motions of intra-plate earthquakes or long-period ground motions of inter-plate earthquakes (Tanaka et al. 1979; Heaton et al. 1995; Hall et al. 1995; Celebi and Sanli 2002). In other words, if the frequency of oscillation of the ground is close to the natural frequency of the building, resonance may cause severe damage. In case of resonance frequency content of the ground motion is close to the building's natural frequency and tends to increase or amplify building response. In this framework, the seismic response of long-period structures (high-rise buildings, long-span bridges, base-isolated structures and others like these) is a scientific curiosity. Therefore, long-period structures can be vulnerable to earthquakes occurring in faraway locations. Some examples of these phenomena are presented below.

First example from the Mexican Earthquake of September 19, 1985, a magnitude of 8.1. These earthquakes resulted from the Cocos Plate pushing (Fig. 1.1) under the North American Plate (Butcher 1988; Smolka and Berz 1988).

Serious shaking lasted for nearly 3 min. And 350 km from the epicenter damage was concentrated in a 25 km^2 area (Fig. 1.2) of Mexico City.

Periods of the most (about 60%) vulnerable buildings (Fig. 1.3) ranged roughly from 0.6 to 2 s (6–20 storey buildings).

Second example (Zhigalin et al. 2013) from the 1977 Vrancea Earthquake, Romania on March 4, 1977, with magnitude 7.2. This earthquake caused the residents' panic and discomfort in high-rise buildings in Moscow, which are not considered as seismic regions far away from the epicenter. The news of this event

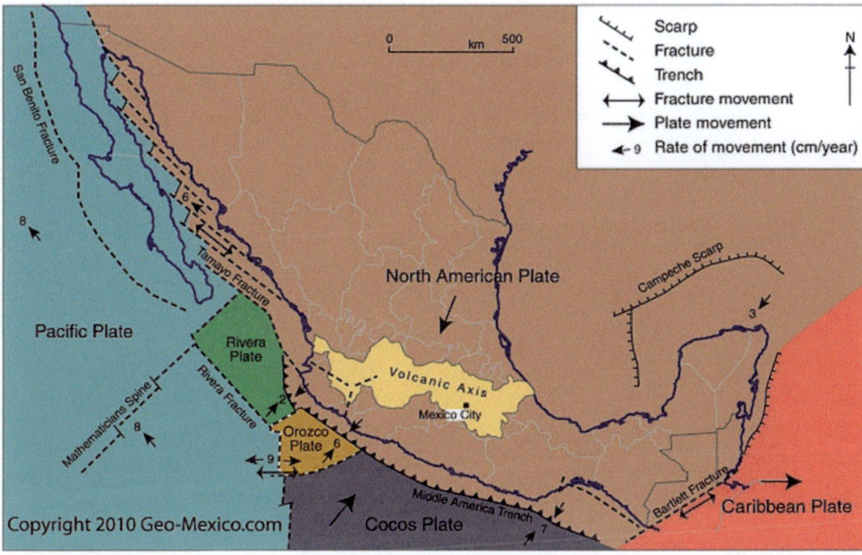

Fig. 1.1 Illustration of the Mexico's Cocos Plate position in relation to tectonic plates on the Mexican Earthquake of September 1985 (Butcher 1988)

fell into Moscow's doubt. The earthquake magnitude on ground level was 3–4, and the fluctuations at the height of the main building of the Moscow State University were about 7.

The third example is (Zhigalin et al. 2013) from the 2013 Okhotsk Sea Earthquake, Russian on May 24, 2013 with magnitude 7.7, depth 600 km (according to the US Geological Survey, the magnitude was Mw = 8.3, and the focal depth was 609 km). It was recorded in Moscow and in a number of satellite cities. For Moscow, which has more than once felt the echoes of strong Carpathian earthquakes from the Vrancea zone (Romania), this is a unique and unexpected event. The new seismic event caused a significant social resonance. The city was surveyed. In the center and southwest of Moscow (predominantly in high-rice buildings), chandeliers swung, the dishes rattled, objects moved, and personnel were evacuated from some offices. In Moscow and St. Petersburg, the intensity of quakes at the ground level was estimated at magnitude 1–2. The effect of a strong deeps focus earthquake at the level of perceptibility in the European part of Russia and in Moscow at a great distance (about 6500 km) from the earthquake focus in the Far East is a new phenomenon.

The Department of Earthquake Engineering of Kandilli Observatory and Earthquake Research Institute of Bogazici University (DEE-KOERI) has designed and been operating a significant number of Structural Health Monitoring (SHM) networks in Istanbul and Near East region. The next fourth and fifth examples are based on records of this center.

Fig. 1.2 Illustration of the 25 km^2 damage area of Mexico City 350 km from the epicenter plates on the Mexican Earthquake of September 1985 (Smolka and Berz 1988)

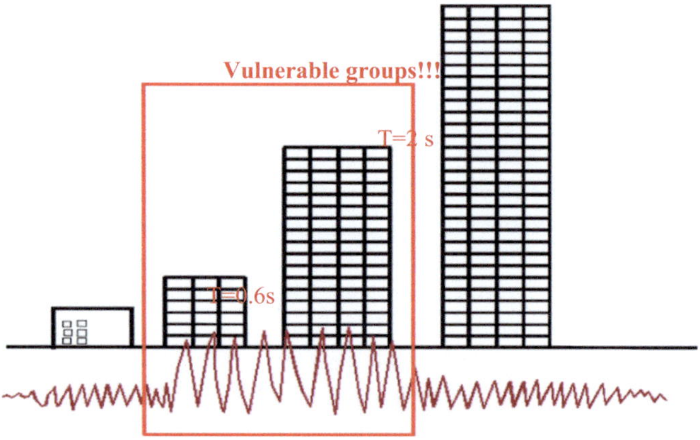

Fig. 1.3 Illustration of the most vulnerable building groups on the Mexican Earthquake of September 1985

The fourth example is from the Northern Aegean Sea Earthquake of May 24, 1914, a magnitude 6.5, which is 300 km away from the 62-storey, 238 m high Sapphire Building in Istanbul. The vibration amplitudes at the top of the building were very large and lasted more than 10 min after the earthquake has stopped. Although the accelerations and displacements were not large enough to present any danger structurally, the long duration of building's vibrations were strongly felt by the occupants and caused panic.

The fifth example is from the Iran–Pakistan border earthquake of 16 April 2013, a magnitude 7.8, which is 900 km away from the 74-storey, 310 m tall building in Abu Dhabi, whose SHM system is designed by the DEE-KOERI. The shaking from the earthquake was felt over a wide area, including the Middle East, causing panic among people in tall buildings in the region.

Since 2000, Japanese codes have required (Taylor 2012) design for velocity response spectrum values of 0.8 m/s up to a period of 10 s, whereas in previous codes this requirement did not exist. Prior to 2000, three ground motions were primarily used for design of high-rise structures (El Centro, Taft, and Hachinohe). This is the first is a group of the ground motions that are popular for evaluation of high-rise or base-isolated building presumed as a very rare case with a return period of 500 years in Japan. Those are adjusted so that the peak ground velocity is 0.50 m/s. Second is a group of ground motions observed in Kobe City (Fujitani et al. 2012), when the Hyogoken Nambu Earthquake occurred (PGA = 8.18 m/s^2, PGV = 0.92 m/s) and Takatori 1995 NS near-fault pulse (PGA = 6.06 m/s^2, PGV = 1.23 m/s). This group is adjusted so that the peak ground velocity in average is 1.075 m/s. During the 2011 Great East Japan Earthquake, the peak ground acceleration at the base of three buildings surrounding Tohoku Institute of Technology were 3.36–354 m/s^2 and the peak acceleration measured at the upper floor of the surrounding five storey building was 8.20 m/s^2. The results of the recent studies (Jangid 2005; Fujitani et al. 2006, 2012; Bonelli 2014; Saito 2014; Warnitchai 2014; Estrada 2014; Kasimzade et al. 2015a, b; Kawabe 2015) also show that an excessively great displacement occurs in the isolation layers of the base-isolated buildings. Active and semi-active control systems have been proposed for response reduction (Jangid 2005; Fujitani et al. 2006, 2012). Another side, in related design references (Eurocode 8 2003) presented design tools bounded with period of 6 s. But in references summarized and reported by (Tanaka et al. 1979; Kawabe 2015) classification of long wave period earthquakes between 1–15 s and 0.1–10 s and also presented examples from the later huge 2011 Tohoku earthquake.

As supported by the above presented and other existing related reference reports seismic isolated structures with period-dependent seismic isolation devices are vulnerable against the near-fault pulse ground motions of intra-plate earthquakes or long-period ground motions of inter-plate earthquakes.

New subjects such as long-period seismic ground motion was one of the main subject of the latest conference 14th Japan Earthquake Engineering Symposium (Osaka, Japan, 2014), International Symposium on Disaster Simulation (DS'15, Osaka University, Japan 2015) and it was discussed in special sessions by the researchers (Saito 2014; Warnitchai 2014; Estrada 2014) and (Kasimzade et al. 2015a, b; Kawabe 2015) respectively.

In summary, existing design tools for seismic isolated structures are bounded with a period of 6 s. LCRB isolators are produced with a period of under 5 s. Long wave period earthquakes between range 1 and 15 s and more. Under long-period earthquakes in the isolation layers excessive deformation occurs and buildings seismically isolated with period-dependent isolation devices (LCRB, LRB and others) may lose safety or the functionality.

In recent works (Kasimzade et al. 2015a, b) presented Spherical Foundation Structural Seismic Isolation—SFSSI system is presented for the first time, which deprived above mentioned deficiency of seismic isolated structures with period-dependent seismic isolation devices and may be used in active seismic zones, including against the near-fault pulse ground motions of intra-plate earthquakes or long-period ground motions of inter-plate earthquakes.

Here is presented SFSSI system's nonlinear dynamic performance, particularities and effects of behavior under long-period ground motions with comparison to traditional rubber bearing base isolation systems and not isolated (fixed base—FB) buildings respectively.

1.2 Governing Equations of Motion of the SFSSI System

SFSSI system's mathematical model was illustrated in Fig. 1.4. With the assumption that system masses were concentrated on storey levels, the SFSSI system was modeled as a system with (n + m) number is the dynamic DOF of the system. (n) is superstructure and (m) is the underground parts' dynamic DOF respectively. SFSSI system's equation of rigid body motion (Fig. 1.4) around the gyration center (g.c.) was derived based on the De Lambert principle and main converting steps was presented as following (Kasimzade et al. 2015a, b):

$$\begin{aligned}&\sum_{i=1,n} I_{\varphi_i}\ddot{\varphi} + \sum_{i=1,n} F_{I_i} h_i - \sum_{i=1,n} F_{g_i} u_{0i} + \left(F_{c,\text{eq}} + F_{\text{cb}} + F_{\text{kb}}\right)\rho_2 - \sum_{i=1,n} F_{\ddot{u}_{g_i}} h_i \\ &+ \sum_{j=1,m} I_{\varphi_{0j}}\ddot{\varphi} + \sum_{j=1,m} F_{I,0j} h_{0j} + \sum_{j=1,m} F_{g,0j} u_{0j} \\ &+ \sum_{j=1,m} F_{\ddot{u}_{g,0j}} h_{0j} = 0,\end{aligned} \tag{1.1}$$

where, φ is the rotation angle of the absolute rigid structure around of the gyration center; $I_{\varphi i} = m_i h_i^2$ is the rotational inertia of the superstructure's i-th mass m_i; $F_{I_i} = m_i\, \ddot{u}_{0i}$, $F_{\ddot{u}_{g_i}} = m_i\, \ddot{u}_g$, i = 1, n are forces in the i-th mass of the superstructure's from the $\ddot{u}_{0i}$, $\ddot{u}_g$ accelerations respectively; $\ddot{u}_g$ is the horizontal component of the earthquake acceleration; $F_{g_i} = m_i g$ is the gravitation force of the superstructure's i-th mass; g is the gravitational acceleration equal to 9.81 m/s^2; $I_{\varphi_{0j}}$, $F_{I,0j}$, $F_{\ddot{u}_{g,0j}}$, $F_{g,0j}$, j = 1, m are similar parameters for the underground part of the SFSSI system; $F_{c0} = c_b \rho_2 \dot{\varphi}, F_{\text{kb}}$ are the total damping and total stiffness forces of the isolator devices installed on spherical foundation of SFSSI system with total damping coefficient c_b and spherical radius ρ_2; $F_{c,\text{eq}} = c_d \rho_2 \dot{\varphi}$ is the total equal damping force in the external dampers with damping coefficient c_d.

Substituting above relation in Eq. 1.1, it becomes

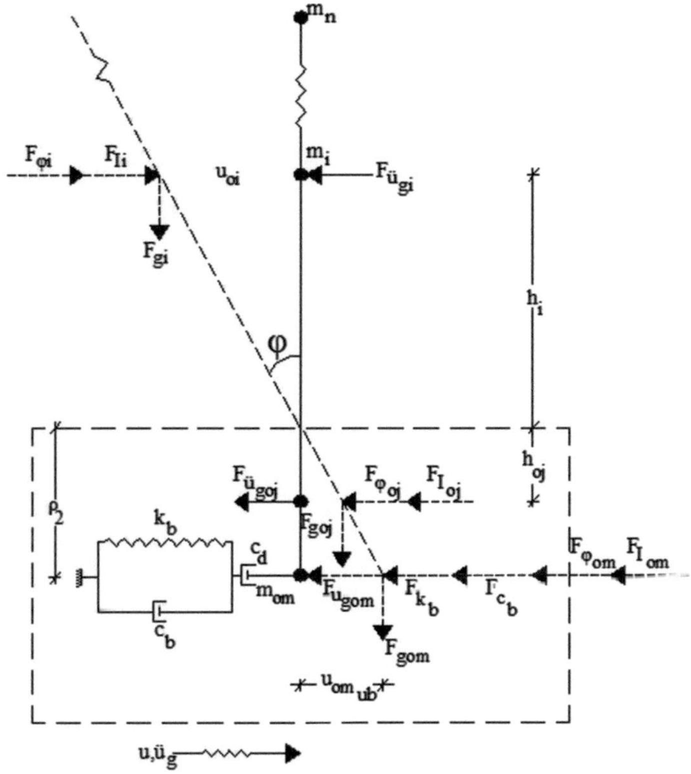

Fig. 1.4 SFSSI system's rigid body behavior—schematic illustration under earthquake excitation

$$
\begin{aligned}
&\ddot{\varphi} \sum_{i=1,n} m_i h_i^2 + \sum_{i=1,n} m_i \ddot{u}_{0i} h_i - \sum_{i=1,n} m_i g u_{0i} + \left(F_{c,\mathrm{eq}} + F_{\mathrm{cb}} + F_{\mathrm{kb}}\right) \rho_2 \\
&\quad - \sum_{i=1,n} m_i \ddot{u}_g h_i + \ddot{\varphi} \sum_{j=1,m} m_{0j} h_{0j}^2 + \sum_{j=1,m} m_{0j} \ddot{u}_{0j} h_{0j} \\
&\quad + \sum_{j=1,m} m_{0j} g u_{0j} + \sum_{j=1,m} m_{0j} \ddot{u}_g h_{0j} = 0
\end{aligned}
\tag{1.2}
$$

Defining terms as follows:

$$
\begin{aligned}
&\tan\varphi = \sim \varphi = \frac{u_{0i}}{h_i}, \quad u_{0i} = \varphi h_i, \quad \ddot{u}_{0i} = \ddot{\varphi} h_i, \\
&F_{c,\mathrm{eq}} + F_{c_b} + F_{k_b} = c_d \rho_2 \dot{\varphi} + c_b \rho_2 \dot{\varphi} + F_{\mathrm{kb}}
\end{aligned}
\tag{1.3}
$$

Equation 1.2, may be rewritten as the following:

$$
\begin{aligned}
&\ddot{\varphi}\sum_{i=1,n} m_i h_i^2 + \ddot{\varphi}\sum_{i=1,n} m_i h_i^2 - \varphi g \sum_{i=1,n} m_i h_i + \dot{\varphi}(c_d + c_b)\rho_2^2 + F_{\mathrm{kb}}\rho_2 \\
&\quad - \ddot{u}_g \sum_{i=1,n} m_i h_i + \ddot{\varphi}\sum_{j=1,m} m_{0j} h_{0j}^2 + \ddot{\varphi}\sum_{j=1,m} m_{0j} h_{0j}^2 \\
&\quad + \varphi g \sum_{j=1,m} m_{0j} h_{0j} + \ddot{u}_g \sum_{j=1,m} m_{0j} h_{0j} = 0
\end{aligned}
\tag{1.4}
$$

or

$$
\begin{aligned}
&\ddot{\varphi}\left[\left(\sum_{i=1,n} m_i h_i^2 + \sum_{j=1,m} m_{0j} h_{0j}^2\right)2\right] + \dot{\varphi}(c_d + c_b)\rho_2^2 \\
&\quad - \varphi\left[\left(\sum_{i=1,n} m_i h_i - \sum_{j=1,m} m_{0j} h_{0j}\right)g\right] + F_{\mathrm{kb}}\rho_2 \\
&\quad = \ddot{u}_g\left[\sum_{i=1,n} m_i h_i - \sum_{j=1,m} m_{0j} h_{0j}\right]
\end{aligned}
\tag{1.5}
$$

SFSSI system's forced vibration covering equation of motion around of the gyration center as a rigid body was obtained as:

$$
\ddot{\varphi}C_2 + \dot{\varphi}C_1 - \varphi C_{01} g + \rho_2 F_{\mathrm{kb}}(\varphi) = \ddot{u}_g C_{02} \tag{1.6}
$$

Here

$$
C_1 = \rho_2^2(c_d + c_b), \tag{1.7}
$$

$$
C_2 = 2\left(\sum_{i=1,n} m_i h_i^2 + \sum_{j=1,m} m_{0j} h_{0j}^2\right) \tag{1.8}
$$

$$
C_{01} = \sum_{i=1,n} m_i h_i - \sum_{j=1,m} m_{0j} h_{0j} \tag{1.9}
$$

$$
C_{02} = \sum_{i=1,n} m_i h_i - \sum_{j=1,m} m_{0j} h_{0j} \tag{1.10}
$$

Accepting that SFSSI system's underground part storey-level forces $F_{\ddot{u}_{g,0j}} = m_{0j}\ddot{u}_g$ will be equilibrated by the opposite forces $F_{\ddot{u}_{g,0j,\mathrm{eq}}} = F_{\ddot{u}_{g,0j}}$ by the outside part of the spherical foundation (Fig. 1.5) in Eqs. (1.2–1.6), the Eq. 1.10 become as

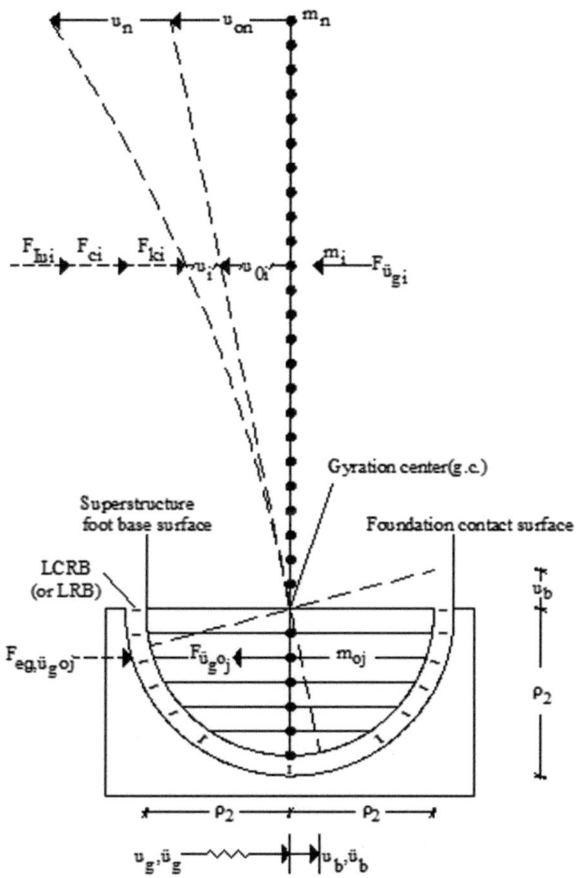

Fig. 1.5 Illustration of the SFSSI system's underground spherical foundation and superstructure forces

$$C_{02} = \sum_{i=1,n} m_i h_i \tag{1.11}$$

The underground part's horizontal deformation and corresponding acceleration are neglected, since it is completely rigid.

SFSSI system's superstructure part deformed states equation of motion (Fig. 1.6) is built and main converting steps are presented as following:

$$F_{I_i} + F_{\text{ci}} + F_{\text{ki}} - F_{\ddot{u}_{g_i}} = 0; \quad i = 1, n \tag{1.12}$$

$$m_i(\ddot{u}_i + \ddot{u}_{0i}) + c_i\,\dot{u}_i + k_i u_i = m_i\,\ddot{u}_g; \quad i = 1, n \tag{1.13}$$

$$m_i \ddot{u}_i + c_i\,\dot{u}_i + k_i u_i = m_i(\ddot{u}_g - \ddot{u}_{0i}); \quad i = 1, n \tag{1.14}$$

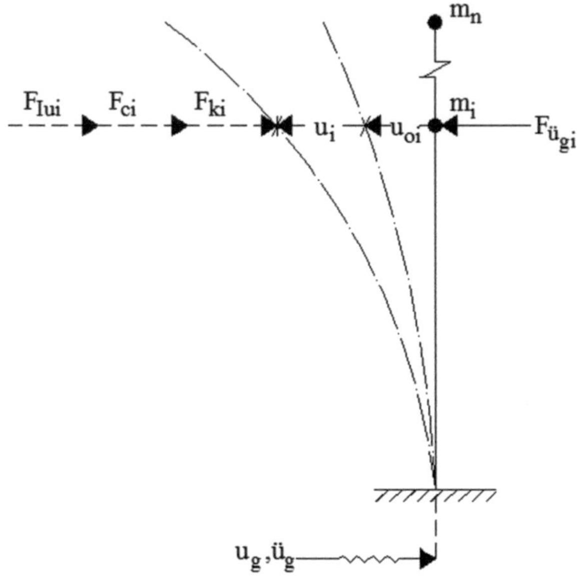

Fig. 1.6 SFSSI system's superstructure part deformed state under earthquake excitation—schematic illustration

SFSSI system's superstructure part deformed states equation of motion becomes.

$$[m]\{\ddot{u}\} + [c]\{\dot{u}\} + [k]\{u\} = \{F_{\ddot{u}_g}\}, \tag{1.15}$$

where $[m]$, $[c]$, $[k]$ are mass, damping and stiffness matrix of the superstructure respectively;

$\{u\}$ is the relative displacement vector of the deformed states superstructure.

$$\{F_{\ddot{u}_g}\} = \begin{Bmatrix} m_1(\ddot{u}_g - \ddot{u}_{01}) \\ m_2(\ddot{u}_g - \ddot{u}_{02}) \\ \cdots \\ m_i(\ddot{u}_g - \ddot{u}_{0i}) \\ \cdots \\ m_n(\ddot{u}_g - \ddot{u}_{0n}) \end{Bmatrix}; \{\ddot{u}_0\} = \begin{Bmatrix} \ddot{u}_{01} \\ \ddot{u}_{02} \\ \cdots \\ \ddot{u}_{0i} \\ \cdots \\ \ddot{u}_{0n} \end{Bmatrix} = \begin{Bmatrix} h_1 \\ h_2 \\ \cdots \\ h_i \\ \cdots \\ h_n \end{Bmatrix} \ddot{\varphi}; \{u\} = \begin{Bmatrix} u_1 \\ u_2 \\ \cdots \\ u_i \\ \cdots \\ u_n \end{Bmatrix}; \tag{1.16}$$

with

$$u_{0i} = \varphi h_{0i}, \ddot{u}_{0i} = \ddot{\varphi} h_i;$$

Here φ is the solution of the Eq. 1.6.

Thus, the behavior of the SFSSI system under lateral earthquake excitation is expressed by a set of Eqs. 1.6 and 1.15.

SFSSI system's absolute displacement and acceleration responses for top and base levels respectively are defined as

$$\begin{aligned} u_{n,\text{abs}} &= \left|-u_n - u_{\text{on}}\right| = u_{\text{top,abs}} \\ \ddot{u}_{n,\text{abs}} &= \left|\ddot{u}_g - \ddot{u}_n - \ddot{u}_{\text{on}}\right| = \ddot{u}_{\text{top,abs}} \\ u_{b,\text{abs}} &= \left|u_g + u_b\right| \\ \ddot{u}_{b,\text{abs}} &= \left|\ddot{u}_g + \ddot{u}_b\right| \end{aligned} \tag{1.17}$$

1.3 Mathematical Model of Base Isolator

For the present study, Wen's model (Wen 1976; Ikhouane and Rodellar 2007) is used to characterize the hysteretic behavior of the Lead-Core Rubber Bearing (LCRB or N-Z) isolator which is shown schematically and graphically in Figs. 1.7, 1.8 and 1.9 respectively. The LCRB isolator was invented in New Zealand (N-Z) in 1975 and have been used extensively in New Zealand, Japan, and the United States (Naeim 1999).

For high (Fig. 1.10) and low (Fig. 1.11) initial isolator stiffness, its nonlinear restoring force F_{kb} can be expressed as following respectively:

$$F_{\text{kb}} = k_b u_b + (1-\alpha)F_y Z \tag{1.18a}$$

$$F_y = Q + k_b u_y \tag{1.18b}$$

$$F_{\text{kb}} = \alpha\, k_b u_b + (1-\alpha)F_y Z \tag{1.19a}$$

$$F_y = Q + \alpha k_b u_y, \tag{1.19b}$$

where u_b is the lateral displacement of the superstructure foot base surface relative to the foundation contact surface; u_y is the yield displacement of bearing; k_b is the total stiffness of the isolators between the superstructure foot base surface and the foundation contact surface; α is the ratio of the post to pre-yielding stiffness and in references (Deb and Paul 1996; Deb et al. 1997), from selected order 0.25 to 0.4

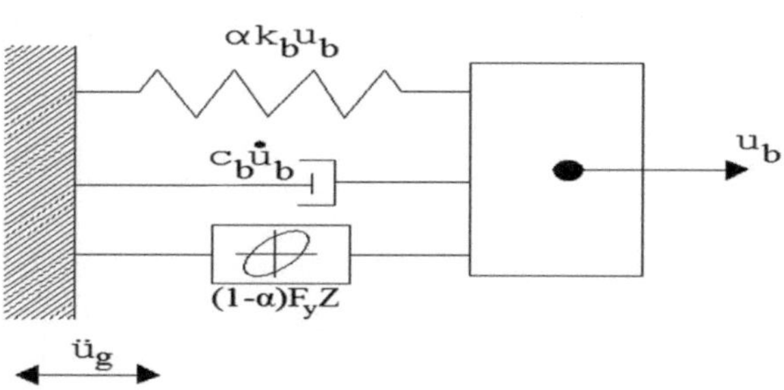

Fig. 1.7 Schematic diagram of the LCRB isolator

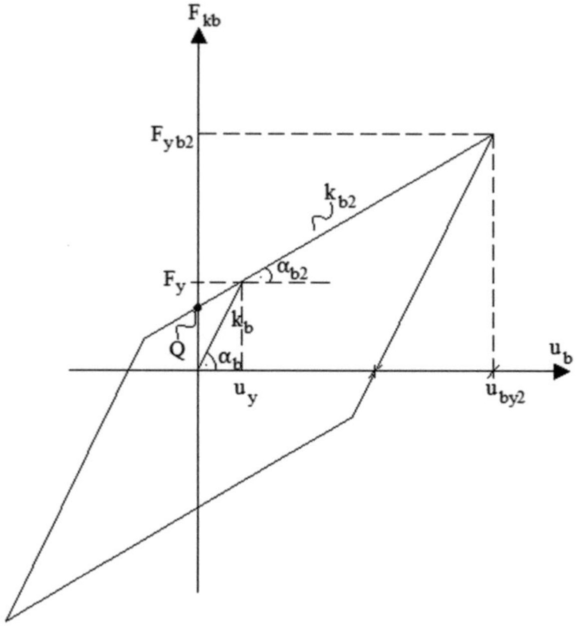

Fig. 1.8 Hysteretic model of the LCRB isolator

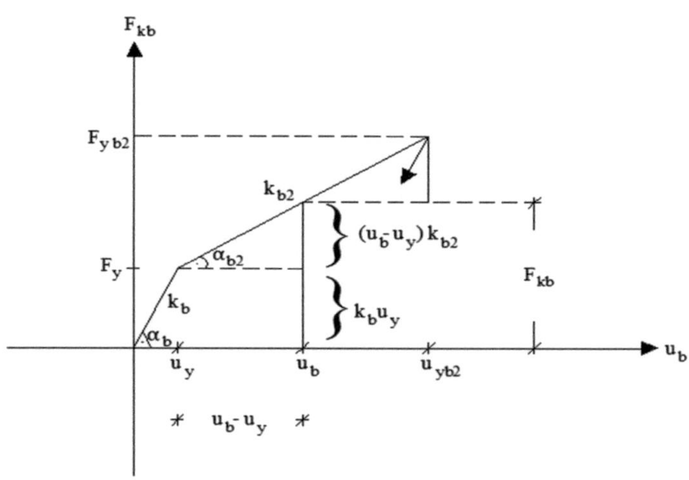

Fig. 1.9 LCRB isolator's hysteretic model main geometric rations—illustration

was used $\alpha = 0.33$; in experimental investigation of low-type LCRB isolators (FIP Industrial 2013) was obtained $\alpha = 0.0882$ and $\alpha = 0.0967$ respectively (roughly $\alpha = 0.1$); in LCRB isolated structural nonlinear vibration control analysis (Li and

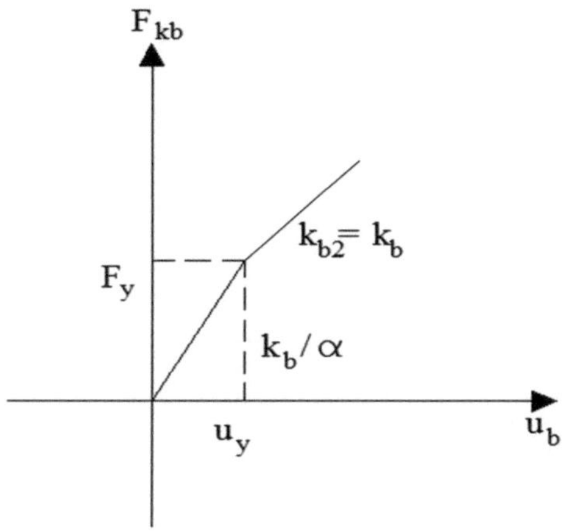

Fig. 1.10 Hysteretic model of the LCRB isolator with high initial stiffness

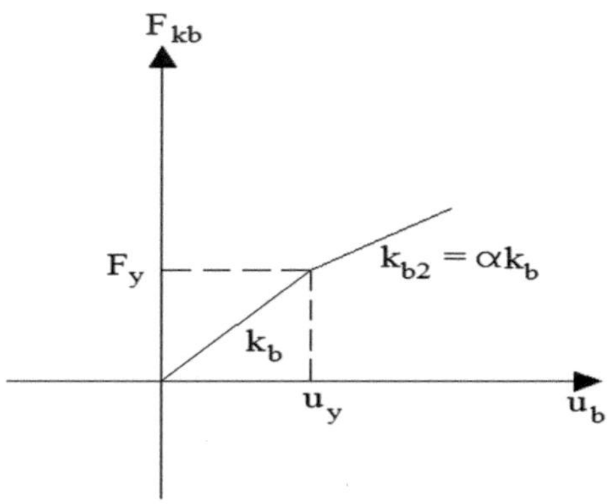

Fig. 1.11 Hysteretic model of the LCRB isolator with low initial stiffness

Ou 2012) also was used $\alpha = 0.1$; in research (Lei and He 2013) aimed to identification nonlinear properties of LCRB isolators, and was used $\alpha = 0.2$;

The ratio (d_r) of maximum bearing displacement (u_{yb2}) and yield displacement (u_y) are found to be in the order of 5 to 7 on (Deb and Paul 1996; Deb et al. 1997), on (FIP Industrial 2013) are found in the range $d_r = 10.387 - 10.523$ (about 10) and $f_r = F_{yb2}/F_y = 1.8279 - 1.92$ (about 2) respectively.

F_y is the yield strength and Q is characteristic strength of the bearing respectively. The yield strength of the bearing is normalized with respect to the total weight (W) of the isolated building and expressed by the parameter, F_0 defined as

$$F_0 = \frac{F_y}{W} \tag{1.20}$$

In the some references normalized stiffness has been expressed as

$$f_0 = \frac{Q}{W} \tag{1.21}$$

$$W = g\, m_t \tag{1.22}$$

$$m_t = m_b + \sum_{i=1}^{n} m_i, \tag{1.23}$$

where m_i, m_b, m_t are the mass of the storey, base slab and total mass of the building respectively.

For high initial stiffness LCRB isolated buildings under near-fault zone earthquake excitation the optimum normalized yield strength (F_0) is found to be in range $F_0 = 0.1-0.15$ in (Jangid 2007). But for low initial stiffness LCRB isolated buildings yield strength is accepted in range $F_0 = 0.05-0.10$ and yield displacement of bearing is acceptance in range $u_y = 0.025-0.05$ m; $\xi_b = 0.1-0.3$; $T_b = 2-3$ s. Generally, for a given performance response bound of the structure, the isolated structure's three characterized parameters (the isolation period (T_b), damping ratio (ξ_b) or damping coefficient (c_b) and normalized yield strength (F_0) bound must be defined (Kasimzade 2002). The parameters T_b and c_b are expressed as

$$T_{0b} = 2\pi\sqrt{\frac{m_t}{k_b}}; \quad \omega_{0b} = \sqrt{\frac{m_t}{k_b}}; \quad \omega_b = \omega_{0b}\sqrt{1-\xi_b^2}; \quad T_b = 2\pi/\omega_b \tag{1.24}$$

$$c_b = 2m_t\omega_b\xi_b \tag{1.25}$$

Z is the non-dimensional hysteretic displacement component satisfying the following nonlinear first order differential equation expressed as (Wen 1976; Ikhouane and Rodellar 2007):

$$\dot{Z} = \left[a\dot{u}_b - \beta|\dot{u}_b|Z|Z|^{n-1} - \gamma\dot{u}_b|Z|^n\right]u_y^{-1} \tag{1.26}$$

The dimensionless parameters a, β, γ, n control the shape of the hysteresis loop and are selected such that the predicted response from the model closely matches the experimental results. It has been shown that: for $n = 2$, $a = 1$ and $+\gamma = 1$, the model of Eq. 1.26, reduces to a model of visco-plasticity and in this case *uy* represents the yield displacement.

In order to compare the results of the analysis in the future, the equations of motion of buildings with traditional LCRB/LRB base isolation and without isolation (fixed base FB) buildings are presented below.

1.4 Governing Equation of Motion of Traditional LCRB/LRB Base-Isolated and Fixed Base Buildings

As known the equation of motion traditional LCRB/LRB base-isolated buildings (Fig. 1.12) are presented as a set of equations

$$m_b \ddot{u}_b + F_{\mathrm{kb}} + c_b \dot{u}_b - k_1 u_1 - c_1 \dot{u}_1 = +m_b \ddot{u}_g \tag{1.27}$$

$$[m]\{\ddot{u}\} + [c]\{\dot{u}\} + [k]\{u\} = +[m]\{r\}(\ddot{u}_g - \ddot{u}_b) \tag{1.28}$$

Here $\{r\}$ is the unit vector of order n: $\{r\} = (111\ldots1\ldots1_n)^{\mathrm{T}}$

LCRB/LRB isolation system's absolute displacement and acceleration responses for top and base levels respectively are defined as

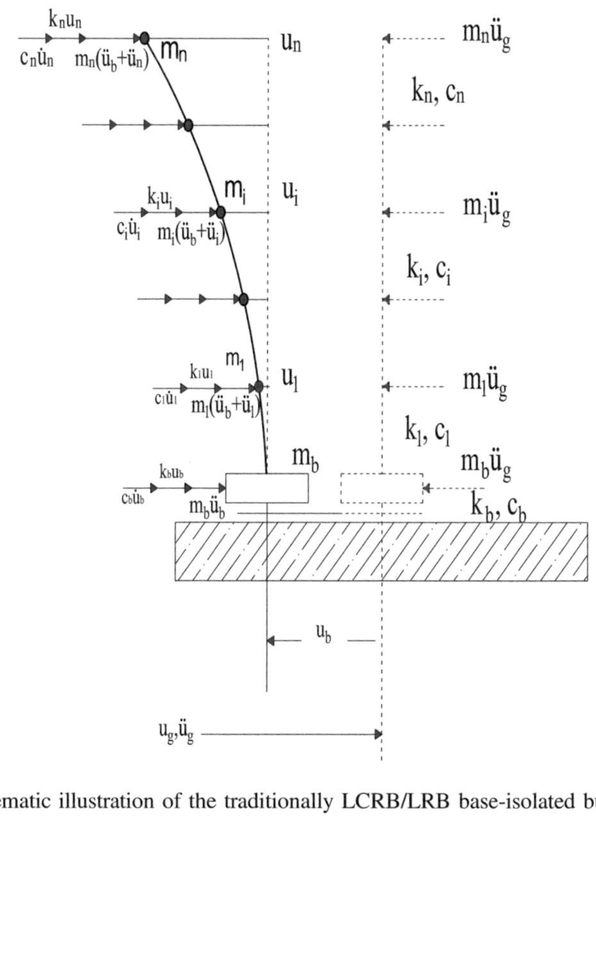

Fig. 1.12 Schematic illustration of the traditionally LCRB/LRB base-isolated building

$$
\begin{aligned}
u_{n,\text{abs}} &= \left|-u_n - u_{\text{ob}}\right| = u_{\text{top,abs}} \\
\ddot{u}_{n,\text{abs}} &= \left|\ddot{u}_g - \ddot{u}_n - \ddot{u}_{\text{ob}}\right| = \ddot{u}_{\text{top,abs}} \\
u_{b,\text{abs}} &= \left|u_g - u_b\right| \\
\ddot{u}_{b,\text{abs}} &= \left|\ddot{u}_g - \ddot{u}_b\right|
\end{aligned} \tag{1.29}
$$

In presented Eqs. 1.27 and 1.28 for the case of $u_b = 0$, the equation of motion of the not isolated-fixed base (FB) structure (Fig. 1.13) is obtained as:

$$[m]\{\ddot{u}\} + [c]\{\dot{u}\} + [k]\{u\} = +[m]\{r\}(\ddot{u}_g) \tag{1.30}$$

Fixed base (FB) system's absolute displacement and acceleration responses for top and base levels respectively are defined as

$$
\begin{aligned}
u_{n,\text{abs}} &= \left|u_g - u_n\right| = u_{\text{top,abs}} \\
\ddot{u}_{n,\text{abs}} &= \left|\ddot{u}_g - \ddot{u}_n\right| = \ddot{u}_{\text{top,abs}} \\
u_{b,\text{abs}} &= \left|u_g\right| \\
\ddot{u}_{b,\text{abs}} &= \left|\ddot{u}_g\right|
\end{aligned}
$$

1.5 SFSSI System's Free Vibration

SFSSI system's undamped free vibration covering equation of motion around the gyration center as a rigid body was obtained from the Eq. 1.6, substituting in it $C_1 = \rho_2^2(c_d + c_b) = 0, \ddot{u}_g = 0$ and assuming $F_{\text{kb}}(\varphi) = k_b u_b = k_b \varphi \rho_2$ as

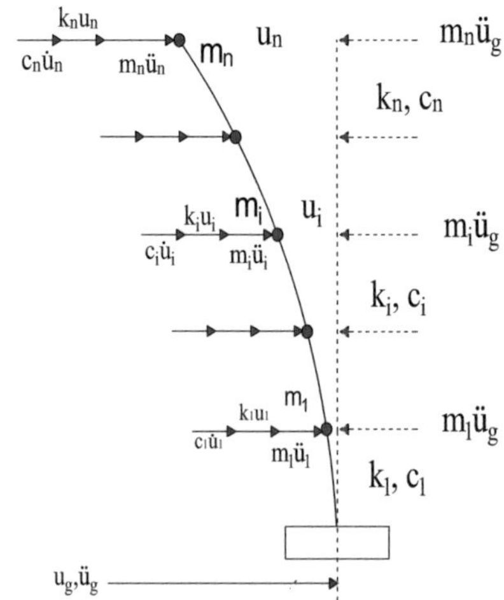

Fig. 1.13 Schematic illustration of the fixed base building

$$\ddot{\varphi}C_2 + \varphi k_b \rho_2^2 - C_{01} g \varphi = 0 \tag{1.31}$$

$$\ddot{\varphi} + \frac{k_b \rho_2^2 - C_{01} g}{C_2} \varphi = 0 \tag{1.32}$$

The embedded system frequency (ω_0) and period (T_0) in this equation are expressed as

$$\omega_0^2 = \frac{k_b \rho_2^2 - c_{01} g}{c_2} \tag{1.33}$$

$$T_0 = 2\pi \sqrt{\frac{C_2}{\mathrm{k_b} \rho_2^2 - C_{01} g}} \tag{1.34}$$

The total stiffness of the isolators (k_{br}) along the foundation contact surface on base required system period (T_{0r}) can be found in the first approach as follows (Kasimzade et al. 2015a, b):

$$k_{\mathrm{br}} = \left[C_2 / (T_{0r}/2\pi)^2 + C_{01} g \right] / \rho_2^2 \tag{1.35}$$

1.6 Hysteretic Damping Model of Isolator

One of the main purposes of employing a LCRB is to dissipate the earthquake energy. The behavior of the base-isolated building is selected in a way to represent the force-deformation behavior of the commonly used isolation systems such as elastomeric systems. The energy dissipated per cycle of the hysteretic loop (A_h) when the lead has yielded, is illustrated in Fig. 1.14: 2^2

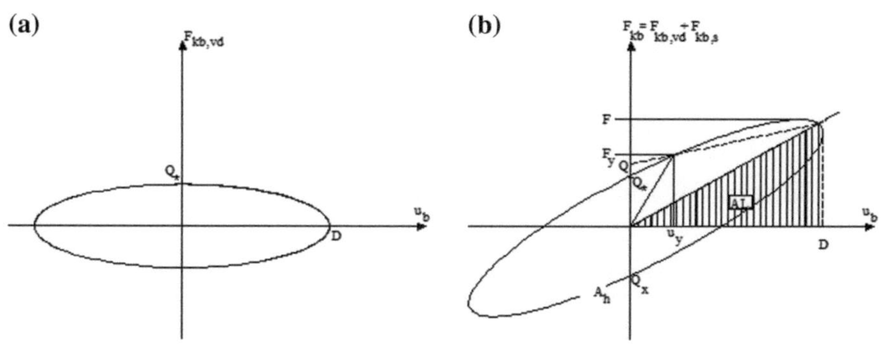

Fig. 1.14 Hysteresis loops for **a** viscous damper (vd); **b** Spring and viscous dumper in parallel

The width of the hysteresis loop is related to the loop area which is in effect a measure of the energy dissipation that can be conveniently described by the equivalent damping ratio (ξ_{eq}). At a specified design isolation displacement (D), for resonance case ($r = \theta/\omega_{0b} = 1$) equivalent damping ratio (ξ_{eq}) for system are expressed as

$$\xi_{\text{eq}} = \frac{1}{4\pi r}\frac{A_h}{A_L}, \tag{1.36}$$

where $A_L = \frac{\text{DF}}{2}$ is stored energy (linear energy)—the energy produced by the linear component of the isolator restoring force for high $F_{\text{kbH}} = k_b u_b$ and low $F_{\text{kbL}} = \alpha k_b u_b$ initial stiffness restoring forces respectively; $r = \theta/\omega_{0b}$ is the ratio of the forced vibration's frequency to natural frequency of the isolator.

$A_h = \pi Q_*(D - u_y)$ (or roughly $A_h \approx \pi Q_* D$) is hysteresis area or dissipated energy (nonlinear energy)—the energy produced by the nonlinear component of the isolator restoring force for high $F_{\text{kbH}} = (1-\alpha)(Q + k_b u_y)Z$ and low $F_{\text{kbL}} = (1-\alpha)(Q + \alpha k_b u_y)Z$ initial stiffness restoring forces respectively;

Q is characteristic strength; D is maximum shear displacement and u_y is yield displacement.

For steady-state harmonic response

$$\xi_{\text{eq}} = D/2u_{\text{st}}$$

Here u_{st} is the static displacement for steady-state harmonic response from the harmonic forces amplitude excitation.

The equivalent stiffness and damping ratio of an isolator can be computed by

$$k_{\text{eq}} = (F^+ - F^-)/(D^+ - D^-) \tag{1.37}$$

$$\xi_{\text{eq}} = (1/2\pi)[A_h/FD], \tag{1.38}$$

where F^+, F^- and D^+, D^- are maximum forces and maximum displacements of the isolator in positive and negative directions respectively; F, D are absolute maximum force and absolute maximum displacement of the isolator (in the LCRB isolators can be accept $Q^* = \sim F_y$).

1.7 Solution Method of Equation of Motion

SFSSI system's set of nonlinear governing equations of motion-Eqs. 1.6, 1.15, 1.18a, b, 1.19a, b and 1.26 is solved by the Runge–Kutta method using MATLAB differential equation tools (Matlab 2017).

1.8 Numerical Study

SFSSI system's parameters. Numerical verifications are presented on the example 26 storey (superstructure is 20 storey and underground part is six-storey) SFSSI steel framed structure (Figs. 1.15 and 1.16) with storey height h = 4 m and total height H = 20 h + 6 h = 102 m; the floor load including self-weight of the steel members is set at 7840 N/m^2. The structure is designed so that the maximum storey angle is less than 1/200. The material of steel members is SN490, so that the yield strength is 357 MPa. The columns have box-section, beam have H-shape section and all members are designed to satisfy FA rank building design grade. Building section plane axial radius is $\rho_{2a} = 22.63\,\text{m}$;

Superstructure foot base surface radius is $\rho_2 = 24\,\text{m}$; Superstructure's each storey area is $A_{\text{st}} = \pi\rho_{2a}^2 = 1608.9\ \text{m}^2$; Superstructure total area is A_{sup} = 20 A_{st} = 32,178 m^2; Underground part space occupancy rate is $\alpha_b = (A_{\text{st}}/2)/(2\rho_2\,2\rho_2/2) = 0.7854$; underground part total area is $A_{\text{ung}} = 6A_{\text{st}}\,\alpha_b = 7581.6\ \text{m}^2$; Total storey area of the 26 storey structure is A_{SFSSI} = 39,759 m^2; Building storey mass m_{st} = 7840 A_{st}/g = 1.2858e6 kg; SFSSI steel framed structure storey masses calculated on base this information are given in Table 1.1. For comparing results classically LCRB isolated and fixed base (FB) 24-storey steel framed structure parameters with same section area are generated respectively and also presented in the Table 1.1. LCRB isolated and fixed base (FB) structure total storey areas are A_{LCRB} = A_{FB} = A_{st} 24 = 38,613 m^2.

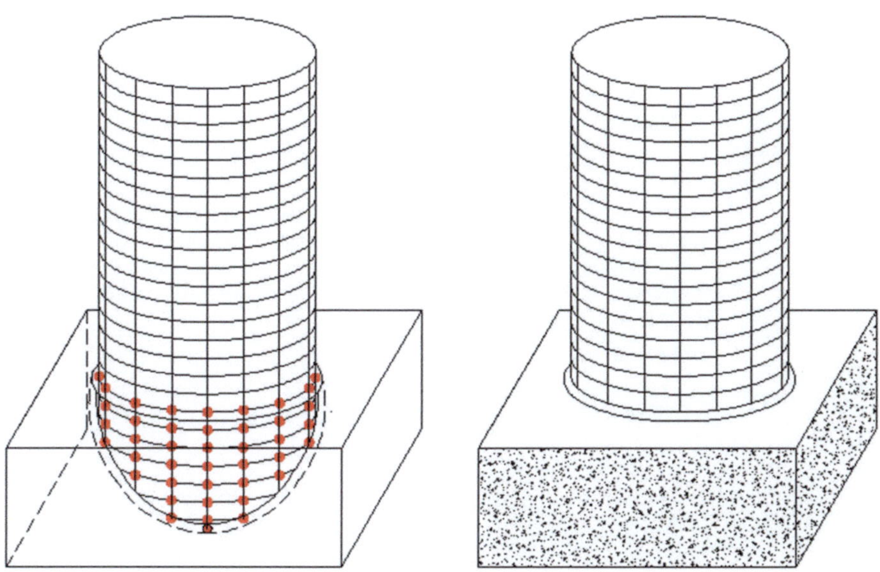

Fig. 1.15 Illustration of the spherical foundation structural seismic isolation—SFSSI system

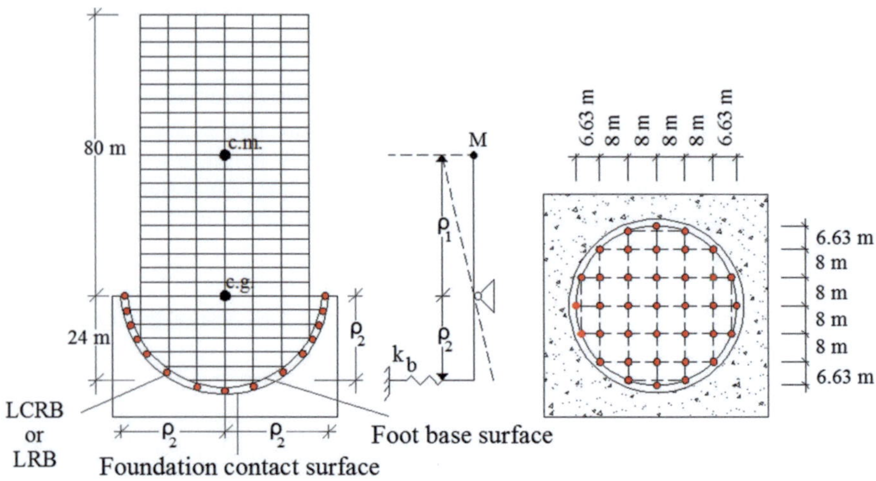

Fig. 1.16 Sectional view of the SFSSI system

Table 1.1 SFSSI, classical LCRB isolated and Fixed Base (FB) steel framed structures masses distributions

Storey number (for LCRB FB structures)	SFSSI steel framed structure storey masses (kg)	Classically LCRB isolated structure storey masses (kg)	Fixed Base (FB) structure storey masses (kg)
Superstructure			
1–19 (1–23) 20 (24)	1.2858e+6 0.91597e+6	1.2858e+6 0.91597e+6	1.2858e+6 0.91597e+6
Total superstructure	25.3460e+6	30.4894e+6	30.4894e+6
Underground part			
−1 −2 −3 −4 −5	1.1192e+6 1.0702e+6 0.9830e+6 0.8460e+6 0.6274e+6	This mass is not presented because it is not reflected in the analysis	This mass is not presented because it is not reflected in the analysis
−6 (isolator)	1.7830e+6	1.2858e+6	–
Total underground part	6.4288+6	–	–
Total structure	31.7750e+6	31.775e+6	30.4894e+6

Assuming that the predominant period of the earthquakes in the area where examined (SFSSI, LCRB, FB) buildings will be build is less than 11 s, SFSSI structure's required total isolator stiffness for the first approximation may be defined on base Eq. 1.35 as $k_{\mathrm{br}} = 8.2455\mathrm{e}+\mathrm{N/m} = k_b$. Other parameters of the LCRB isolator defined by the Eqs. 1.24 and 1.25 respectively as $T_b = 3.9451$ s and $c_b = 1.5182\mathrm{e}+07$ Ns/m for assumed isolator damping ratio $\xi_b = 0.15$ and stiffness $k_b = 8.2455\mathrm{e}+7$ N/m respectively. For possibility comparing SFSSI system with the classically LCRB isolated structure, in presented SFSSI system example additional dampers are not used, so $c_d = 0$ and related parameter defined by Eq. (1.7) as $C_1 = \rho_2^2(c_d + c_b) = 8.7449\mathrm{e}+09$ Nsm. Hysteretic parameters in Eqs. (1.18a, b or 1.19a, b and 1.26) was accepted as $\alpha = 0.1$, $\beta = 0.5$, $\gamma = 0.5$, $a = 1$, $(\frac{a}{\gamma+\beta} = 1)$, u_y = 0.0601 m. Yielding strength F_y is defined on LCRB's experimental investigations on base empirical relation Eq. 1.20. In presented example it was assumed that F_0 = 0.0159, so $F_y = 0.0159\ W = 49562\mathrm{e}+6N$.

SFSSI system's superstructure stiffness $[k_s]$ and mass matrices $[m_s]$ was obtained as

```
[ks]=kss*[ 2 -1 0 0 0 0 0 0 0 0 0 0 0 0 0 0 0 0 0 0
          -1 2 -1 0 0 0 0 0 0 0 0 0 0 0 0 0 0 0 0 0
          0 -1 2 -1 0 0 0 0 0 0 0 0 0 0 0 0 0 0 0 0
          0 0 -1 2 -1 0 0 0 0 0 0 0 0 0 0 0 0 0 0 0
          0 0 0 -1 2 -1 0 0 0 0 0 0 0 0 0 0 0 0 0 0
          0 0 0 0 -1 2 -1 0 0 0 0 0 0 0 0 0 0 0 0 0
          0 0 0 0 0 -1 2 -1 0 0 0 0 0 0 0 0 0 0 0 0
          0 0 0 0 0 0 -1 2 -1 0 0 0 0 0 0 0 0 0 0 0
          0 0 0 0 0 0 0 -1 2 -1 0 0 0 0 0 0 0 0 0 0
          0 0 0 0 0 0 0 0 -1 2 -1 0 0 0 0 0 0 0 0 0
          0 0 0 0 0 0 0 0 0 -1 2 -1 0 0 0 0 0 0 0 0
          0 0 0 0 0 0 0 0 0 0 -1 2 -1 0 0 0 0 0 0 0
          0 0 0 0 0 0 0 0 0 0 0 -1 2 -1 0 0 0 0 0 0
          0 0 0 0 0 0 0 0 0 0 0 0 -1 2 -1 0 0 0 0 0
          0 0 0 0 0 0 0 0 0 0 0 0 0 -1 2 -1 0 0 0 0
          0 0 0 0 0 0 0 0 0 0 0 0 0 0 -1 2 -1 0 0 0
          0 0 0 0 0 0 0 0 0 0 0 0 0 0 0 -1 2 -1 0 0
          0 0 0 0 0 0 0 0 0 0 0 0 0 0 0 0 -1 2 -1 0
          0 0 0 0 0 0 0 0 0 0 0 0 0 0 0 0 0 -1 2 -1
          0 0 0 0 0 0 0 0 0 0 0 0 0 0 0 0 0 0 -1 1];
```

$$k_{ss} = 1.8428\text{e}+9\,\text{N/m};$$

SFSSI system's superstructure part predominant period roughly was defined depending on the superstructure total height 80 m (or storey number n = 20) as:

$$T_{\_sfssi\_supers} = 0.08 * (h * n)^{(3/4)} = 2.14s\,(\text{or}\,0.1n = 0.1\,20 = 2\,s)$$

The correctness of the matrix $[k_s]$ is evidenced by the solution of the eigenvalue problem of the matrices $[k_s]$, $[m_s]$ and comparing with predominant period (2.1362 ~ = 2.14).

Damping matrix $[c_s]$ for the superstructure with damping ratio $\xi_s = 0.05$ are defined by the classical-Rayleigh mass $[m_s]$ and stiffness $[k_s]$ proportional damping as

$$[c_s]=c_{ss}\,[c_1 + c_2 \quad -c_2$$
$$\dots$$
$$-c_{i-1} \quad c_{i-1} + c_i \quad -c_i$$
$$\dots$$
$$-c_{n-1} \quad c_{n-1} + c_n \quad -c_n$$
$$-c_n \quad c_n\,];$$

c_{ss}= 1.0496e+7 Ns/m;

LCRB isolated building parameters. To be able to compare the SFSSI system with the classically LCRB isolated system, the n = 24 storey classically LCRB isolated building system with total mass is equal to the SFSSI system total mass and with same geometry and section will be examined. LCRB isolator parameters and numbers (37) are same. Another word LCRB parameters used in SFSSI and classical LCRB isolated systems are same and summarized as following:

$T_b = 3.9451\,s$, $c_b = 1.5182\text{e}+07$ Ns/m, $\xi_b = 0.15$, $k_b = 8.2455\text{e}+7\,\text{N/m}$, $\alpha = 0.1$, $\beta = 0.5$, $\gamma = 0.5$, $a = 1$, u_y = 0.0601 m, F_0 = 0.0159

LCRB isolated system's superstructure part predominant period roughly was defined depending to superstructure total height 104 m (or storey number n = 24) as

$$T_{\_lcrb\_supers} = 0.08 * (h * n)^{(3/4)} = 2.4535\,\text{s}(\text{or}\,0.1n = 0.1\,24 = 2.4\,s)$$

LCRB isolated system's superstructure stiffness $[k_s]$ and mass matrices $[m_s]$ was obtained as

```
[kI]=kIs*[ 2 -1 0 0 0 0 0 0 0 0 0 0 0 0 0 0 0 0 0 0 0 0 0 0
          -1 2 -1 0 0 0 0 0 0 0 0 0 0 0 0 0 0 0 0 0 0 0 0 0
          0 -1 2 -1 0 0 0 0 0 0 0 0 0 0 0 0 0 0 0 0 0 0 0 0
          0 0 -1 2 -1 0 0 0 0 0 0 0 0 0 0 0 0 0 0 0 0 0 0 0
          0 0 0 -1 2 -1 0 0 0 0 0 0 0 0 0 0 0 0 0 0 0 0 0 0
          0 0 0 0 -1 2 -1 0 0 0 0 0 0 0 0 0 0 0 0 0 0 0 0 0
          0 0 0 0 0 -1 2 -1 0 0 0 0 0 0 0 0 0 0 0 0 0 0 0 0
          0 0 0 0 0 0 -1 2 -1 0 0 0 0 0 0 0 0 0 0 0 0 0 0 0
          0 0 0 0 0 0 0 -1 2 -1 0 0 0 0 0 0 0 0 0 0 0 0 0 0
          0 0 0 0 0 0 0 0 -1 2 -1 0 0 0 0 0 0 0 0 0 0 0 0 0
          0 0 0 0 0 0 0 0 0 -1 2 -1 0 0 0 0 0 0 0 0 0 0 0 0
          0 0 0 0 0 0 0 0 0 0 -1 2 -1 0 0 0 0 0 0 0 0 0 0 0
          0 0 0 0 0 0 0 0 0 0 0 -1 2 -1 0 0 0 0 0 0 0 0 0 0
          0 0 0 0 0 0 0 0 0 0 0 0 -1 2 -1 0 0 0 0 0 0 0 0 0
          0 0 0 0 0 0 0 0 0 0 0 0 0 -1 2 -1 0 0 0 0 0 0 0 0
          0 0 0 0 0 0 0 0 0 0 0 0 0 0 -1 2 -1 0 0 0 0 0 0 0
          0 0 0 0 0 0 0 0 0 0 0 0 0 0 0 -1 2 -1 0 0 0 0 0 0
          0 0 0 0 0 0 0 0 0 0 0 0 0 0 0 0 -1 2 -1 0 0 0 0 0
          0 0 0 0 0 0 0 0 0 0 0 0 0 0 0 0 0 -1 2 -1 0 0 0 0
          0 0 0 0 0 0 0 0 0 0 0 0 0 0 0 0 0 0 -1 2 -1 0 0 0
          0 0 0 0 0 0 0 0 0 0 0 0 0 0 0 0 0 0 0 -1 2 -1 0 0
          0 0 0 0 0 0 0 0 0 0 0 0 0 0 0 0 0 0 0 0 -1 2 -1 0
          0 0 0 0 0 0 0 0 0 0 0 0 0 0 0 0 0 0 0 0 0 -1 2 -1
          0 0 0 0 0 0 0 0 0 0 0 0 0 0 0 0 0 0 0 0 0 0 -1  1];
```

$$k_{\mathrm{Is}} = 2.0925\mathrm{e}+9\,\mathrm{N/m}$$

The correctness of the matrix $[k_{\mathrm{Is}}]$ is evidenced by the solution of the eigenvalue problem of the matrices $[k_{\mathrm{Is}}]$, $[m_{\mathrm{Is}}]$ and comparing with predominant period (2.4335 s $\sim$ = 2.4 s).

Damping matrix $[c_{\mathrm{Is}}]$ for the superstructure with damping ratio $\xi_s = 0.05$ are defined by the classical-Rayleigh mass $[m_{\mathrm{Is}}]$ and stiffness $[k_{\mathrm{Is}}]$ proportional damping as

$$[c_I]=c_{Is}\,[c_1+c_2 \quad -c_2$$
$$\ldots$$
$$-c_{i-1} \quad c_{i-1}+c_i \quad -c_i$$
$$\ldots$$
$$-c_{n-1} \quad c_{n-1}+c_n \quad -c_n$$
$$-c_n \quad c_n\,];$$

$c_{Is}=1.3375e+7$ Ns/m

FB building parameters. To be able to compare the SFSSI system to the fixed base (FB) system, An n = 24 storey FB building system with total mass equal to the SFSSI system total mass and with the same geometry and section will be examined.

FB building system's stiffness matrix $[k_{FB}]$ was defined depending to its total height 104 m (or storey number n = 24), mass matrix $[m_{FB}]$ and predominant period which roughly defined as

$$T_{\_FB\_supers} = 0.08 * (h * n)(3/4) = 2.4535\,\text{s}\,(\text{or}\,0.1n = 0.1\,24 = 2.4\,\text{s}),$$

and it is the same as the LCRB isolated building system' superstructure part:

$$[k_{FB}] = [k_{Is}]$$

Damping matrix $[c_{FB}]$ for the superstructure with damping ratio $\xi_s = 0.05$ are defined by the classical-Rayleigh mass $[m_{FB}]$ and $[k_{FB}]$ stiffness proportional damping and it is same with the LCRB isolated building system's superstructure part

$$[c_{FB}] = [c_{Is}]$$

Earthquake specifications. The earthquake ground motions selected for the study are illustrated in Table 1.2 with their specifications.

The response quantities of interest are the top and bearing behavior of the SFSSI system building, classically LCRB isolated building and FB building and are presented below in Table 1.3 and Figs. 1.17, 1.18, 1.19, 1.20, 1.21, 1.22, 1.23, 1.24 and 1.25 respectively from the earthquakes specified in Table 1.2. On the Figures resonance causes was marked by the red lines.

For all nine earthquake excitations (Figs. 1.17a, 1.18a, 1.19a, 1.20a, 1.21a, 1.22a, 1.23a, 1.24a and 1.25a) fixed base (FB) building's top (24-th storey) level three time (in Figs. 1.17b, 1.18b, 1.19b, 1.20b, 1.21b, 1.22b, 1.23b, 1.24b and 1.25b), classically LCRB isolated building base (Figs. 1.17c, 1.18c, 1.19c, 1.20c,

1.21c, 1.22c, 1.23c, 1.24c and 1.25c) and top (Figs. 1.17d, 1.18d, 1.19d, 1.20d, 1.21d, 1.22d, 1.23d, 1.24d and 1.25d) levels twice in average were caused by resonant vibrations. The SFSSI building base (Figs. 1.17e, 1.18e, 1.19e, 1.20e, 1.21e, 1.22e, 1.23e, 1.24e and 1.25e) and top (Figs. 1.17f, 1.18f, 1.19f, 1.20f, 1.21f, 1.22f, 1.23f, 1.24f and 1.25f) levels is not caused resonant vibrations.

To withstand the same strong earthquake (Fig. 1.17a) excitation, classically isolated LCRB building's base isolator dissipates energy more than twice (Fig. 1.17h) in comparison with SFSSI system (Fig. 1.17). Base isolator displacement in classically LCRB isolated building is mostly more than SFSSI building's base displacement (Table 1.2). Lower deformation than the LCRB isolator of the SFSSI building is due to being protected from the brake or buckle damages the laminated rubber layer of LCRB isolators compering with the classically LCRB isolated building LCRB isolators. These particulars indicate that the base isolators of the SFSSI building remain on the safer side against both major and long-period earthquakes.

As can be seen from Ls-Dyna software modeling results of the SFSSI building under Kobe/Japan, 1955 earthquake (Appendix, Figs. 1.26, 1.27, 1.28, 1.29, 1.30, 1.31, 1.32, 1.33 and 1.34): 1) The normal and bending moment forces' orders approximately are same; 2) The torsional moment's order compeering with bending moment and shear forces, two order is lower; 3) Averaged absolute maximum of the base and top levels accelerations was decreased 87% and 67% reactively (Fig. 1.32); 4) Averaged absolute maximum of the base and top levels velocities under 0.51 m/s (Fig. 1.33) and this velocity is much lower than the velocity limit (0.7^{-1} m/s) that can

Table 1.2 Earthquake ground motion specifications

N	Earthquake name, year, station name, horizontal acceleration record Id	PGA (m/s^2)	PGV (m/s)	PGD (m)	T_d (s) duration
1	Kobe/Japan, 1995, KJMA RSN1106_KOBE_KJM000.AT2	8.1838	0.9111	0.2108	49.98
2	Darfield (a)/New Zealand, 2010, Christchurch, RSN6889_DARFIELD_CHHCN01 W.AT2	2.0541	0.672	0.5992	110
3	Duzce/Turkey, 1999, Bolu, RSN1602_DUZCE_BOL000.AT2	7.2520	0.5593	0.2559	56
4	Tottori/Japan, 2000, Tottori, RSN3942_TOTTORI_SMN010NS.AT2	0.182	0.0363	0.0426	120
5	Imperial Valley-02, California/USA, 1940, El Centro Array #9 RSN6_IMPVALL.I_I-ELC270.AT2	2.0674	0.3133	0.2416	54
6	El Centro/USA, 1940, California, elcentro_NS	3.1276	0.3615	0.2135	32
7	Darfield(b)/New Zealand, 2010, Christchurch Cathedral College, RSN6888_DARFIELD_CCCCN26 W.AT2	1.9079	0.5917	0.4909	137
8	El Mayor-Cucapah/Mexico, 2010, Chihuahua, RSN5823_SIERRA.MEX_CHI000.AT2	2.4377	0.3835	0.4823	130
9	Kocaeli/Turkey, 1999, Bursa Sivil, RSN1154_KOCAELI_BSI090.AT2	0.4415	0.0811	0.0355	141

Table 1.3 Response of the fixed base (FB) building, classically LCRB isolated building and the SFSSI system building and resonance states at the base and top levels respectively from the earthquakes specified in Table 1.2

Maximum response	Isolation type			Isolation type			Isolation type		
	FB	LCRB	SFSSI	FB	LCRB	SFSSI	FB	LCRB	SFSSI
Earthquake no.	#1			# 2			#3		
a									
u_b		0.1678	0.0517		0.5461	0.2124		0.2022	0.1028
$u_t + u_{0t}$	0.5365	0.3354	0.3595	0.9312	0.8407	0.8102	0.4091	0.3142	0.4087
Resonanse state: Yes +/No−	3+	2+	–	3+	2+	–	3+	2+	–
Earthquake no.	#4			#5			#6		
b									
u_b		0.0288	0.0184		0.1539	0.0509		0.169	0.055
$u_t + u_{0t}$	0.0154	0.044	0.0607	0.2557	0.2352	02806	0.3412	0.2286	0.2565
Resonanse state: Yes +/No−	3+	2+	–	2+	3+	–	1+	1+	–
Earthquake no.	#7			#8			#9		
c									
u_b		0.4739	0.1844		0.2656	0.3883		0.0362	0.0166
$u_t + u_{0t}$	0.9361	0.7435	0.77	0.2623	0.4068	1.3682	0.0569	0.0538	0.0783
Resonanse state: Yes +/No−	5+	3+	–	3+	1+	–	4+	2+	–

Fig. 1.17 Response of the fixed base (FB) building, classically LCRB isolated building and the SFSSI building under Kobe/Japan, 1955 earthquake **a** Kobe/Japan, 1995 earthquake acceleration **b** FB building top floor displacement **c** LCRB isolated building bearing displacement **d** LCRB isolated building top floor displacement **e** SFSSI building bearing displacement **f** SFSSI building top floor displacement **g** Force-deformation behavior of the SFSSI building's base LCRB isolator for hysteretic model expressed by Eqs. (18, 19, 29) **h** Force-deformation behavior of the LCRB isolated building's base LCRB isolator for hysteretic model expressed by Eqs. (18, 19, 29)

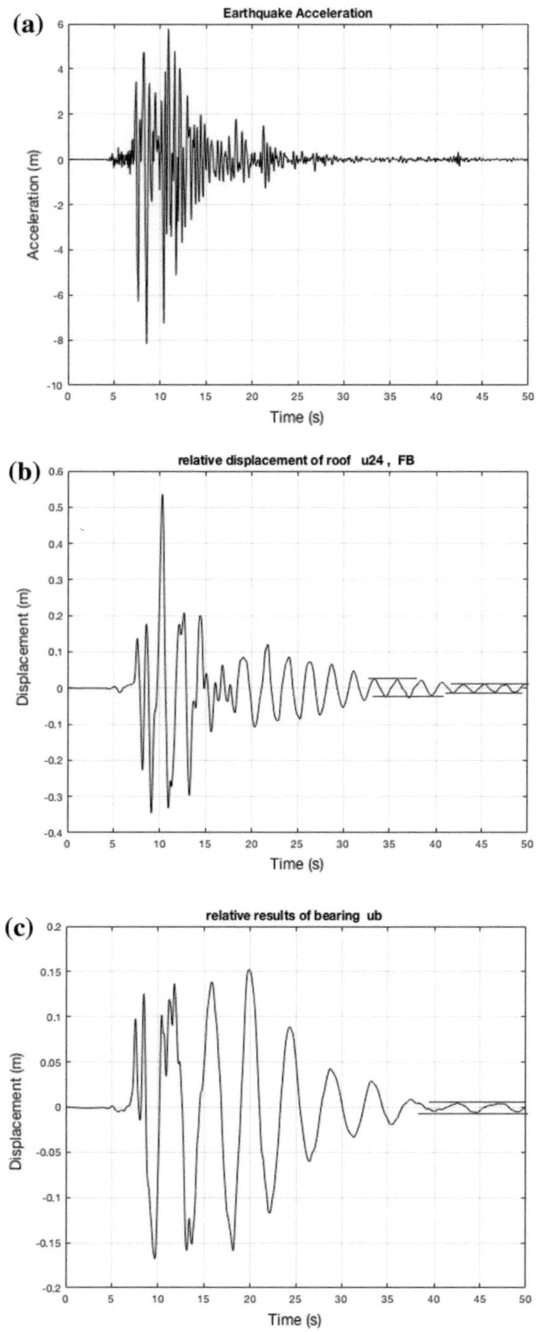

Fig. 1.17 (continued)

(d)

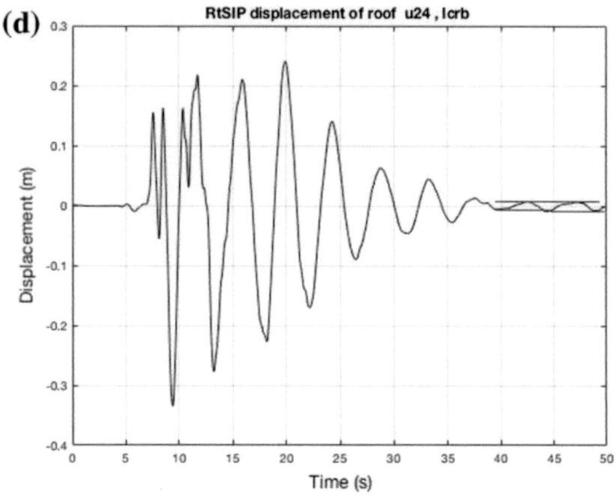

(e)

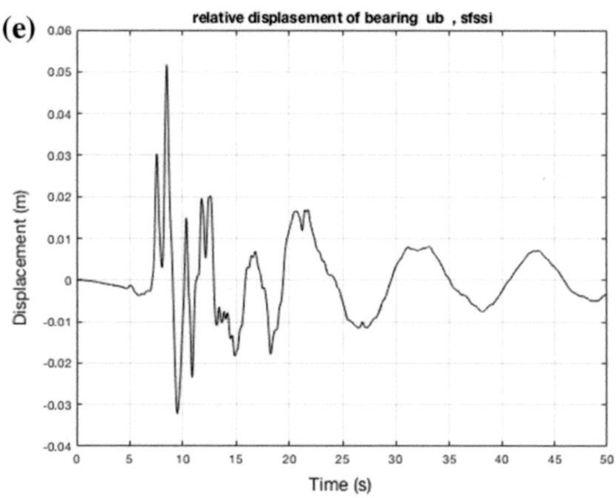

(f)

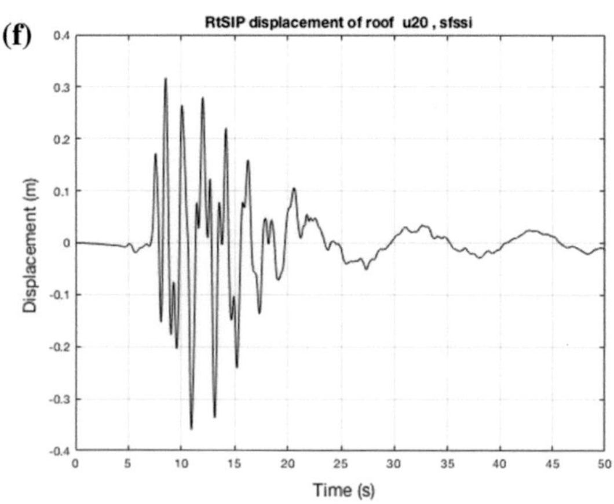

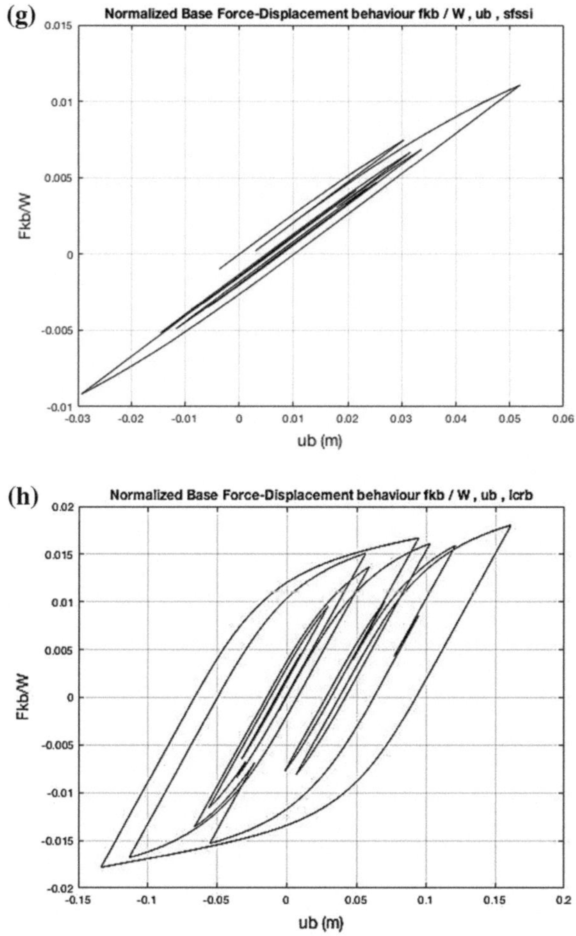

Fig. 1.17 (continued)

damage sensitive equipment involved inside of the building; 5) Averaged absolute maximum of the base and top levels displacements (Fig. 1.34) under acceptable limits for such type building structures. Obtained this results allows the SFSSI building to reduce the cross-sections of the current elements, compared with the fixed base (FB) and classically LCRB isolated buildings. In this study, the cross sections and characteristics of the current system for SFSSI, FB and conventional LCRB isolated structures were taken same for possibility its comparison.

The isolator damping values calculated from the decay of the vibration (response) displacements of the bearings, illustrate low damping of SFSSI building's LCRB isolators compared to the classically LCRB isolated building's LCRB isolators.

By the installing dampers SFSSI building's parameters may be improved and system reliability can be increased even more.

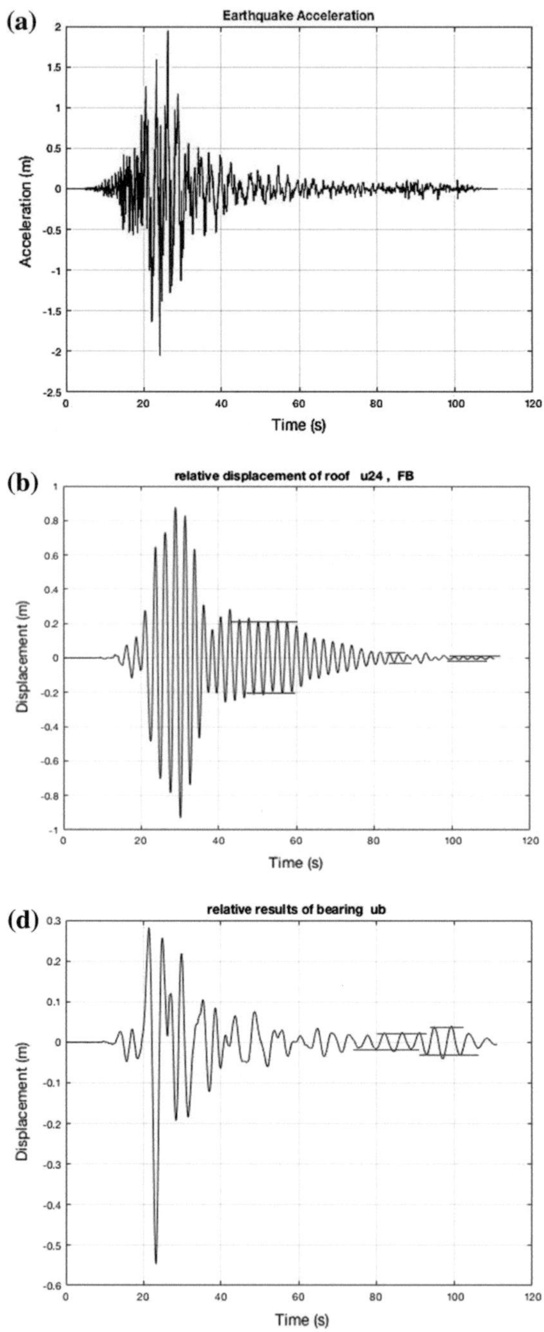

Fig. 1.18 Response of the fixed base (FB) building, classically LCRB isolated building and the SFSSI building under Darfield (a)/New Zealand, 2010 earthquake **a** Darfield(a)/New Zealand, 2010 earthquake acceleration **b** FB building top floor displacement **c** LCRB isolated building bearing displacement **d** LCRB isolated building top floor displacement **e** SFSSI building bearing displacement **f** SFSSI building top floor displacement

(d)

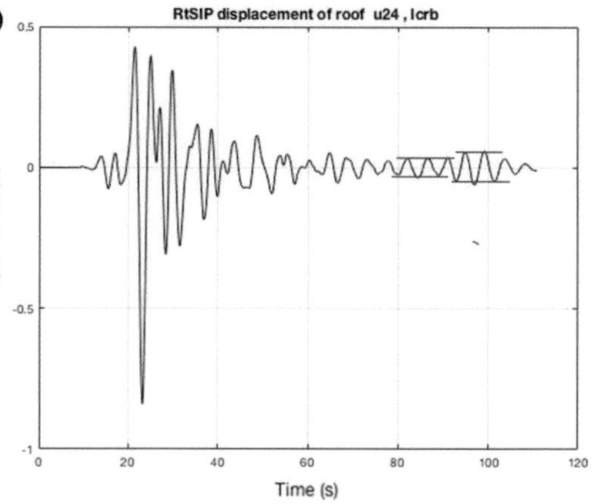

(e)

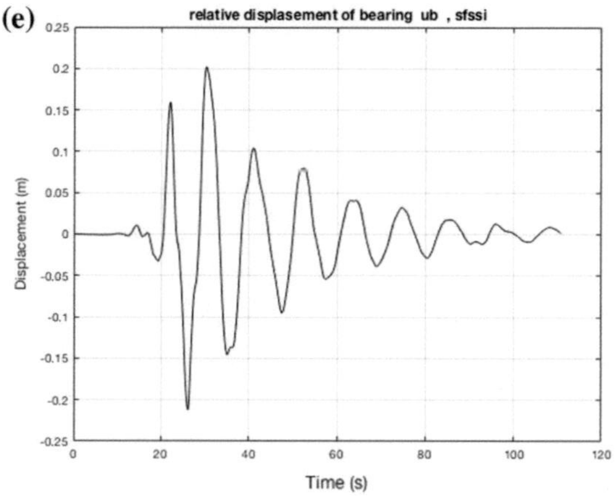

(f)

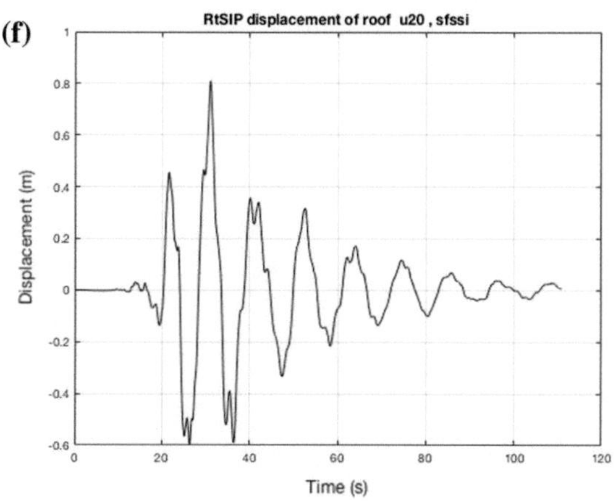

Fig. 1.18 (continued)

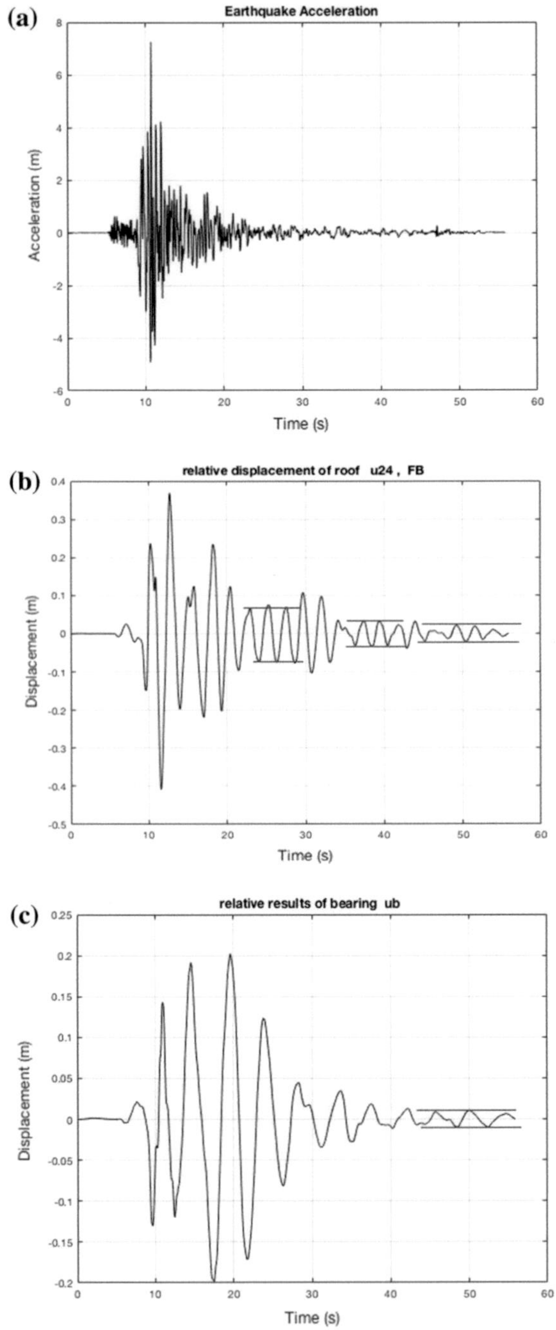

Fig. 1.19 Response of the fixed base (FB) building, classically LCRB isolated building and the SFSSI building under Duzce/Turkey, 1999 earthquake **a** Duzce/Turkey, 1999 earthquake acceleration **b** FB building top floor displacement **c** LCRB isolated building bearing displacement **d** LCRB isolated building top floor displacement **e** SFSSI building bearing displacement **f** SFSSI building top floor displacement

(d)

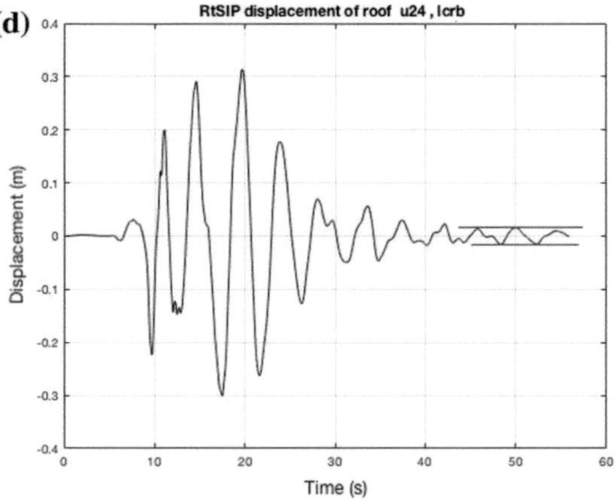

(e)

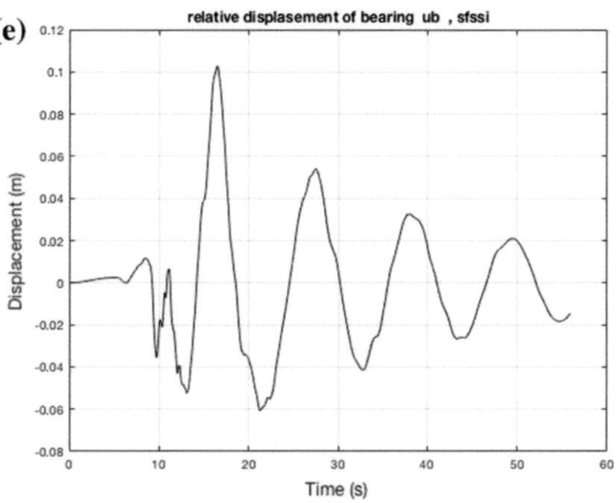

(f)

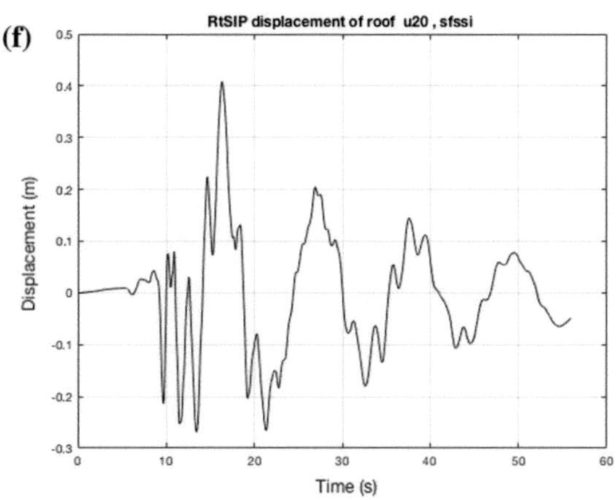

Fig. 1.19 (continued)

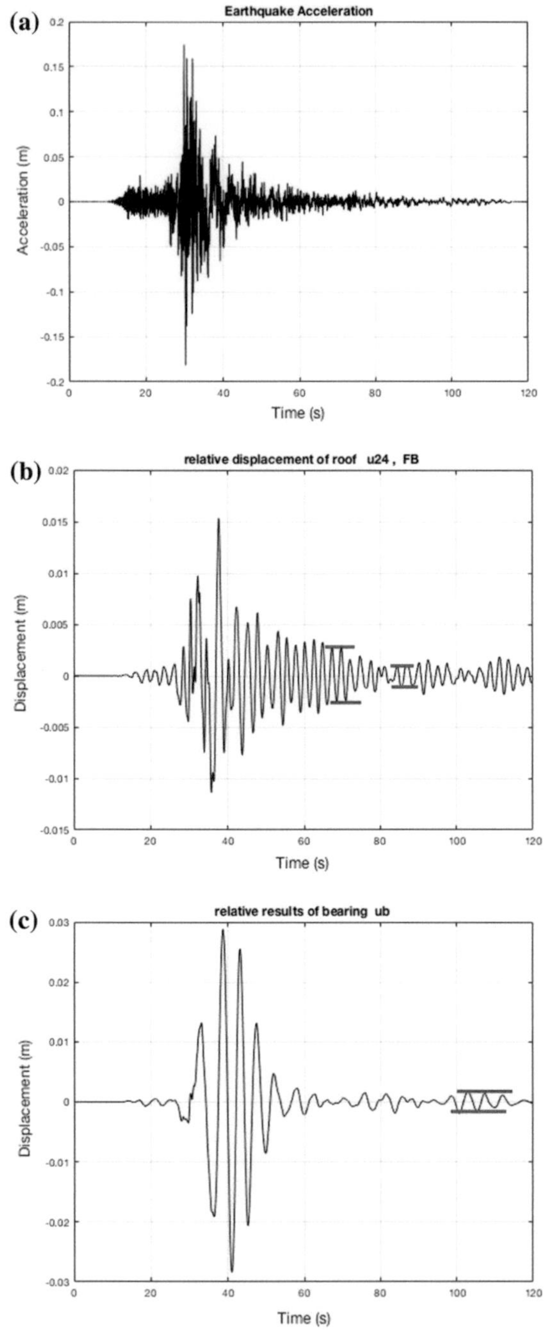

Fig. 1.20 Response of the fixed base (FB) building, classically LCRB isolated building and the SFSSI building under Tottori/Japan, 2000 earthquake **a** Tottori/Japan, 2000 earthquake acceleration **b** FB building top floor displacement **c** LCRB isolated building bearing displacement **d** LCRB isolated building top floor displacement **e** SFSSI building bearing displacement **f** SFSSI building top floor displacement

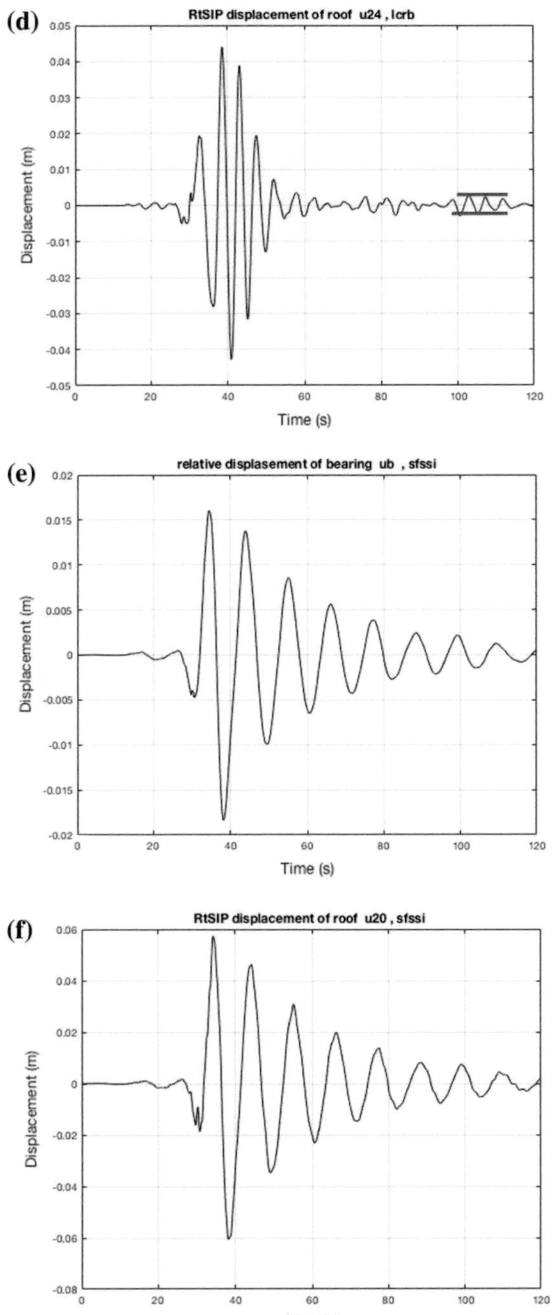

Fig. 1.20 (continued)

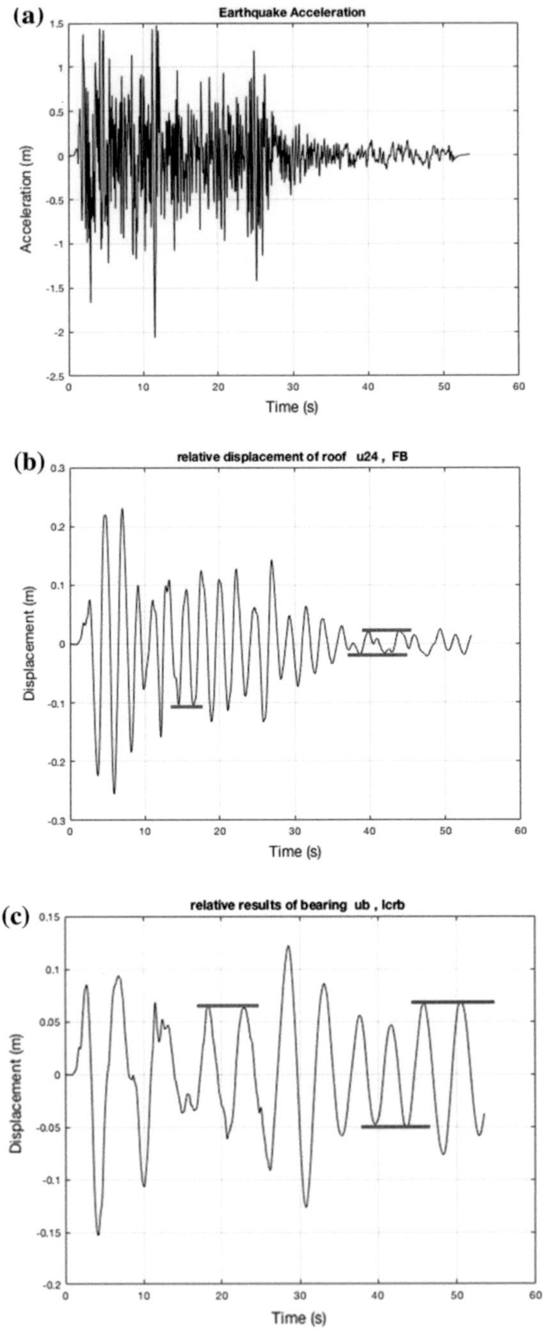

Fig. 1.21 Response of the fixed base (FB) building, classically LCRB isolated building and the SFSSI building under Imperial Valley-02, California/USA, 1940 earthquake **a** Imperial Valley-02, California/USA, 1940 earthquake acceleration **b** FB building top floor displacement **c** LCRB isolated building bearing displacement **d** LCRB isolated building top floor displacement **e** SFSSI building bearing displacement **f** SFSSI building top floor displacement

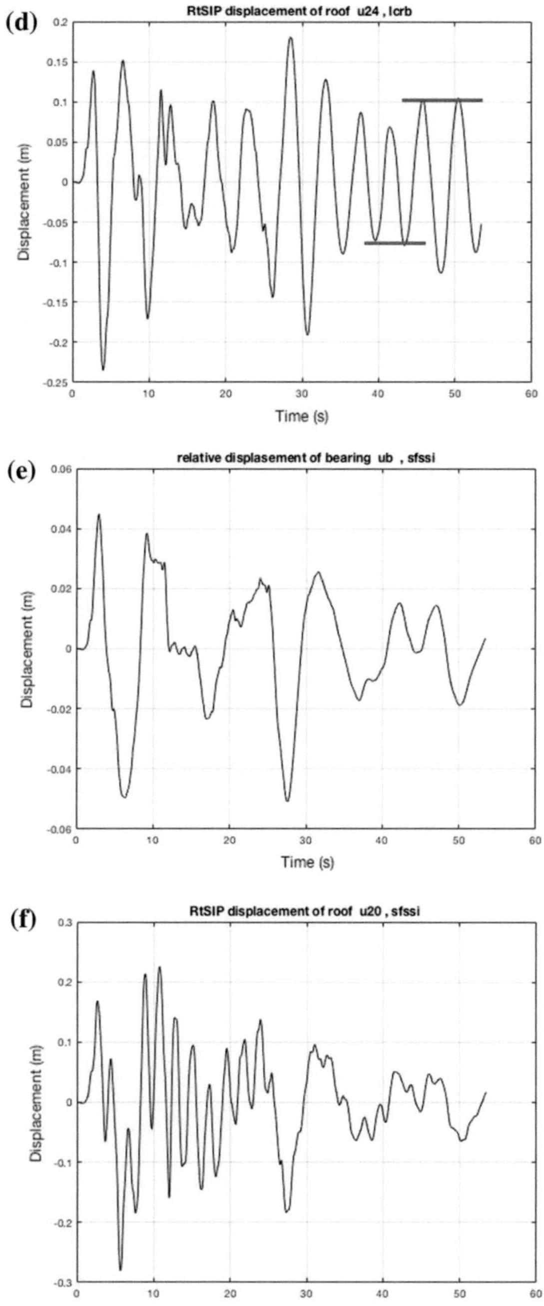

Fig. 1.21 (continued)

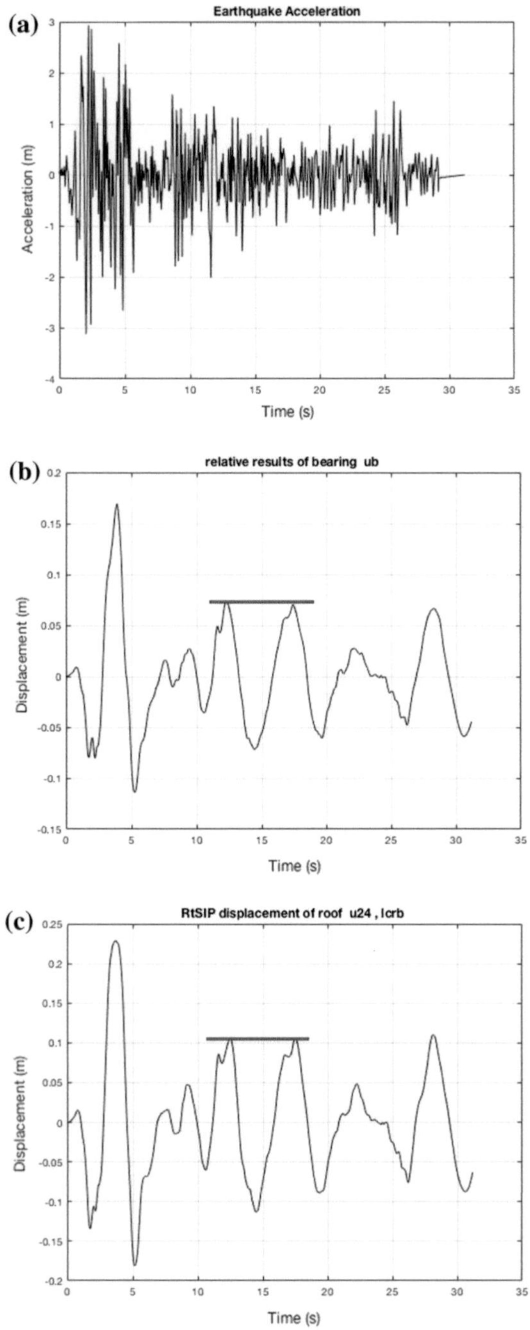

Fig. 1.22 Response of the fixed base (FB) building, classically LCRB isolated building and the SFSSI building under El Centro/USA, 1940 earthquake **a** El Centro/USA, 1940 earthquake acceleration **b** LCRB isolated building bearing displacement **c** LCRB isolated building top floor displacement **d** SFSSI building bearing displacement **e** SFSSI building top floor displacement

(d)
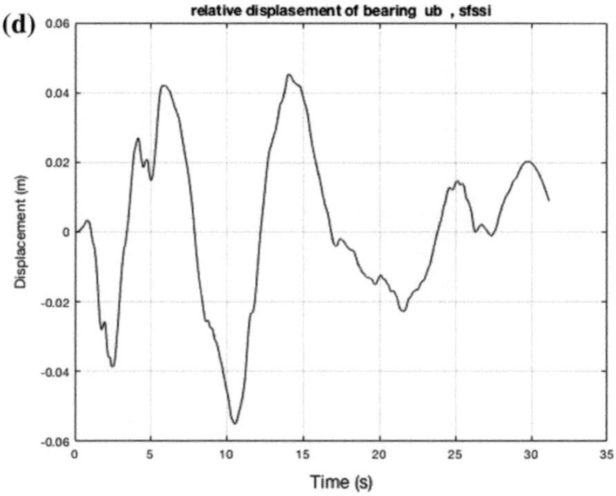

(e)
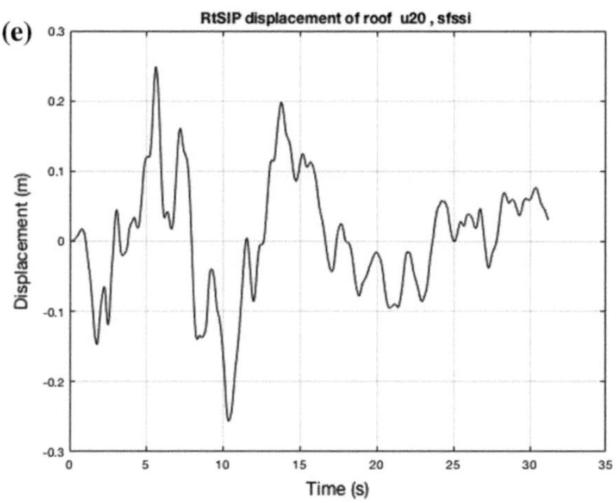

Fig. 1.22 (continued)

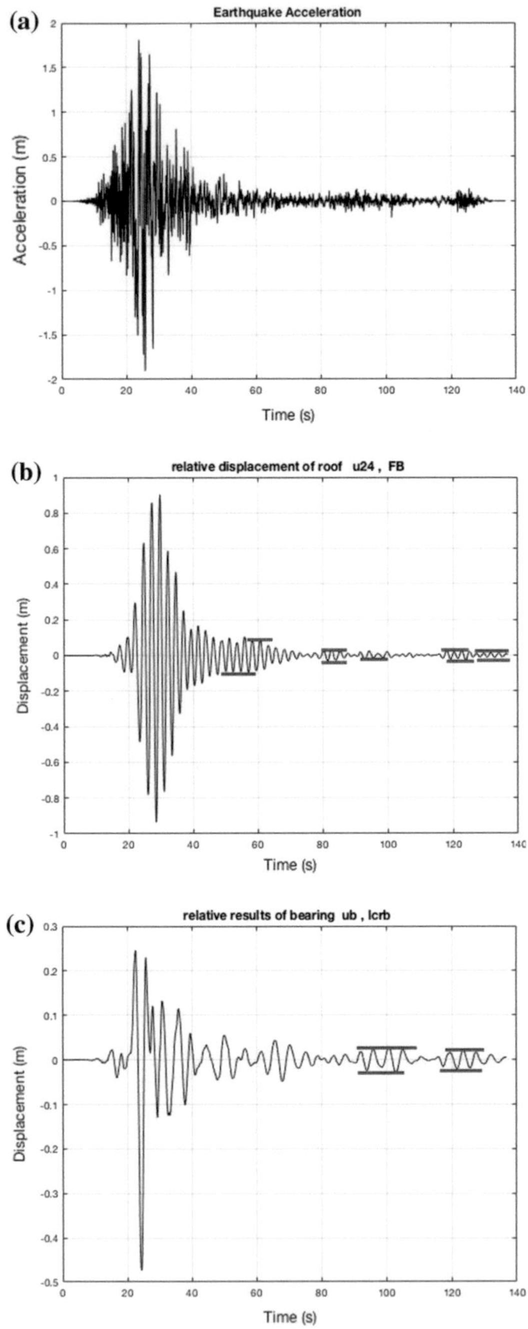

Fig. 1.23 Response of the fixed base (FB) building, classically LCRB isolated building and the SFSSI building under Darfield(b)/New Zealand, 2010 earthquake **a** Darfield(b), 2010 earthquake acceleration **b** FB building top floor displacement **c** LCRB isolated building bearing displacement **d** LCRB isolated building top floor displacement **e** SFSSI building bearing displacement **f** SFSSI building top floor displacement

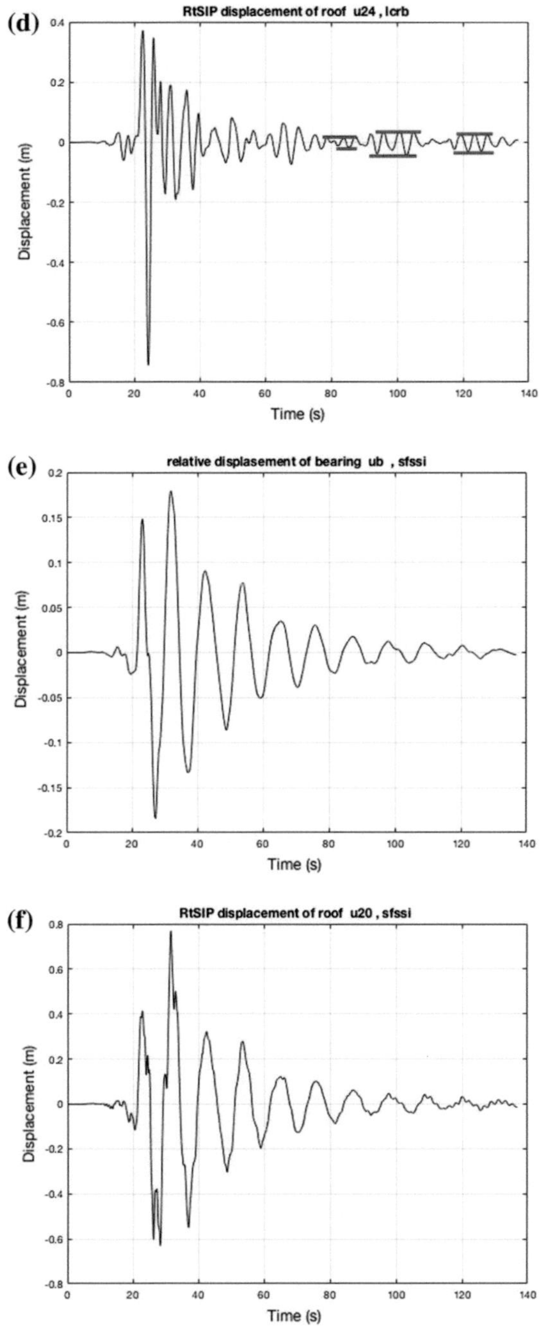

Fig. 1.23 (continued)

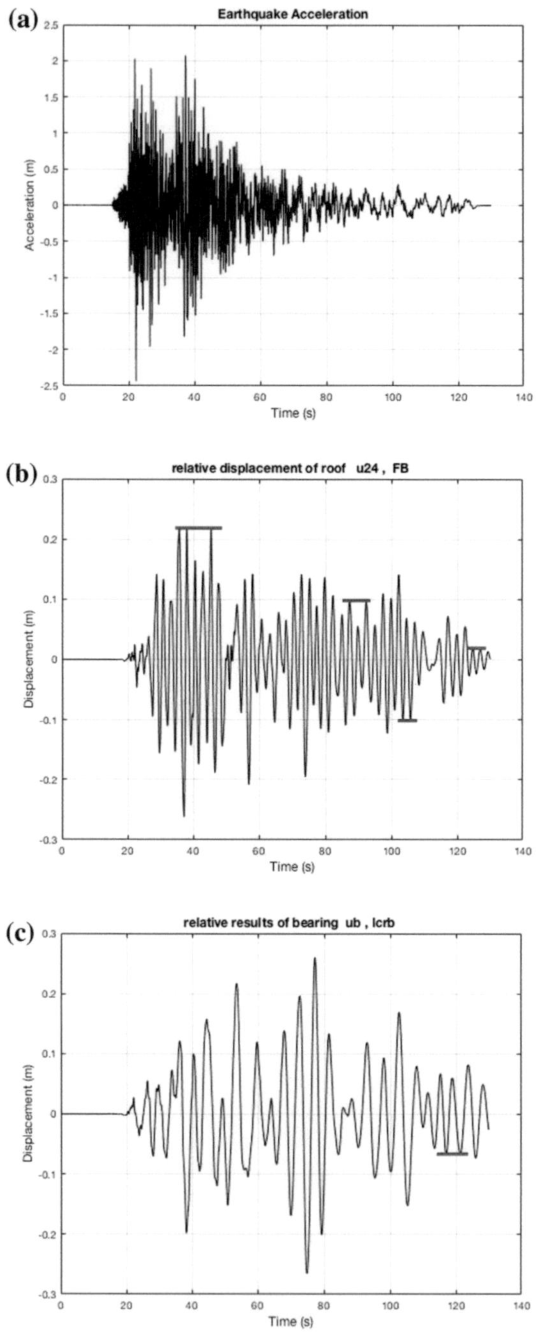

Fig. 1.24 Response of the fixed base (FB) building, classically LCRB isolated building and the SFSSI building under El Mayor-Cucapah/Mexico, 2010 earthquake **a** El Mayor-Cucapah/Mexico, 2010 earthquake acceleration **b** FB building top floor displacement **c** LCRB isolated building bearing displacement **d** LCRB isolated building top floor displacement **e** SFSSI building bearing displacement **f** SFSSI building top floor displacement

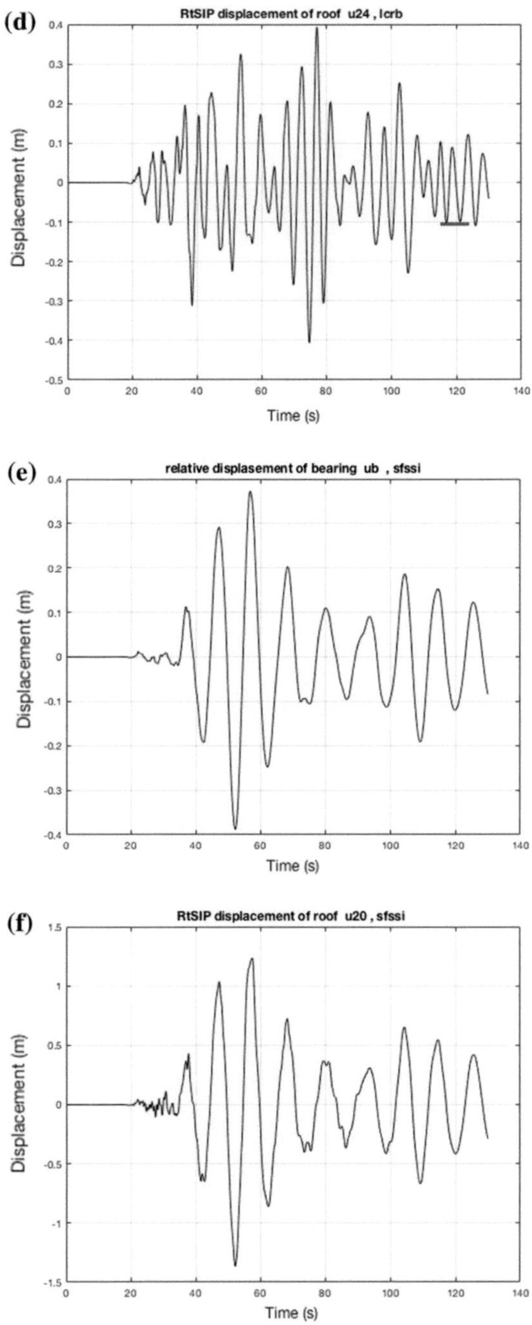

Fig. 1.24 (continued)

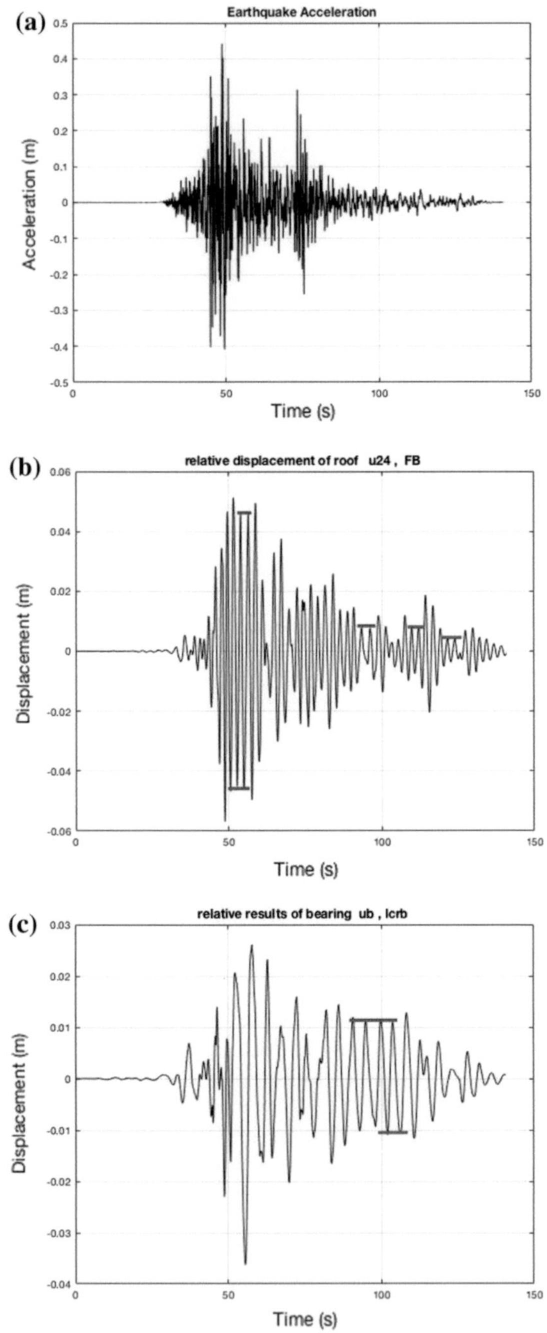

Fig. 1.25 Response of the fixed base (FB) building, classically LCRB isolated building and the SFSSI building under Kocaeli/Turkey, 1999 earthquake **a** Kocaeli/Turkey, 1999 earthquake acceleration **b** FB building top floor displacement **c** LCRB isolated building bearing displacement **d** LCRB isolated building top floor displacement **e** SFSSI building bearing displacement **f** SFSSI building top floor displacement

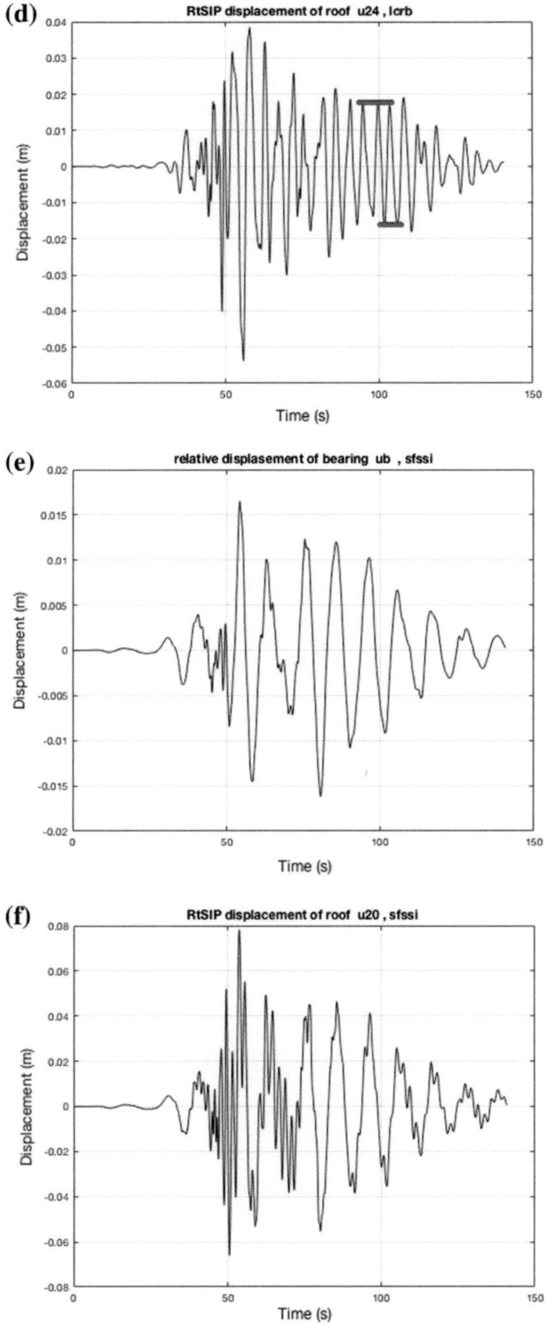

Fig. 1.25 (continued)

As mentioned in the introduction in summary, existing design tools for seismic isolated structures are bounded with a period of 6 s. LCRB isolators are produced with a period of under 5 s. Long wave period earthquakes between range 1–15 s and more. Under long-period earthquakes in the isolation layers excessive deformation occurs and buildings seismically isolated with period-dependent isolation devices (LCRB, LRB and an others) may lose safety or the functionality.

The presented SFSSI system's dominant period is adjustable and must be kept under mentioned above limits.

Under projection above information and exhibited performance of the presented example of SFSSI building, the SFSSI system is a good option for reliable seismic protection of the structures.

1.9 Conclusions

Presented new structural seismic isolation—SFSSI system remain on the safer side against both major and long-period earthquakes compering with fixed base(FB) and classically LCRB isolated buildings and it is without resonance.

Presented SFSSI system's dominant period is adjustable and must be kept under mentioned above(produced today LCRB isolators period 5 s; long wave period earthquakes periods 1–15 s) limits.

On the other hand along with the same number of LCRB isolators SSFSI system displays a more secure behavior than the rather traditional LCRB isolated buildings.

Having the same mass and number of isolators a 3% more total building usable areas and 8% more number of storeys, as is presented in the example under study.

Obtained stress-strain state results (Appendix 1) give the opportunity to design cross sections of the SFSSI building system much lower than existing ones, compared with the fixed base (FB) and classically LCRB isolated buildings (In this study, the cross sections and characteristics of the current system for SFSSI, FB and conventional LCRB isolated structures were taken same, only for possibility its comparison).

As a consequence of the information and exhibited performance of the presented example of SFSSI building, the SFSSI system is a good option for reliable seismic protection of the structures and it is a novel type and progression of the earthquake-resistant structures.

Acknowledgements This study is supported by Scientific and Technological Research Council of Turkey (TUBITAK). The support is gratefully acknowledged.

Appendix

(Appendix, Figs. 1.26, 1.27, 1.28, 1.29, 1.30, 1.31, 1.32, 1.33 and 1.34): 1). Ls-Dyna software modeling results of the SFSSI building under Kobe/Japan, 1955 earthquake for time 8.73 s (Fig. 1.26) Axial forces and its details (Fig. 1.27) Torsional forces and its details (Fig. 1.28) Bending moments in section S direction and its details (Fig. 1.29) Bending moments in section T direction and its details (Fig. 1.30) Shear forces in section S direction and its details (Fig. 1.31) Shear forces in section T direction and its details (Fig. 1.32) Averaged foundation (red), base level (green) and top level (navy blue) accelerations (Fig. 1.33) Averaged foundation (red), base level (green) and top level (navy blue) velocities (Fig. 1.34) Averaged foundation (red), base level (green) and top level (navy blue) displacements.

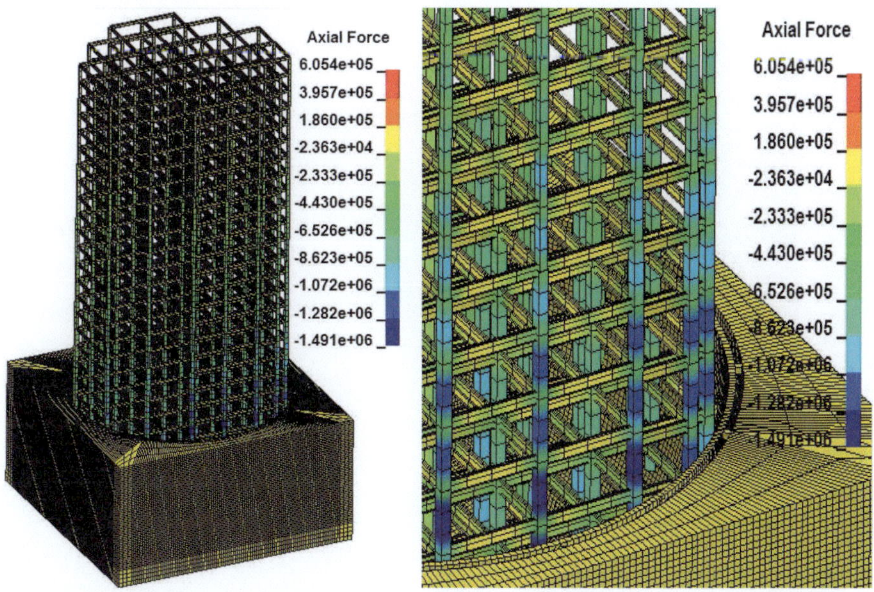

Fig. 1.26 Axial forces and its details

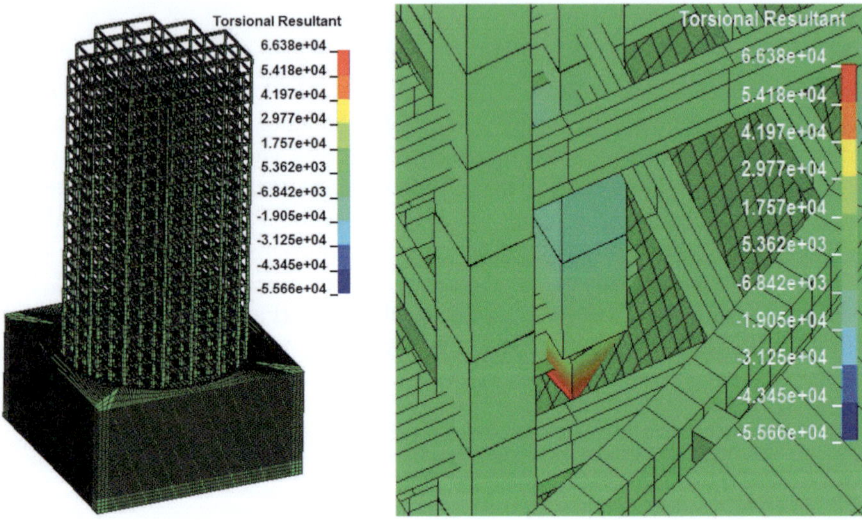

Fig. 1.27 Torsional forces and its details

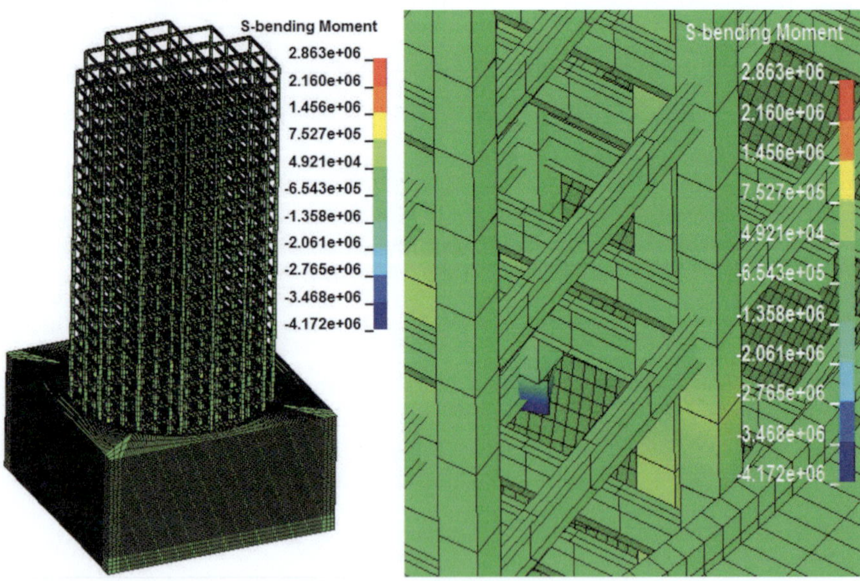

Fig. 1.28 Bending moments in section S direction and its details

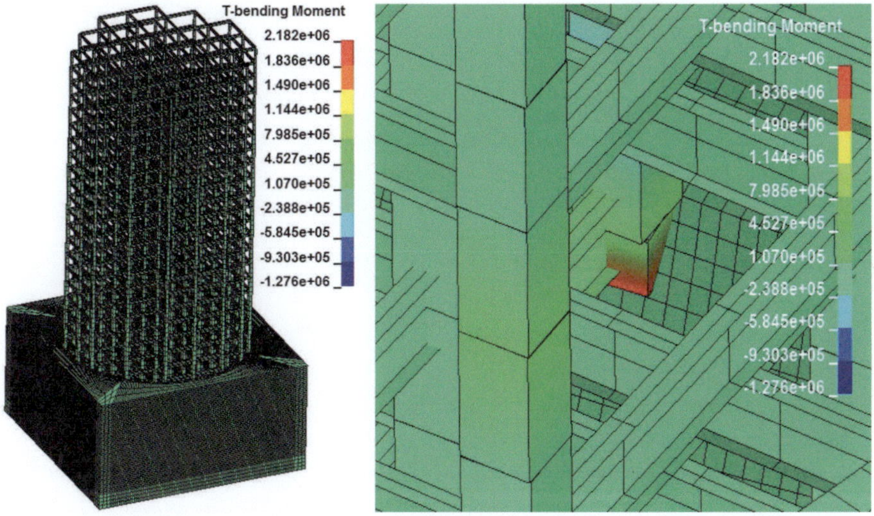

Fig. 1.29 Bending moments in section T direction and its details

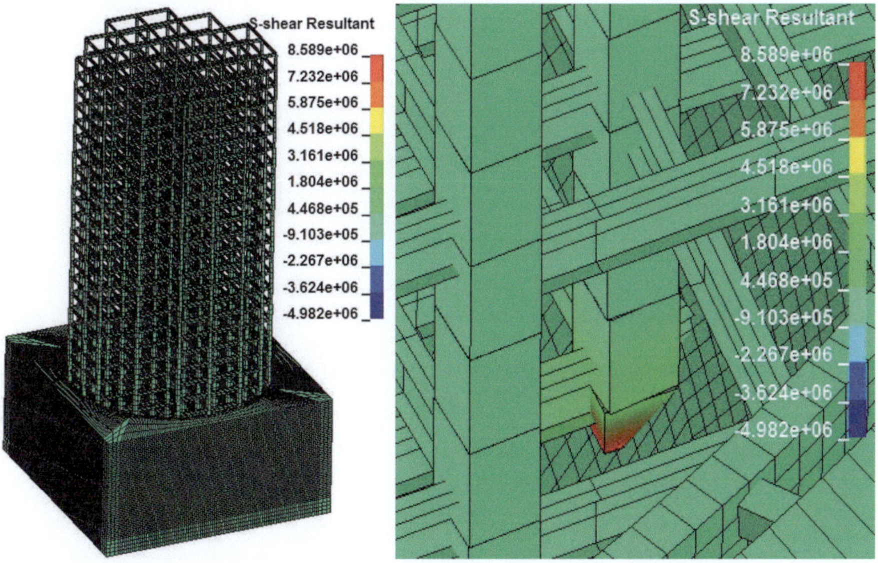

Fig. 1.30 Shear forces in section S direction and its details

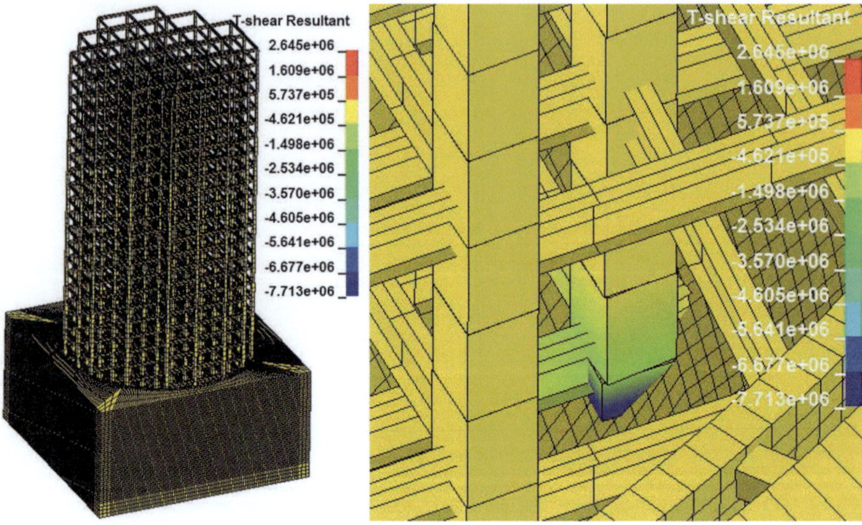

Fig. 1.31 Shear forces in section T direction and its details

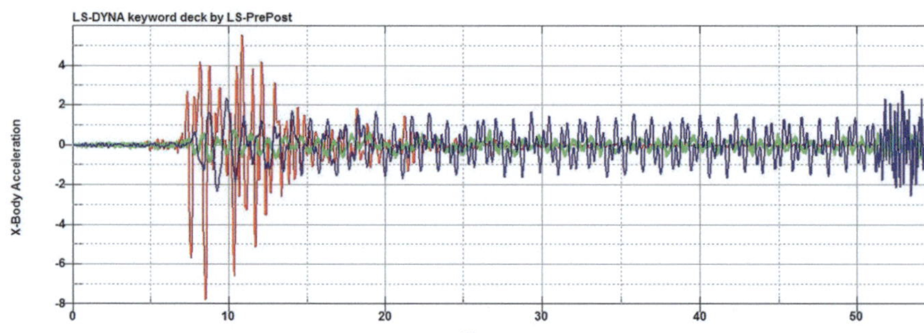

Fig. 1.32 Averaged foundation (red), base level (green) and top level (navy blue) accelerations

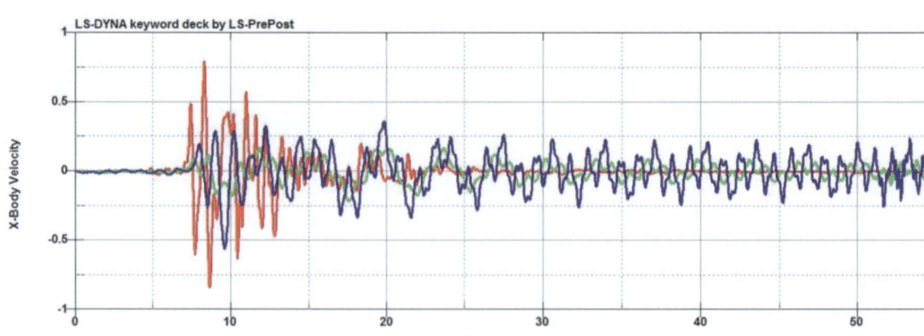

Fig. 1.33 Averaged foundation (red), base level (green) and top level (navy blue) velocities

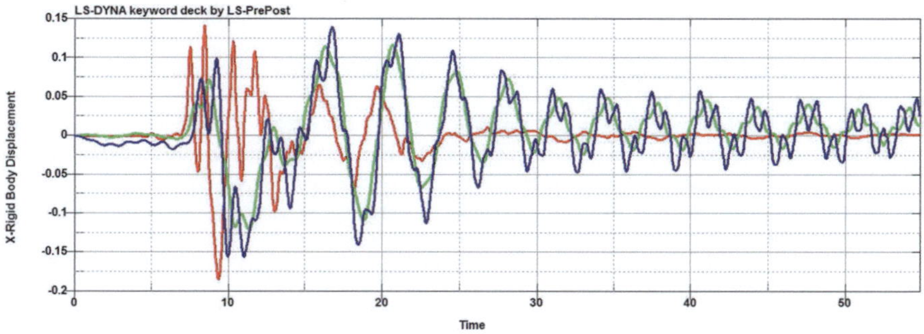

Fig. 1.34 Averaged foundation (red), base level (green) and top level (navy blue) displacements

References

Bonelli CP (2014) Surface ground motions, structural damage and emerging technologies in R/C buildings during and after the 2010 Mw 8.8 Maule earthquake. In: Proceedings of the 14th Japan earthquake engineering symposium, Osaka, Japan 2014

Butcher G (1988) The September 1985 Mexico earthquakes: final report of the New Zealand reconnaissance team. New Zealand Nat Soc Earthq Eng 21(1)

Celebi M, Sanli A (2002) GPS in pioneering dynamic monitoring of long period structures. Earthquake Spectra 18(1):47–61

Deb SK, Paul DK (1996) Simplified non-linear analysis of base isolated building. Eleventh World conference on earthquake engineering, 1344. ISBN 0 08 042822 3

Deb SK, Paul DK, Thakkar SK (1997) Simplified non-linear dynamic analysis of base isolated buildings subjected to general plane motion. Eng Comput 14(5):542–557

Estrada M (2014) Evaluation of damage and recovery process in Pisco City after the 2007 Pisco Earthquake. In: Proceedings of the 14th Japan earthquake engineering symposium, Osaka, Japan 2014

Eurocode 8 (2003) Design of structures for earthquake resistance, part I: general rules, seismic actions and rules for buildings. Pre-ENV 1998-1

FIP Industrial (2013) Development of the full scale prototypes of lead rubber bearings and experimental validation. European Commission 7th framework programme, 2007–2013, EP7-295485–SILER

Fujitani H, Kawabe H, Inoue N, Furuta T (2006) Response control of base-isolated building against strong ground motions of future Nanaki and Yamasaki fault earthquake (Part 1) Research objective and prediction of ground motions. Summaries of Technical papers of annual meeting, architectural Institute of Japan (in Japanese) B-2, pp 693–694

Fujitani H, Mukai Y, Tomizawa T, Hirata K, Mazuka Y, Fujii H (2012) Response reduction of base-isolation system against near-fault pulse and long-period ground motions. In: Proceedings of 15th world conference on earthquake engineering, Lisbon, Portugal 2012

Hall JF, Heaton TH, Halling MW, Wald DJ (1995) Near-source ground motion and its effects on flexible buildings. Earthquake Spectra 11:569–605

Heaton TH, Hall JF, Wald DJ, Halling MW (1995) Response of high-rise and base-isolated buildings to a hypothetical MW 7.0 blind thrust earthquake. Science 267:206–211

Ikhouane F, Rodellar J (2007) Systems with hysteresis analysis, identification and control using the Bouc-Wen model. Wiley Interscience

Jangid RS (2005) Optimum friction pendulum system for near fault motions. Eng Struct 27 (4):349–359

Jangid RS (2007) Optimum lead–rubber isolation bearings for near-fault motions. Eng Struct 29:2503–2513

Kasımzade AA (2002) Bounds of the structural response imprecisely-defined systems under earthquake action. PJAS 2(10):96

Kasımzade AA, Tachibana E, Mukai Y, Tuhta S, Atmaca G (2015a) Spherical foundation base isolation system on base ancient architecture inherence. In: International symposium on disaster simulation, DS'15, Osaka University, Japan, pp 127–133, 13–14 June 2015

Kasımzade AA, Tachibana E, Mukai Y, Tuhta S, Atmaca G (2015b) Spherical foundation structural seismic isolation system: development of the new type earthquake resistant structures. In: 6th international conference on theoretical and applied mechanics (TAM '15), Salerno, Italy, 27–29 June 2015

Kawabe H (2015) Long period ground motion simulation of huge plate boundary earthquakes. In: International symposium on disaster simulation (DS'15), Osaka University, Japan 2015

Lei Y, He MY (2013) Identification of the nonlinear properties of rubber- bearings in base isolated buildings with limited seismic response data. Sci China, Technol Sci 56(5):1224–1231

Li LY, Ou JP (2012) Seismic base isolation analysis for the control of structural nonlinear vibration, 15 WCEE, Lisboa

MATLAB R2017b Documentation (2017) The MathWorks Inc., Natick, MA. https://www.mathworks.com/help/matlab/index.html. Accessed 1 Dec 2017

Naeim F, Kelly JM (1999) Design of seismic isolated structures. Wiley, New York

Saito T (2014) Response of high-rise buildings to long period earthquake ground motions. In: Proceedings of the 14th Japan earthquake engineering symposium, Osaka, Japan 2014

Smolka A, Berz G (1988) The Mexican earthquake of September 1985: Damage statistics and implications for risk assessment. In: Proceedings of ninth world conference on earthquake engineering, Tokyo-Kyoto, Japan, 2–9 August 1988

Tanaka T, Yoshizawa S, Osawa Y (1979) Characteristics of strong earthquake ground motion in the period range from 1 to 15 seconds: analysis of low -magnification seismograph records. Bull Earthq Res Inst, Univ Tokyo 54:629–655

Taylor WA (2012) Response control systems in the United States and lessons learned from the Tohoku earthquake. In: Proceedings of the international symposium on engineering lessons learned from the 2011 great east Japan Earthquake, Tokyo, Japan, 1–4 March 2012, Tokyo

Warnitchai P (2014) Estimating the potential impacts of large distant earthquakes on tall buildings in Bangkok (Asian Institute of Technology). In: Proceedings of the 14th Japan earthquake engineering symposium, Osaka, Japan 2014

Wen YK (1976) Method for random vibratio of hysteretic systems. J Eng Mech (ASCE) 102:249–263

Zhigalin AD, Zavyalov AD, Mindel IG, Nikonov AA, Popova OG, Rogozhin EA, Ruzaikin AI, Sevostyanov VV (2013) The phenomenon of the sea of Okhotsk earthquake of 24 May 2013, in Moscow. Herald Russ Acad Sci 84(4):283–291

Chapter 2
Development of Resilient Seismic Response Control with a Semi-active System

Yoichi Mukai and Hideo Fujitani

Abstract An approach to develop resilient structures using seismic response control with a semi-active system is introduced. Actual system performance of control device installing to the practical building structure is evaluated with a RTHS (real-time hybrid simulation) test and an online simulator with the shaking table of Kobe University is used. Specifications of this online simulator system are explained in detail. As one example operating the RTHS tests, performance evaluation for the mid-story isolated building structure controlled by a semi-active device is focused. First, mechanical properties of the testing apparatus of the MR (magneto-rheological) rotary inertia mass damper used for semi-active control are quantified by the device tests, and then numerical simulation model of this damper is provided. Next, the RTHS test studies are operated and the test results are compared with the pure numerical simulation results by the fully modeled analyses. Validity of this online simulator and exactness of numerical model of the semi-active damper are confirmed. Moreover, reproducibility and reliability of the operation for the RTHS tests are evaluated by observing the displacement and the acceleration generated on the shaking table. As a result, it is assured that both the target displacement and acceleration of the shaking table were accurately reproduced under the real-time calculation in this online simulator. And it is confirmed that performance evaluations for the actual test specimen of the semi-active control device were also accurately done through the RTHS test method by using the shaking table.

Keywords Shaking table · Real-time hybrid simulation · Semi-active control
Mid-story isolated structure

Y. Mukai (✉) · H. Fujitani
Kobe University, Hyogo, Japan
e-mail: ymukai@port.kobe-u.ac.jp

A. A. Kasimzade et al. (eds.), *Seismic Isolation, Structural Health Monitoring, and Performance Based Seismic Design in Earthquake Engineering*,
https://doi.org/10.1007/978-3-319-93157-9_2

2.1 Introduction

Huge ground vibration of the Hyogo-ken Nambu Earthquake in 1995 in Japan had brought numerous damages and destruction in building structures and infrastructures, and it had killed or injured lots of people. On the other hand, this earthquake has triggered lots of construction of seismic isolation buildings after the earthquake. It is worried that near future's earthquake disasters could be caused by huge pulse-type ground motions of an inland earthquake or by long-period-type ground motions of an oceanic trench zone earthquake. Actually, the Nankai Trough Earthquake or the other undersea earthquakes are predicted to occur in quite near future in Japan (Architectural Institute of Japan 2007, 2013a; Taga et al. 2011). Practical concerns about seismic isolation buildings under these kinds of earthquakes are pointed out as follows: (1) Extra-large deformation of isolators as exceeding the design criteria due to these types of earthquakes must make the laminated rubber layer of isolators break or buckle, or must lead other impacting damage on bearing walls, elevators, stairways or pipes, etc., (2) Degradation of seismic isolation effects as being unable to reduce acceleration responses due to the continuation of the long-time vibration must cause equipment inside of buildings to move or topple, or must lose the safety or the functionality of the buildings (Architectural Institute of Japan 2013b, 2016).

As a general design solution, lots of dampers are installed to prevent excessive deformation of the isolation layer because dampers can work effectively for reducing deformation of isolation layer. However, the floor acceleration responses upper than the isolation layer are often increased as losing the isolation effects by dampers' actions. To overcome this problem, authors focus on introducing semi-active control systems on the isolation layer. A semi-active control is well-known system available to vary the vibration characteristics of the structural system by online operation as necessary. Some previous research results have reported effective performance improvements using semi-active controls to reduce deformation of seismic isolation layer without exciting accelerations (Fujii et al. 2007; Nagashima et al. 2010; Yoshida and Fujio 2001).

Recently, testing method with a "real-time hybrid simulation" (it is called as RTHS in the following) is actively adopted to evaluate a semi-active control system (Ito et al. 2011). Christenson has researched the semi-active vibration controlled structure equipping the MR (magneto-rheological) dampers in the work of the sub-theme research on NEES (the project for Network for Earthquake Engineering Simulation, which was funded by the NSF from 2004 to 2014) (Christenson et al. 2008), and Asai has performed the RTHS tests on semi-active control system for a seismic-isolated structure (Asai et al. 2015). Shao has operated the RTHS test by combination of the shaking table and the auxiliary actuators (Shao et al. 2011). In this study, the RTHS test is used for performance evaluation of the mid-story isolated structures installing a newly designed semi-active device. This semi-active device is designed as the rotary inertia mass damper encapsulating the MR fluid (Aoyama et al. 2014; Ito et al. 2015; Yoshida et al. 2016).

The RTHS test for the mid-story isolated structure is operated on the shaking table by providing the experimental part and the numerical part, respectively. The experimental parts are actually produced as the test specimen of the isolation layer and the upper structure, and the numerical part is simulated in a computer as the lumped mass system model for the lower structure. The RTHS test can always secure reproducing the actual behaviors for the "actual experimental parts". Thus, even though the experimental parts include the unclarified sub-system, the performance of the other experimental parts and the numerical part of the entire system can be exactly simulated, only to generate interactions at the boundary between the experimental part and simulation part. The interaction force for the MR rotary inertia mass damper is generated through the shaking table motion on the real time.

In the followings, the system component for the RTHS tests is explained in detail. Then, the testing model of the mid-story isolated building system is evaluated by the RTHS tests using the shaking table. Reliability of this simulator system and reproducibility of the shaking table is verified through this study.

2.2 Special Feature and Function of Shaking Table

A shaking table is one of key items on studies for developing and evaluating advanced vibration control systems. In Kobe University, the new shaking table was constructed in 2013 and has started its operations since 2014. The control system function to perform the RTHS tests has been provided in 2015.

2.2.1 Basic Performance of Shaking Table

Performance specifications of the Kobe University's shaking table are listed in Table 2.1. This shaking table provides allowable stroke of 550 mm in whole movable range available to operate tests for long period and long displacement seismic motions. At the same time, it is designed as to provide accurate reproducibility for various input motion targets containing both high-frequency and low-frequency vibrations. To actualize these requirements, as seen in Fig. 2.1, a unique mechanism using a dual driving system combining the electro-dynamic motor and the AC servo motor is installed for the Kobe University's shaking table.

Figure 2.2 depicts dimension of the shaking table and the whole view. As seen in this figure, the electro-dynamic motor is placed on the liner guide and its position can be displaced by the AC servo motor. Thus, long stroke motion of the shaking table is mostly generated by the AC servo motor drive and minute motion of the shaking table is compensated by the electro-dynamic motor drive at the same time. This cooperating mechanism with the dual driving system can reproduce accurately

Table 2.1 Specification of the shaking table in Kobe University

Detail		Capacity of performance
Maximum acceleration[a]	Harmonic[b]	10 m/s^2
	Random[c]	20 m/s^2
Maximum velocity[a]	Harmonic[b]	1.0 m/s
	Random[c]	1.5 m/s
Maximum stroke	Harmonic[b]	±275 mm (550 mm p-p)
	Random[c]	±275 mm (550 mm p-p)
Maximum driving force	Harmonic[b]	70 kN
	Random[c]	140 kN
Frequency range for driving		From 0.1 to 20 Hz
Allowable loading weight		10,000 kg
Movable direction		1 axis
Table size		3000 mm in vibrating direction, 2000 mm in cross direction

[a]In a case of placing mass of 5000 kg on the shaking table
[b]Harmonic means a case of continuous harmonic input motion
[c]Random means a case of random input like earthquake motions

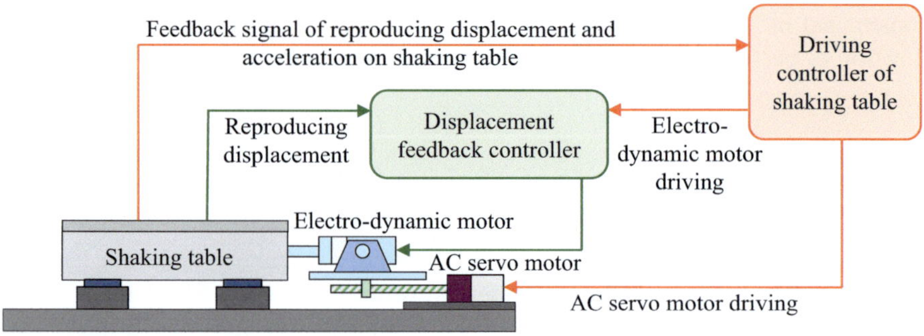

Fig. 2.1 Dual driving system of the shaking table

target motion on this shaking table even if the input stroke is long or short. The target motion is divided into two parts in the frequency domain, those two parts are low-frequency motion for the AC servo motor drive and high-frequency motion for the electro-dynamic motor drive. A band division filter for each drive is designed as shown in Fig. 2.3. The target input motion is reproduced by the superposition of two band-limited motions through a low pass filter (LPF) and a high pass filter (HPF).

Figure 2.4 shows reproducing responses on the shaking table. The target input is the full scale of "JMA-Kobe" wave record which was observed at the Kobe station of the JMA (Japan Meteorological Agency) during the Hyogo-ken Nambu

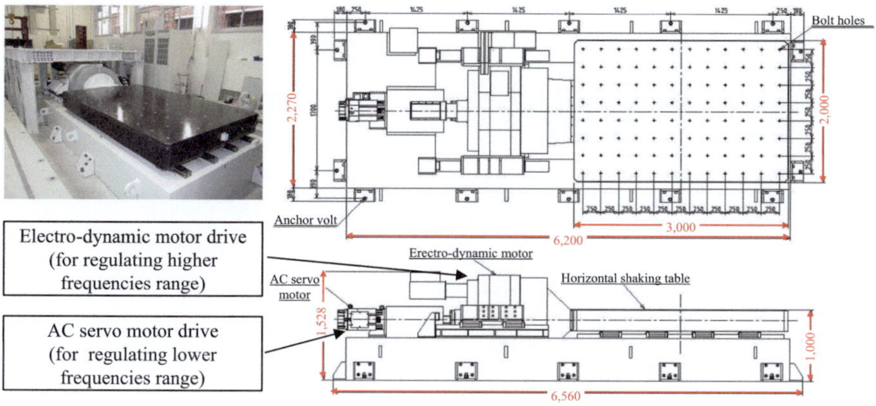

Fig. 2.2 Dimension of shaking table

Fig. 2.3 Band division filter for controlling dual motion drive

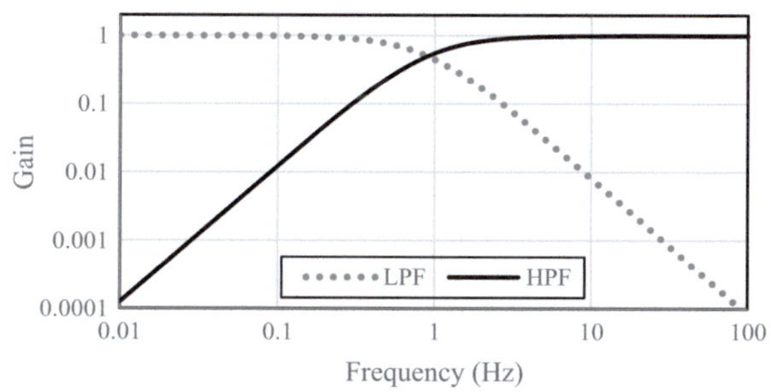

Earthquake in 1995 in Japan. The shaking table is driven in the condition of no weight placed on the shaking table (in which, motors need to bear 2000 kg weight according to the permanent load for the shaking table own part). In Fig. 2.4a, the reproduced displacements of the AC servo motor and the electro-dynamic motor are shown, and these motions are generated by using the band division filters mentioned in Fig. 2.3. Figure 2.4b, c, d show comparisons between the reproduced response and the target input for displacement, velocity and acceleration on the shaking table, respectively. Figure 2.4e shows errors of the reproduced acceleration from the target input. As seen these figures, it is considered that this shaking table can perform accurate reproducibility of the target input motion. Figure 2.5 shows the spectrum ratio (ratio of the transfer functions) of the reproduced acceleration response for the target acceleration input. As seen in this figure, it is observed that accurate reproducibility is gained in the frequency range from about 0.3 to about 10 Hz. On the other hand, in the high-frequency range over 10 Hz, reproducibility seems to decrease, and controllability cannot be gained in the higher range than about 20 Hz.

This shaking table can be used not only as a vibration generator but also as a dynamic load generator for the other kind of way to use. Namely, in general way to use as seen in Fig. 2.1, this shaking table can reproduce any requested vibration

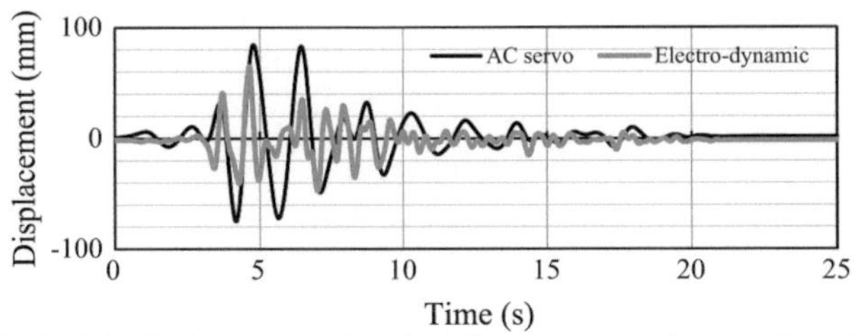

(**a**) Divided displacement motions for AC servo motor and erectro-dynamic motor.

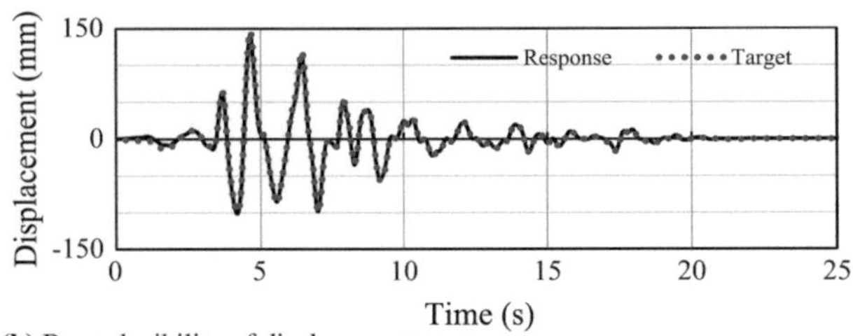

(**b**) Reproducibility of displacement.

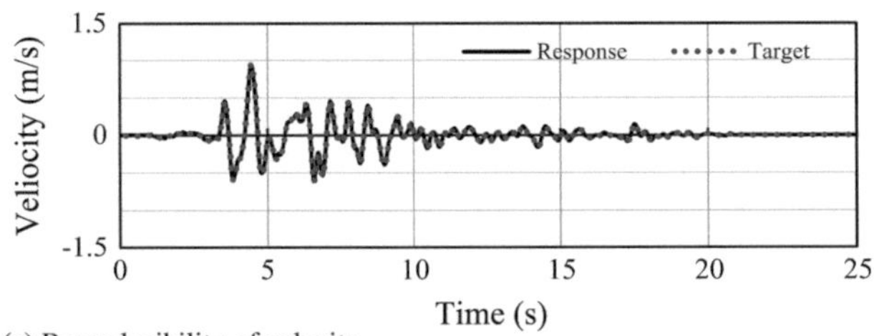

(**c**) Reproducibility of velocity.

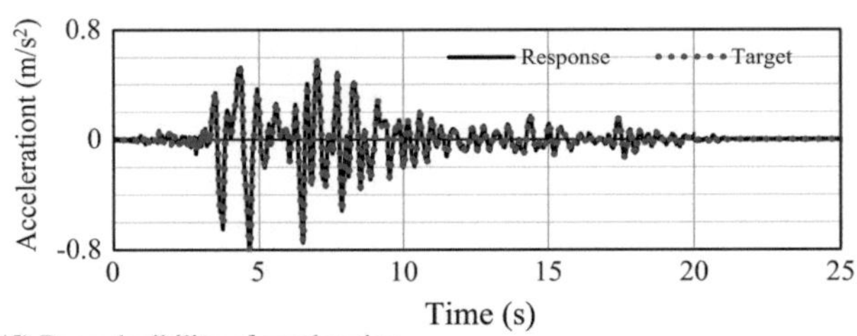

(**d**) Reproducibility of acceleration.

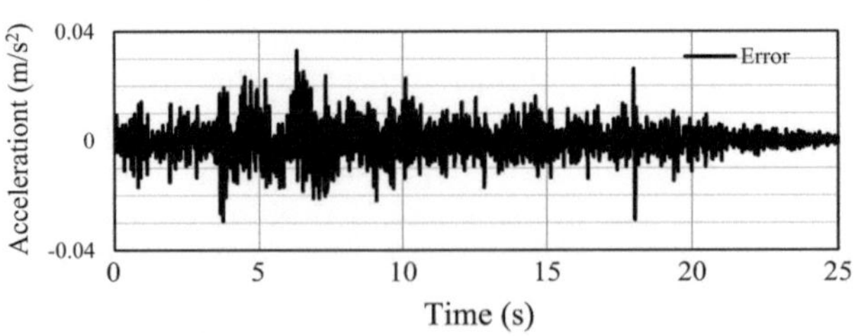

(**e**) Errors of acceleration response from the target input.

Fig. 2.4 Reproducing motion on the shaking table

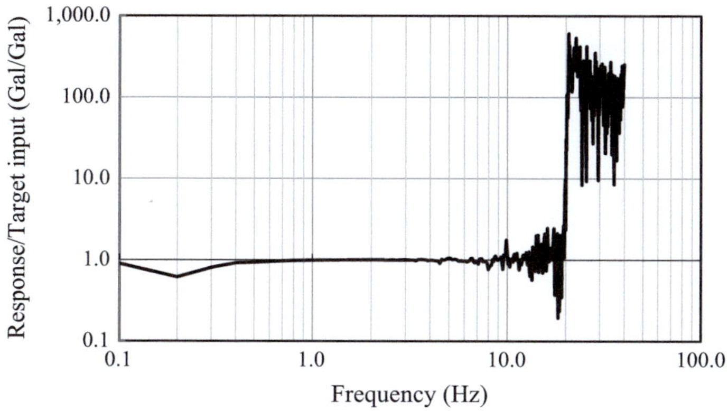

Fig. 2.5 Spectrum ratio of reproduced response for the target input

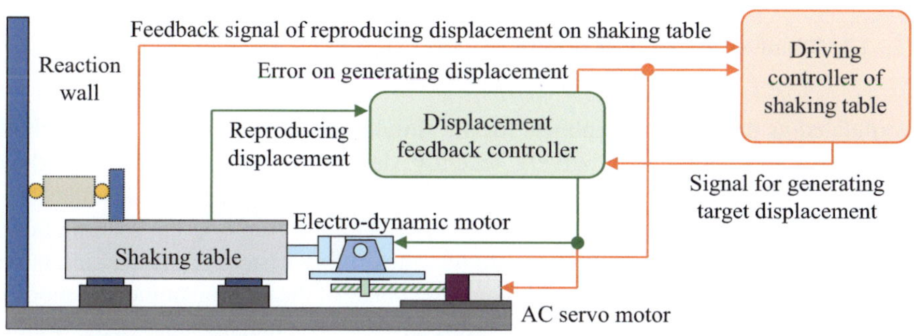

Fig. 2.6 Load generator mode of generating programmed displacement with shaking table and reaction wall

motion. On the other hand, it can be used as the compulsory deforming device by using a reaction wall. System composition and controlling flow for the load generator mode is depicted in Fig. 2.6.

2.2.2 *Function of Shaking Table as an Real-Time Hybrid Simulator*

The shaking table of Kobe University has been customized for executing various kinds of online tests by hybrid operation of experiments and simulations. This kind of operations is usually called "real-time hybrid simulations" (RTHS) test. The RTHS test is remarked as an effective way to operate online simulation for huge system by using only a specified part within the entire system. The sub-system to examine is provided as the actual testing specimen and the remaining parts are

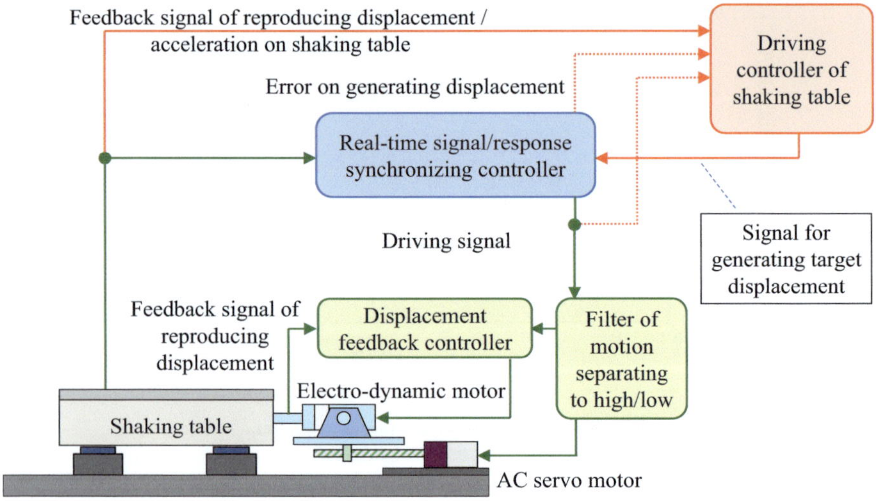

Fig. 2.7 Control system block of the shaking table to operate RTHS test

formulated as the numerical models for the simulations in a computer. This method for system evaluation is relatively recently developed but it has been widely introduced in the various field on the scientific and the engineering researches.

Figure 2.7 depicts the control system block diagram for driving the shaking table to operate RTHS tests. As seen in this figure, the driving system of the shaking table forms online simulator to calculate dynamic response of the internal numerical model. At the same time, this driving system can directly get the actual response of the testing specimen for the specified part in all the target system. And the numerical model's responses are recalculated while considering interactions with the test specimen's responses. The shaking table is designed to reproduce the interactive behavior of the test specimen as if it was set upon the practical entire system while synchronizing both responses of the test specimen and the numerical model in real time.

2.3 Feature of Semi-active Control Device and Test Specimen of Building Model

To aim for developing advanced structural control systems, a new type of semi-active control device with a unique mechanism and properties is proposed and a prototype specimen is actuary produced. In this study, a rotary inertia mass damper that can generate inertia mass effect is used and this damper also functions semi-active control device by using the property for MR (magneto-rheological) fluid encapsulating in the space around the flywheel. This new type of semi-active damper is called as "MR rotary inertia mass damper" (MR-RIM damper) in the followings.

2.3.1 Mechanical Design of Newly Proposed Semi-active Damper

Figure 2.8 depicts the structural and mechanical view of the MR-RIM damper. Figure 2.9 explains schematic diagrams of generating power with this damper according to its motion and supplying current to MR fluid. Figure 2.10 explains schematic diagrams of the load vs. deformation relationships corresponding to the motion at the part of rotary inertia mass and the part of encapsulating MR fluid. Table 2.2 lists the design specifications of the MR-RIM damper for this study. Table 2.3 lists the design specifications of the electro-magnets introduced as a mechanism for generating the magnetic field in the MR-RIM damper.

As seen in Fig. 2.8, rotational motion of the flywheel connected to the ball screw generates rotating inertia force. Ball screw and ball nut can change rotational motion of the flywheel into linear motion of the piston rod. Accordingly, rotating inertia force generated by the rotary inertia mass is proportionally transformed into the acceleration along the axial direction of the piston rod. At the same time, since shear velocity occurs resulting from the relative sliding motions between the flywheel and the baring case, damping force according to the resistant force, which is resulted from the shear flow of the MR fluid filling in the space around the flywheel, is generated. MR fluid encapsulated in the bearing case is placed in a magnetic field and electro-magnet mechanism can make strength of the magnetic field in the MR fluid variable. And then, the resistant force applied via MR fluid becomes controllable within a certain range and this property can be used for a semi-active control to adjust damping force as needed.

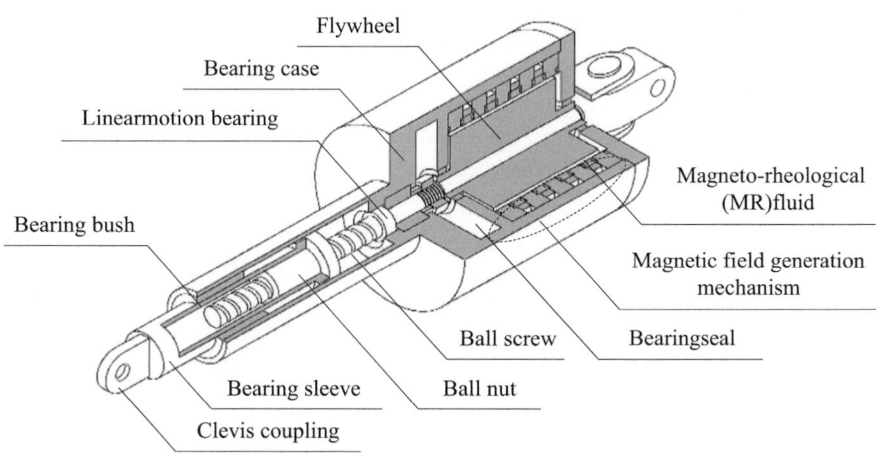

Fig. 2.8 Structural view and explanation of particle of the MR-RIM damper

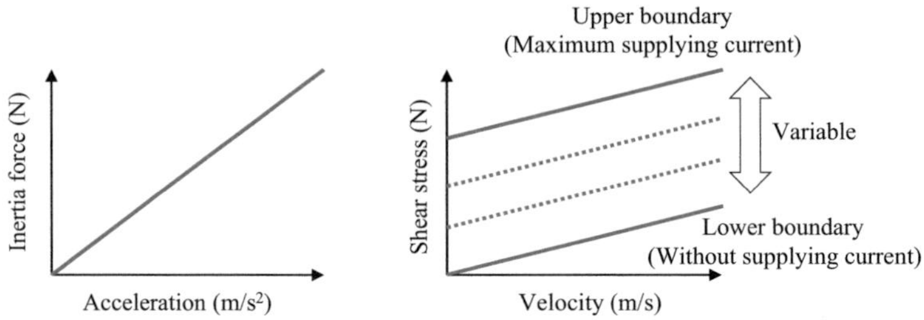

(a) Inertia force vs. acceleration. **(b)** Shear stress of MR fluid vs. velocity.

Fig. 2.9 Generating force in the MR-RIM damper

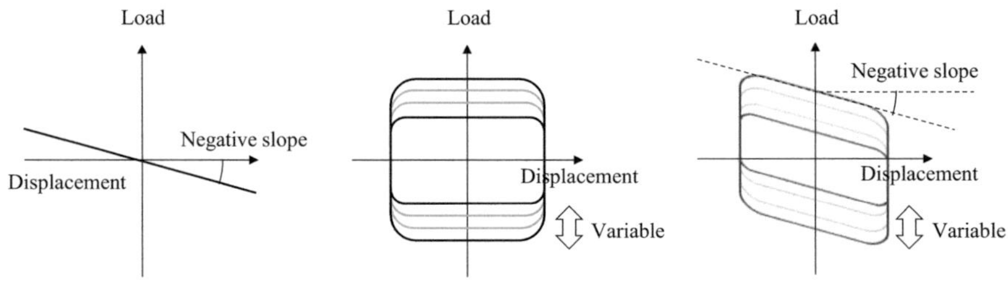

(a) Inertia mass force. **(b)** Viscous damping and Coulomb's friction force. **(c)** Total force.

Fig. 2.10 Hysteresis loop of the MR-RIM damper

Table 2.2 Design specifications of the MR-RIM damper

Detail	Capacity of performance
Maximum stroke	±300 mm (600 mm p-p)
Maximum loading	2.45 kN
Relief loading	250 kg
Maximum driving force	2.45 kN
Maximum acceleration	9.8 m/s^2
Maximum velocity	1.0 m/s

Table 2.3 Design specifications of the electro-magnets for the magnetic field generating mechanism

Item	Design specification
Number of coils	5 pieces
Connecting type of coils	Direct current
Number of winding of coils	280 windings/unit
Allowable supplying current	1.25 A

2.3.2 Performance Test of the Designed MR-RIM Damper

To evaluate performance of the MR-RIM damper, the element test by using a single device was carried out under sinusoidal displacement inputs for all combinations of parameters about frequency, amplitude and supplying current. These are three kinds of excitation frequencies, 0.2, 0.333 and 0.5 Hz; two kinds of amplitudes, 100 and 200 mm; six kinds of supplying current, 0, 0.25, 0.50, 0.75, 1.00 and 1.25 A. The numerical model of the MR-RIM damper is considered as the combination of three types of structural elements as depicted in Fig. 2.11. The generating force F with this damper can be expressed as Eq. (2.1) and it is consisted of inertia mass, viscous damping and Coulomb's friction forces. In which, Coulomb's friction force F_{MR} is corresponded to the yield load of the MR fluid.

$$F = m'\ddot{x} + C\dot{x}^{\alpha} + F_{\mathrm{MR}} \tag{2.1}$$

In which,

- F total output force in the MR-RIM damper,
- m' equivalent mass of the inertia mass part of the MR-RIM damper,
- C viscous damping coefficient of the MR fluid,
- α mechanical constant ($0 < \alpha < 1$),
- F_{MR} yield load of the MR fluid according to supplying current,
- x, $\dot{x}$, $\ddot{x}$ displacement, velocity, and acceleration of the MR-RIM damper.

The values of numerical model parameters of the MR-RIM damper is determined from the results obtained by the element test for a single device under sinusoidal displacement inputs. First, the equivalent mass m' correlated to the rotary inertia mass effect is calculated according to the value of negative slope gradient on the load hysteresis curve (mentioned as a dotted line in Fig. 2.10c). In which, the negative slope part can be observed on the load-deformation hysteresis loop diagrams while the deformation in the same direction is caused by monotonic loading.

The equivalent mass m' is evaluated as the average value for all cases. The calculated value of the equivalent mass m' is 241 kg. This MR-RIM damper could be well produced because the original design value of the equivalent mass was

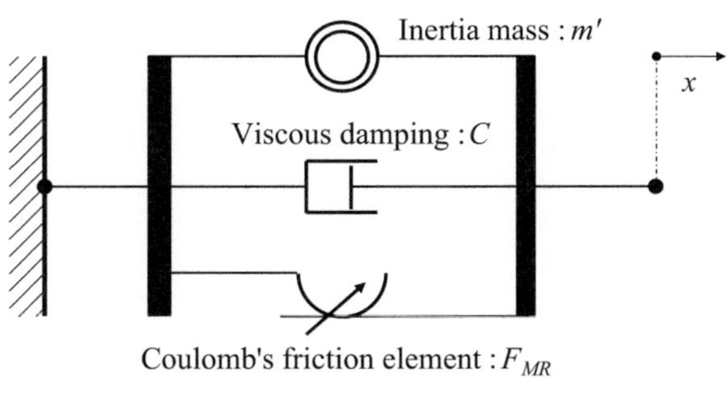

Fig. 2.11 Numerical model for the MR-RIM damper

250 kg. Next, the yield load of the MR-RIM damper according to the resistant force in the MR fluid is obtained from the loading values when the piston rod's velocity of damper is equal to zero in the relationship diagrams between load and velocity. Figure 2.12 shows the relationship between yield load and supplying current. The different kinds of plot points are corresponding to the different parameters' combinations of loading frequency and amplitude. As seen in this figure, the yield load of the MR-RIM damper depends less on the loading frequency and amplitude, but mostly on the supplying current. Accordingly, the yield load of the MR fluid F_{MR} can be approximately expressed as the quadratic equation of the current value I [A] as mentioned in Eq. (2.2), which is evaluated by the method of least squares.

$$F_{\mathrm{MR}} = 614I^2 + 316I + 172 \quad [\mathrm{N}] \tag{2.2}$$

Finally, C and α relating to the viscous damping in the MR fluid are obtained in the following way. Trial values of constant α ($0 < \alpha < 1$) for the viscous damping force definition expressed as the right side second term in Eq. (2.1) are assumed beforehand, and then the square residuals sum between the least squares regression curve and the actually measured characteristic curve for the viscous damping force vs. damper velocity is calculated. The optimal value for α is selected one from the trial values to minimize the square residuals sum. The final value of viscous damping coefficient C is determined as the average value for the set of calculated values C which are obtained from the inverse calculation using the actually measured viscous damping force and the right side second term in Eq. (2.1) applying the final value of α. After these data processing, the estimating values of C and α are gained as 541 Ns/m and 0.679, respectively. The approximate equation of the output force of the MR-RIM damper is expressed as Eq. (2.3). In which, the sign of the term of F_{MR} depends on the direction of the piston rod's velocity of the damper.

$$F = 241\ddot{x} + 541\dot{x}^{0.679} + (614I^2 + 316I + 172) \cdot sign(\dot{x}) \quad [\mathrm{N}] \tag{2.3}$$

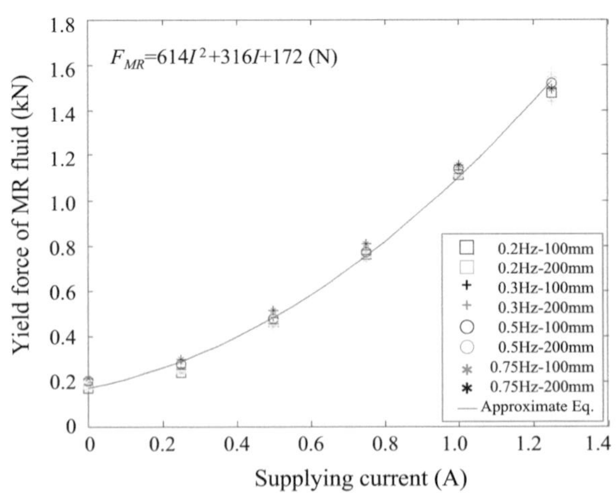

Fig. 2.12 Relationship between yield load and supplying current in the MR-RIM damper

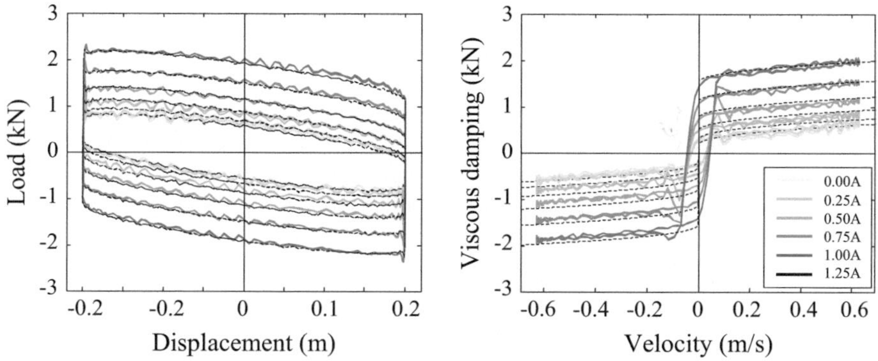

Fig. 2.13 Comparison of experimental results with the approximate equation of Eq. (2.3) (Relationship between output force and displacement or velocity, in the case of the stroke 200 mm and the input frequency 0.5 Hz)

The first, second, and third terms are inertia mass effect, viscous damping effect and yield load of the MR fluid, respectively. The last term for yield load of the MR fluid is controllable by changing supplying current to the MR fluid. Figure 2.13 shows comparison of output forces of damper between the actually measured results of the element tests and the approximately calculated values with Eq. (2.3) (where, these results are corresponding to the case for the stroke of 200 mm and the input frequency of 0.5 Hz). In these figures, the solid lines show the test results and the dotted lines show the approximate values. Since both lines are well corresponding, the validity of this approximate equation of Eq. (2.3) can be confirmed.

2.3.3 Test Specimen of Building Model for Shaking Table Experiments

Figure 2.14 depicts details of the test specimen of the four-story building model. Every floor part of this specimen is supported by the flat type roller bearings (produced by THK Co.) and restoring force on each story is given by coil springs connected adjacent floors each other. Compression springs are placed for the upper three stories from the top and the tension springs are used at the lowest story. Because this building model supposes the seismic isolation layer at the lowest story and less stiffness than the compression springs is given by the tension springs. Material of the compression springs and the tension springs are made from Steel Wire of Oil-tempered Silicon Chromium (SWOSC: Japanese Industrial Standards No. G 3560; 1994), and the tension spring was made from Steel Wire of Piano on A-grade (SWP-A: Japanese Industrial Standards No. G 3522; 1991), respectively. Those springs are considered as having linearity within their designed usable range.

During experimental tests, displacement of each story and acceleration of each floor on the test specimen are measured by inductance type displacement sensors

(a) Test specimen of building model set on the shaking table.

Acceleration sensors : placing at every floor
Displacement sensors : placing at every story
Load cell : setting at the end of the piston rod of the damper placing at the isolation layer

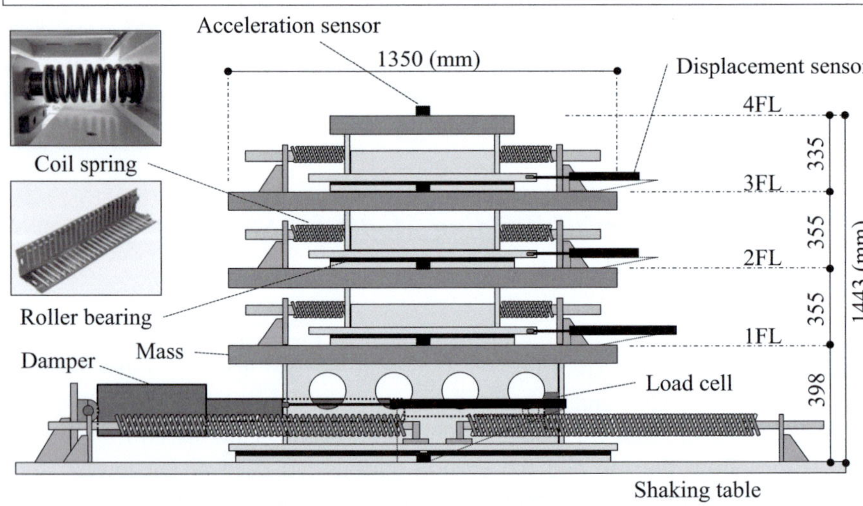

(b) Detail of the test specimen of building model.

Fig. 2.14 Test specimen of 4-story building model

(DLAS series; product of Kyowa Electric Instruments Co.) and servo type accelerometers (LS-10C; product of RION Co.). The output force of the MR-RIM damper is directly measured by the load cell equipped on the end of the piston rod. The load cell used for measuring this damper's force is selected as the type of

Table 2.4 Identified values of dynamic parameters of the test specimen

Movable story	h	f (Hz)	ω (rad/s)	m (kg)	M (kg)	$k = M\omega^2$ (N/m)	$c = 2h^*(Mk)^{1/2}$ (Ns/m)
4	0.013	2.45	15.7	386	386	95.2×10^3	158
3	0.018	1.81	11.4	475	861	111.9×10^3	353
2	0.126	1.62	10.2	471	1331	138.5×10^3	706
1	0.019	0.35	22	695	2027	9.81×10^3	169

capable for both tension and compression (LUK-A-20KN; Kyowa Electric Instruments Co.).

Mass of each floor m of the test specimen is once separated to each part and these weights are directly measured by the scale. Stiffness k and damping coefficient c are identified by every story with the vibrating test on the shaking table using impulse input and random wave (band-limited white noise) input. The parameter identification of the test specimen is operated under the condition to set up the specified single mass movable state. Namely, by making the specified single story movable and making the other three stories fix, this condition is implemented. Natural frequencies f [Hz] for every single mass condition are estimated from the dominant frequency appearing on the Fourier Spectrum of the displacement response under the random input. Damping ratios h are identified from logarithmic decrement of the RD (Random decrement) waves that are gained as the autocorrelation function for the displacement response under the random input. Table 2.4 shows these system identification results about the dynamic properties of the test specimen. In which, M [kg] mentioned in Table 2.4 means the sum of the upper mass supported by the movable story.

2.4 RTHS Test for Semi-active Control System by Using Shaking Table

Dynamic performance of a mid-story isolated building system is evaluated under installation of the semi-active device on the isolation story. The MR-RIM damper is used for the semi-active control operation in this study. The RTHS test system is designed as the online simulator for reproducing seismic response of entire parts of the target mid-story isolated building structure virtually by combining an actual test specimen of the upper structural part than the isolation layer and a numerical model of the structural part below the isolation layer. Numerical model part is expressed as a multi-mass model and online simulations for this part are done in the computer of the RTHS test system.

2.4.1 Outline of RTHS Test

Figure 2.15 shows the controlling flow diagram for the RTHS test system applying the mid-story isolated building. The actual test specimen of building model consists

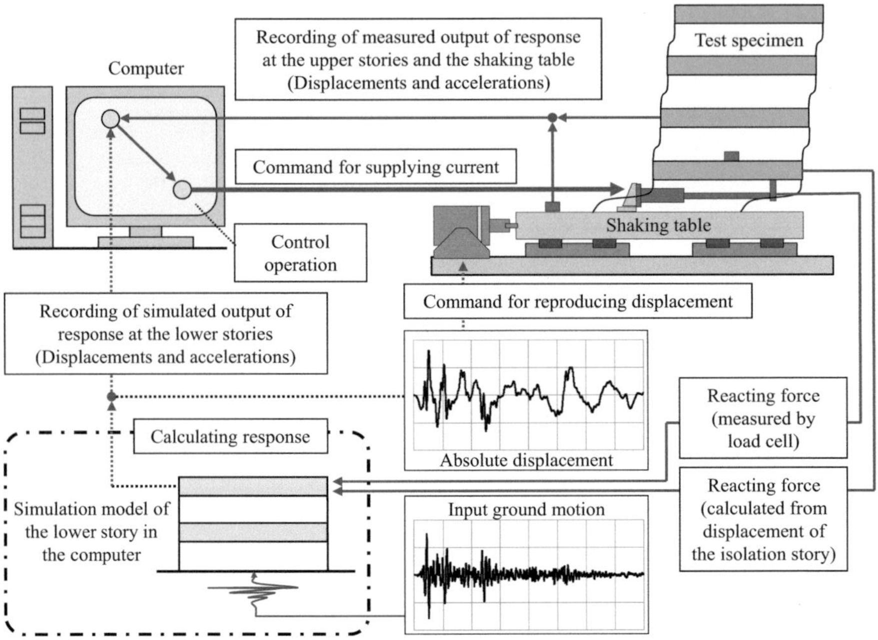

Fig. 2.15 Flow diagram of the RTHS test for the mid-story isolated system

of a single story of isolation layer and the three more stories upper than the isolation layer. The structural part below the isolation layer is provided as a numerical model for computer calculation. Online calculation for the numerical model responses is executed on the DSP (AD5436-I7; product of A&D Co.), while reproducing the interactions between the experimental specimen and the numerical model in the real time. In this study, two-mass system model is used as the numerical model part in the computer.

As displayed in Fig. 2.15, at first, evaluative input earthquake motion is supposed to act on the two-mass system model inside the computer and the computer calculates the top floor's displacement of this numerical model. The simulator of the RTHS test system reproduces the absolute displacement at the top floor of the numerical model as interactive motion on the shaking table. This operation can directly generate the floor acceleration response right under the isolation layer on the shaking table, thus the seismic force inputs supposed for the isolation layer and the upper structural part above the isolation layer can be properly given for the actual test specimen.

Next, actual sensors placed on the isolated layer measure real responses of the test specimen. The RTHS test system calculates the interacting forces through the isolation layer to the lower floor placed right under the isolation layer. The interacting force is considered as the reaction force of the coil springs, the mechanical damping force in the isolation story and the output force generated by the MR-RIM damper. In which, the reaction force is calculated by multiplication of the stiffness of the coil springs placed at the first story on the test specimen and the inter-story displacement at this story measured by sensor. The stiffness of the coil springs has been identified as the inter-story stiffness through the previous test as listed in Table 2.4. The output force of the MR-RIM damper is directly measured by the load cell, thus the output force from the MR-RIM damper at any moment can be observed accurately in this system whether the semi-active controls were operated or not. In this study, the effect of the mechanical damping force on the isolation story is excluded, because this force is regarded much less than the viscous damping force on the MR-RIM damper.

Semi-active control operation of the MR-RIM damper is determined and manipulated according to the control law by considering both the observed responses of the test specimen and the calculated responses at the numerical model in the computer. The input forces for the numerical model of the two-mass system are given as the interacting forces from the isolation layer and the evaluative input earthquake motion at every controlling time step, and the computer calculates seismic responses of the numerical model for the next time step every time. Every control time step on the RTHS test system is continuously processed while synchronizing the actual part and the numerical part of the entire system on the real-time scale. In this RTHS test system, time interval between every control time steps is set to 0.002 s (the simulator operates online calculation by 500 Hz).

To operate the RTHS test for the mid-story isolated building in this study, the shaking table is the key item to actualize the interacting motions between the real part with the test specimen and the ideal part with the numerical model. The RTHS test can give lots of advantages to assure the practical problems on the mid-story isolated buildings. For example, verification of the impulsive behavior on the floor response acceleration at the upper parts than the isolation layer due to the control force of the damper is available through this testing way. Moreover, various kinds of models for the lower structure can be supposed easily on the tests of the mid-story isolated buildings, because this operation can be executed with nothing but replacing the part of the numerical model.

2.4.2 Structural Design of Testing Model for Mid-Story Isolated Building

Figure 2.16 shows a schematic diagram to consider modeling of the target mid-story isolated building from the practical scale of structural design to the test specimen scale justifying for the capacity of the shaking table. As depicted in Fig. 2.16, the upper structure and the lower structure are expressed to the apart multi-mass models and are contracted to the four-mass model and the two-mass model, respectively. The structural properties of the upper part of the contraction model is adjusted to the actual properties of the test specimen as mentioned in Table 2.4. These structural parameters of the four-stories test specimen are

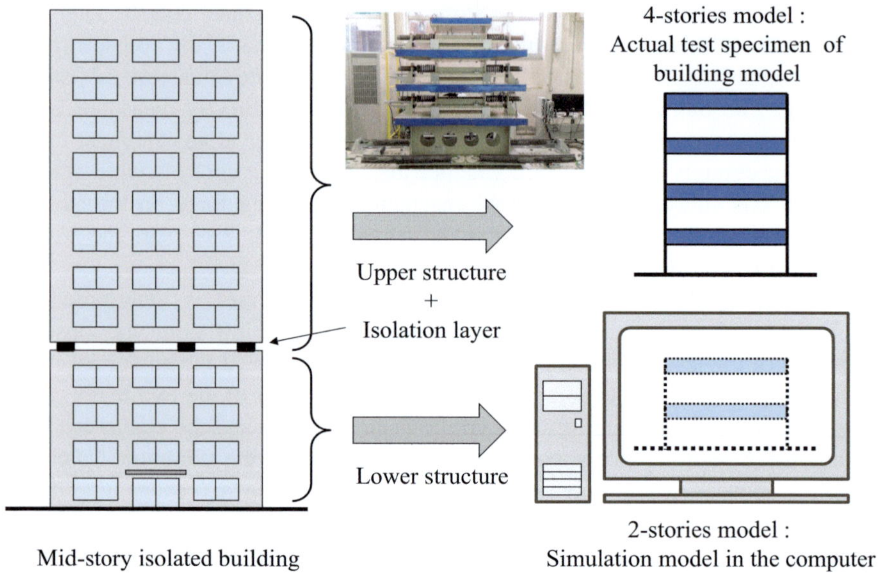

Fig. 2.16 Model contraction of the mid-story isolated building structure

Table 2.5 Structural property of the entire model of the target mid-story isolated building for RTHS test

Story	Mass: m (kg)	Stiffness: k (N/m)
(a) Upper structure part (actual test specimen of building model)		
6	386	95.2×10^3
5	475	111.9×10^3
4	471	138.5×10^3
3	695	9.81×10^3
(b) Lower structure part (simulation model in the computer)		
2	750	250.0×10^3
1	800	250.0×10^3

corresponded to the upper structure's part from third to sixth-stories for the entire model of the target mid-story isolated building of six-stories. This mid-story isolated building of six-stories is designed as that the upper structure's part has about 56.7% mass ratio for the total weight of building.

Structural properties of the entire model of the target mid-story isolated building for the RTHS test are mentioned in Table 2.5. Natural periods of the entire model of the mid-story isolated building of six-stories, the part of the structural model of four-stories upper than the isolation layer and the part of the structural model of two-stories below the isolation layer are listed in Table 2.6, respectively. Table 2.7

Table 2.6 Natural period of the testing model of the mid-story isolated building

Mode	1st	2nd	3rd	4th	5th	6th
(a) Entire model of 6-stories						
Natural period (s)	3.03	0.544	0.504	0.290	0.220	0.212
Mode	1st	2nd	3rd	4th		
(b) Upper structural part of 4-stories						
Natural period (s)	2.93	0.537	0.290	0.220		
Mode	1st	2nd				
(c) Lower structural part of 2-stories						
Natural period (s)	0.527	0.212				

Table 2.7 Modal participation factors of the target mid-story isolated building of 6-stories

Mode	Natural period (s)	Modal participation factor
6	3.03	0.03109
5	0.544	0.1025
4	0.504	0.5545
3	0.290	0.0001627
2	0.220	0.0004064
1	0.212	0.3114

shows the modal participation factors corresponding to every natural frequency of the entire model of the mid-story isolated building of six-stories. Figure 2.17 show the modal vector shapes (which are normalized by the value of the 1st story's component) and the modal participation functions of the target mid-story isolated building of six-stories. As seen in Fig. 2.17b, the first, second, and third modes have lots of participation for the upper structure's responses and the third and sixth modes have lots of participation for the lower structure's responses.

2.5 Semi-active Control Law

Recently, effectiveness of installing a rotary inertia mass damper in the isolation layer is reported to reduce deformation of isolators of seismic isolation buildings. On the other hand, it is known that inertia mass effects may occur to increase acceleration responses on the isolation layer (Isoda et al. 2009). Authors have been reported effectiveness of semi-active controls to overcome this problem through the previous researches with the numerical simulations. In the followings, semi-active control laws are explained to reduce displacement at the isolation layer of the mid-story isolated building model effectively.

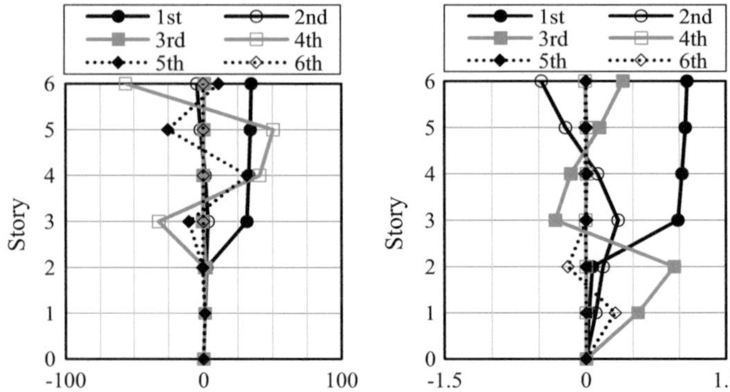

(a) Modal vector shapes. **(b)** Modal participation functions.

Fig. 2.17 Modal shapes of the target mid-story isolated building

2.5.1 *Basis of Proposed Control Law*

As a control law of semi-active controls for the MR-RIM damper, simple operation rule to control the output force generating at the MR fluid is proposed. This rule is based on the virtual on/off method for the inertia mass effects by the rotary inertia mass part. From the previous studies about installation of the rotary inertia mass device for the seismic isolation system, the rotary inertia mass can effectively reduce growth of deformation of the isolation layer, but it makes the absolute velocity response of the isolation layer increase and the seismic isolation effects deteriorate.

Such deterioration of the seismic isolation effects is known to be induced under the phase state that the rotary inertia force effects acts on the same direction with the absolute velocity of the isolation layer. The output force along the piston rod axis, which is transformed from the inertia force, has the opposite direction for the acceleration of the damper's piston rod at all the time, thus the absolute velocity of the isolation layer increases when the acceleration of the damper and the absolute velocity of the isolation layer have the opposite direction each other (Aoyama et al. 2013). An effective strategy for controlling the MR-RIM damper is proposed to cancel the inertia mass effects during the phase state that the rotary inertia mass damper increases the absolute velocity of the isolation layer. To control the additional force on the MR fluid according to this strategy, the supplying current for the MR-RIM damper is calculated considering the modeling formula of the MR-RIM damper of Eq. (2.3). The following condition in Eq. (2.4) is required for the semi-active control rule to cancel the inertia mass effect in the MR-RIM damper.

$$241\ddot{x} + (614I^2 + 316I + 172) \cdot sign(\dot{x}) = 0 \tag{2.4}$$

In which, allowable value for the supplying current I is limited in the range $0 \leq I \leq 1.25$ [A] by the mechanical capacity of this MR-RIM damper using in this study. To actualize the state of Eq. (2.4) is only available under the phase state that the relative acceleration of the damper $\ddot{x}$ and the relative velocity of the damper $\dot{x}$ have the opposite sign. Under this constraint, the semi-active control law of the MR-RIM damper is mentioned as Eq. (2.5) and the required current to supply on the MR fluid is expressed as Eq. (2.6). This equation is given as the expression by solving Eq. (2.5) for the supplying current I.

$$-241|\ddot{x}| + (614I^2 + 316I + 172) = 0 \tag{2.5}$$

$$I = -0.246 + \frac{\sqrt{1545|\ddot{x}| - 853}}{61} \quad [\mathrm{A}] \tag{2.6}$$

In which, both $\ddot{x} \cdot \dot{x} < 0$ and $\ddot{x} \cdot (\dot{x} + \dot{z}) < 0$ are the constraint condition to let this semi-active control operation be available. $\dot{x} + \dot{z}$ means the absolute velocity at the isolation layer.

2.5.2 Rule to Adjust Supplying Current for the MR Fluid According to Ground Velocity

The value of the supplying current is also adjusted considering the earthquake motion level by every control time step. The earthquake motion level for each instant is evaluated by referring the ground velocity. Eq. (2.7) is proposed as the strategy to control the sensitivity of the damping effect according to the earthquake motion level. This rule expressed by Eq. (2.7) is given to determine the supplying current according to the damper's velocity.

$$I = G_0 \cdot |\dot{x}|^{(\dot{z} + 0.5)} \quad [\mathrm{A}] \tag{2.7}$$

As seen in Fig. 2.18, the constant of 0.5 m/s is considered as the boundary of the ground velocity level, and two kind states of relational curves of the supplying current for the damper's velocity $\dot{x}$ are used properly according to the ground velocity $\dot{z}$. This rule gives a large gradient on the relational curve of the damping force for the damper's velocity while the ground motion level $\dot{z}$ is less than 0.5 m/s. On the other hand, a small gradient on the relational curve of the damping force for the damper's velocity while the ground motion level $\dot{z}$ is more than 0.5 m/s. As mentioned in the previous consideration in the Sect. 2.5.1, the supplying current for the MR fluid is manipulated by the Eq. (2.6) or (2.7) according to the state determined by the combination of the directions of the ground velocity, the damper's velocity and the relative acceleration of the rotary inertia mass. In this study, the value of the output gain of the supplying current G_0 is set to unit value (=1.0) in Eqs. (2.7)–(2.9).

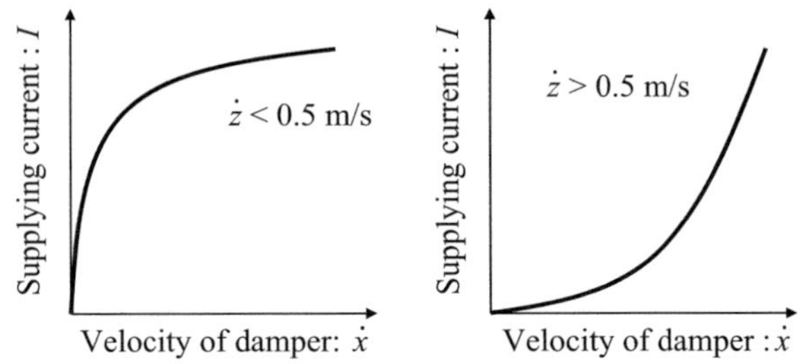

(a) Case for $\dot{z}$ < 0.5 m/s. **(b) Case for $\dot{z}$ > 0.5 m/s.**

Fig. 2.18 Determinative rules for supplying current according to ground velocity

When $\dot{x} \cdot \ddot{x} < 0$ and $\ddot{x} \cdot (\dot{x} + \dot{z}) < 0$:

$$I = I_1, I_1 = G_0 \cdot |\dot{x}|^{(\dot{z}+0.5)} - 0.246 + \frac{\sqrt{1545|\ddot{x}| - 853}}{61} \quad [\mathrm{A}] \tag{2.8}$$

In other states

$$I = I_2, I_2 = G_0 \cdot |\dot{x}|^{(\dot{z}+0.5)} \quad [\mathrm{A}] \tag{2.9}$$

2.5.3 *Reproducibility and Validity of RTHS Test*

To assure accuracy of the RTHS test by using the shaking table of Kobe University, the experimental test results are compared to the results of the pure simulations for the entire numerical model. As mentioned in the previous section, the structural model of the mid-story isolated building of six-stories is used in this study. The isolation layer is placed at the third story and the MR-RIM damper is installed on this isolation layer. The semi-active control operation is carried out and control performance is investigated. Reproducibility of this RTHS test system is also verified.

2.5.3.1 Comparing the Results from RTHS Test and Pure Numerical Analysis

Based on the semi-active control law expressed in the Sects. 2.5.1 and 2.5.2, the RTHS tests were conducted. Table 2.8 lists the input earthquake motions. Level of each earthquake motion is adjusted to a moderate value considering the allowable stroke of the shaking table and the allowable deformation of the test specimen of building model.

First, the actually observed responses of the test specimen for the upper structure and the isolation layer and the simulated responses of the two-mass system model for the lower structure during the RTHS test operations are compared with the responses calculated by pure simulations for the entire numerical model of the

Table 2.8 Input ground motion data using for the RTHS test

Wave data of earthquakes for input motion	Maximum velocity (m/s)	Maximum acceleration (m/s^2)
El Centro 1940 NS (90%)[a]	0.30	3.08
Hachinohe 1968 NS (147%)[a]	0.50	3.33
JMA Kobe 1995 NS (30%)[b]	0.27	2.46
JR Takatori 1995 NS (40%)[b]	0.49	2.42
Tomakomai 2003 NS (60%)[c]	0.30	1.16

[a]Seismic motion for general structural design
[b]Pulse-type seismic motion
[c]Long-period-type seismic motion

target mid-story isolated building. Figure 2.19 shows the time history of the inter-story displacements of every story. Figure 2.20 shows the time history of the accelerations of every floor. Figure 2.21 shows the control force generated by the MR-RIM damper and the supplying current for the MR fluid. These results are corresponding to the case of input with 90% scale El Centro 1940 NS.

Figures 2.19, 2.20 and 2.21 make comparison of the results between the RTHS test and the pure simulation, the solid lines and the dotted lines are corresponded to the time history for the RTHS test and the pure simulation, respectively. In Figs. 2.19 and 2.20, the simulated values of the two-mass system model for the lower structure in the computer are placed at the first and second story/floor and the observed values of the four-mass test specimen for the upper structure are placed from the third to sixth story/floor in the target mid-story isolated building. As seen in Fig. 2.19 and 2.20, it is regarded that both displacement and acceleration responses at every story/floor can be accurately reproduced by this RTHS test system because of well agreement to the pure simulation results. As seen in Fig. 2.21, both the control force of the MR-RIM damper and the supplying current for the MR fluid also show very well agreements between the RTHS test results and the pure simulation results.

In this study, both the test specimen and the MR-RIM damper are exactly quantified and accurately modeled for the pure numerical simulation of the entire model, thus the agreement between both outputs from the RTHS tests and from the pure simulations can prove the validity of this RTHS test system and its test results.

2.5.4 Reproducibility of Shaking Table Motion on RTHS Test

The interaction between the really-tested system part (the test specimen corresponding to the upper structure and the isolation layer) and the simulated system part (the numerical model in the computer corresponding to the lower structure below the isolation layer) is reproduced by the shaking table motion in this RTHS

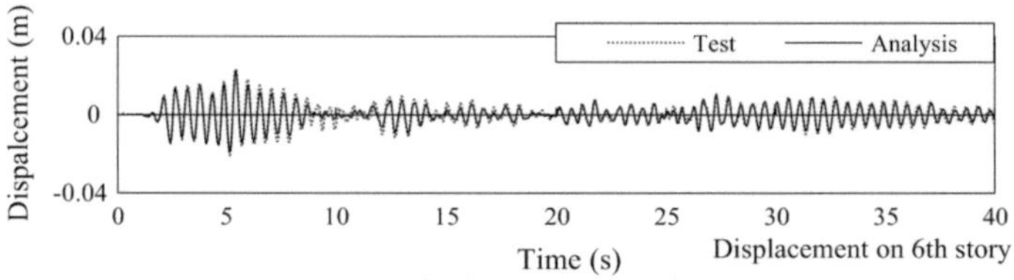

(a) Displacement on 6th story.

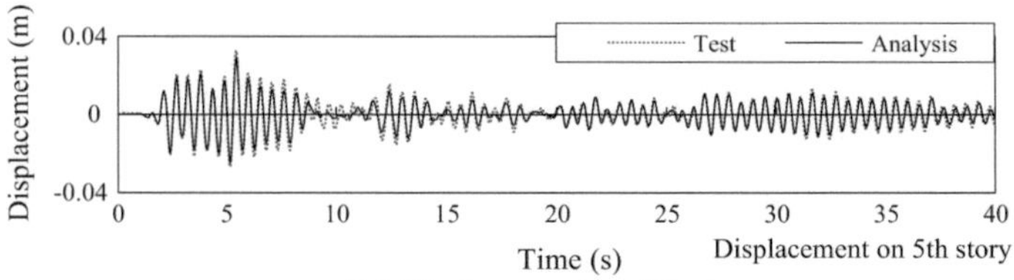

(b) Displacement on 5th story.

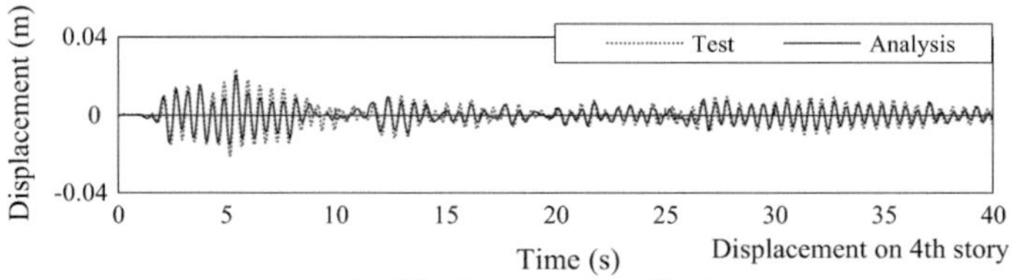

(c) Displacement on 4th story.

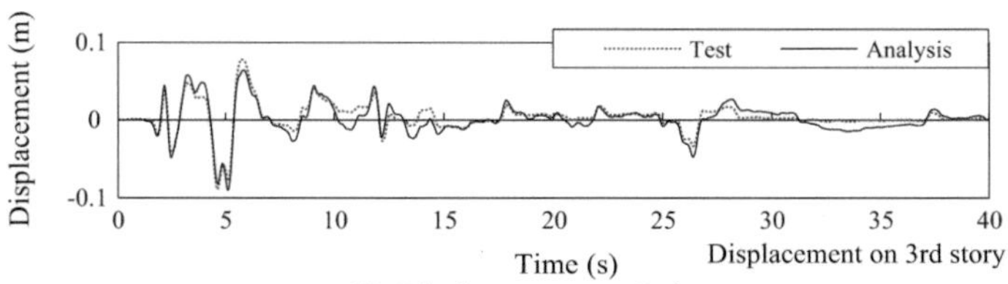

(d) Displacement on 3rd story.

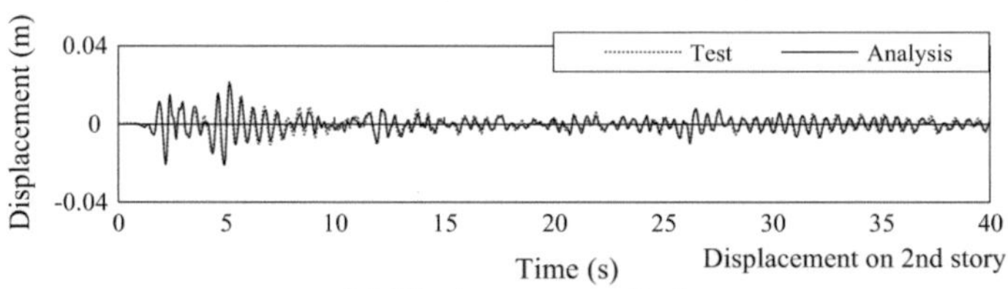

(e) Displacement on 2nd story.

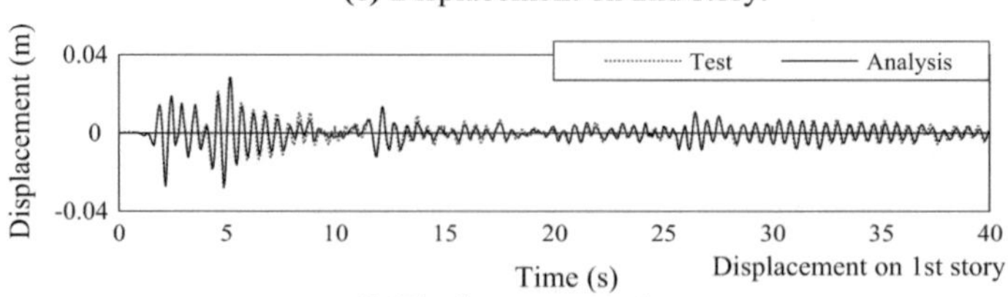

(f) Displacement on 1st story.

Fig. 2.19 Comparison between results of RTHS test and pure simulation (displacements; El Centro)

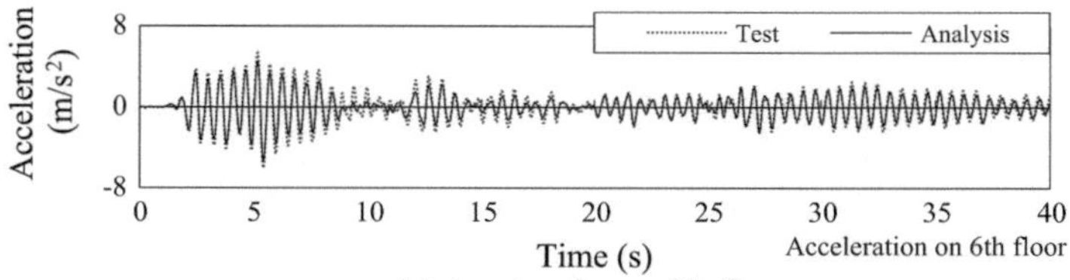

(a) Acceleration on 6th floor.

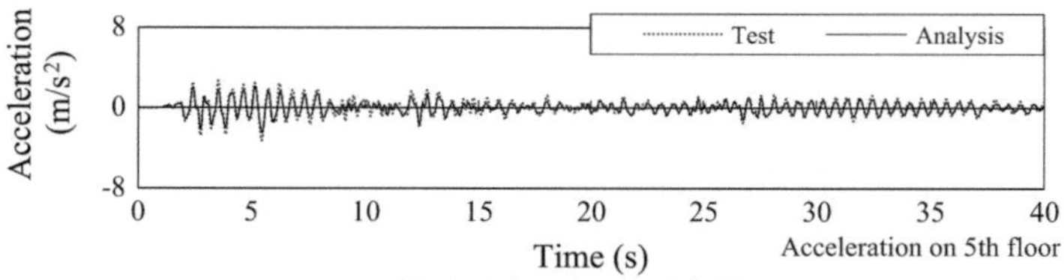

(b) Acceleration on 5th floor.

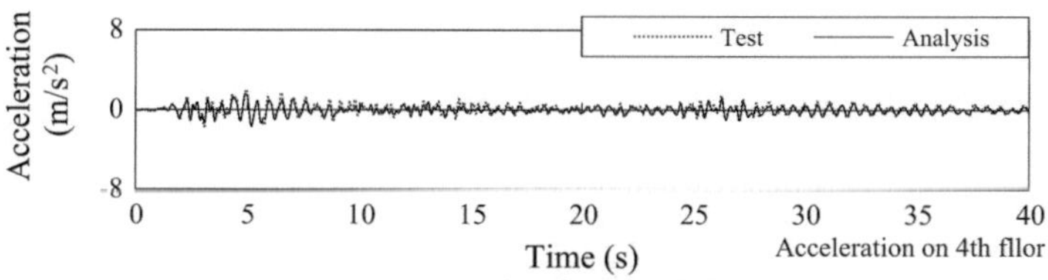

(c) Acceleration on 4th floor.

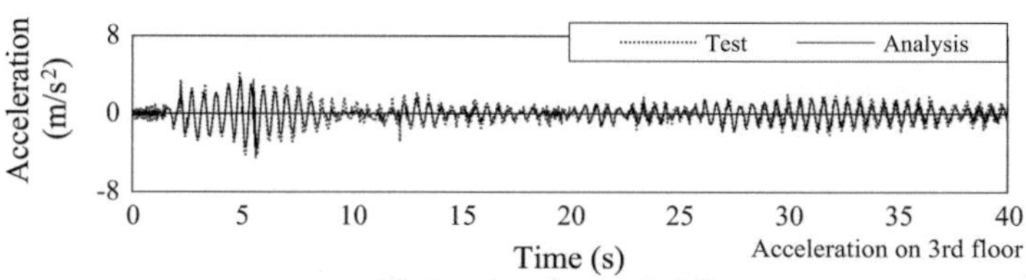

(d) Acceleration on 3rd floor.

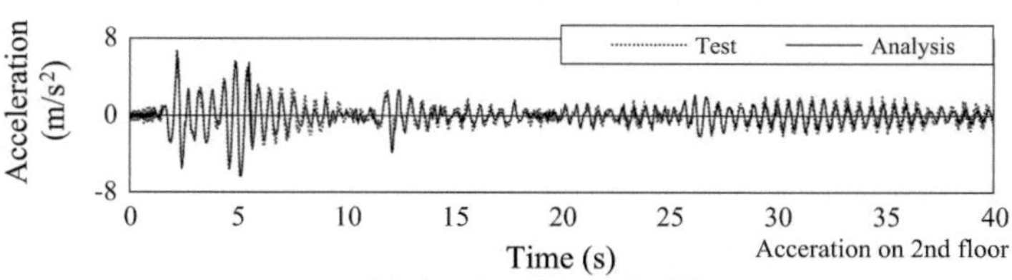

(e) Acceleration on 2nd floor.

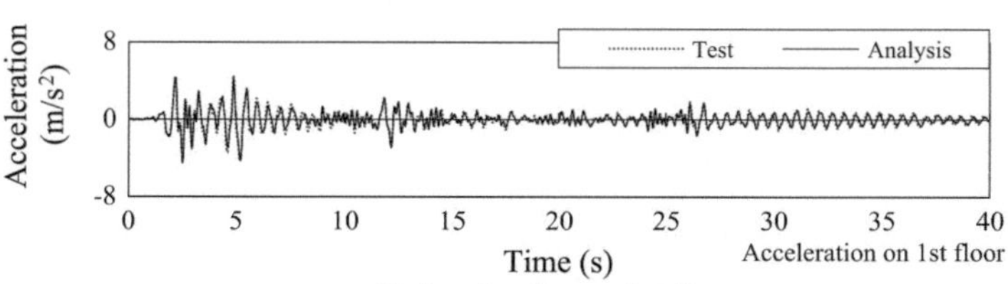

(f) Acceleration on 1st floor.

Fig. 2.20 Comparison between results of RTHS test and pure simulation (accelerations; El Centro)

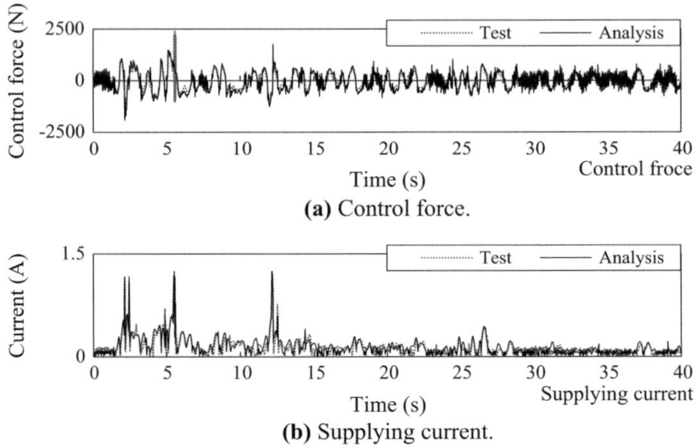

Fig. 2.21 Comparison between results of RTHS test and pure simulation (Control force and supplying current; El Centro)

test system. According to the computer calculation for the two-mass system model of the lower structure conducted in real time, the RTHS test system commands the shaking table to reproduce the absolute displacement at the top floor of the lower structure part. The commanded displacement and the actually reproduced displacement on the shaking table are compared in Fig. 2.22a. The target (simulated) acceleration and the actually reproduced acceleration on the shaking table are compared in Fig. 2.22b. Well agreements of these reproduced responses for the commanded/simulated target can be confirmed. Since the shaking table can be exactly moved according to the commanded displacement as observed in Fig. 2.22a, the reproducibility of this RTHS test system can be regarded to be quite accurate. At the same time, the reproduced acceleration can be resulted to generate the simulated target acceleration quite accurately for the numerical model in the computer.

These results are also evaluated from the correlation factor ρ and the RMS ratio (the ratio of the RMS value of the reproduced outputs for the RMS value of the commanded inputs). The correlation factor ρ and the RMS value are defined by Eqs. (2.10) and (2.11), respectively.

$$\rho = \frac{\frac{1}{N}\sum_{i=1}^{N}(A_i-\mu_A)(B_i-\mu_B)}{\sqrt{\frac{1}{N}\sum_{i=1}^{N}(A_i-\mu_A)^2}\sqrt{\frac{1}{N}\sum_{i=1}^{N}(B_i-\mu_B)^2}} \tag{2.10}$$

$$\mathrm{RMS}\,[\mathrm{A}] = \sqrt{\frac{1}{N}\sum_{i=1}^{N}A_i^2} \tag{2.11}$$

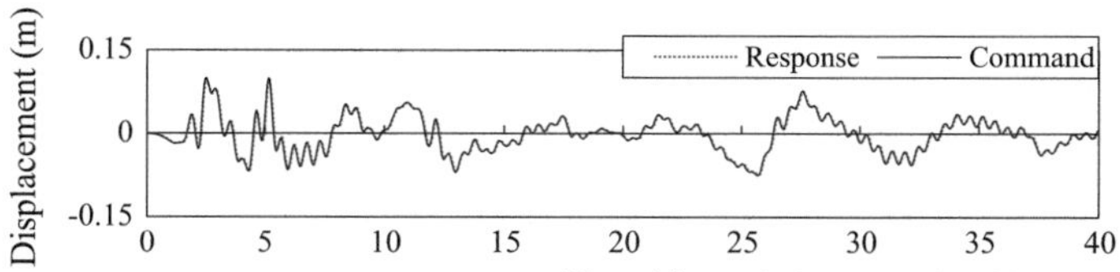

(a) Displacement of shaking table.

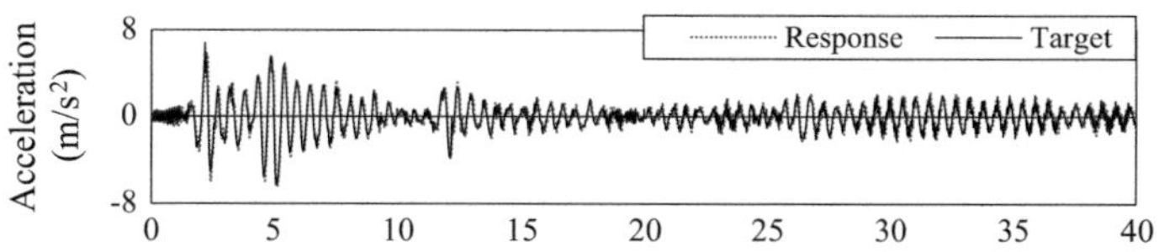

(b) Acceleration of shaking table.

Fig. 2.22 Comparison between command (target) input and reproduced response on the shaking table (El Centro)

N number of data,
A_i, B_i time history data (corresponding to displacement or acceleration for the command input values or the reproduced response values),
μ_A, μ_B average values of A_i and B_i.

The correlation factor takes the value from −1 to 1 and it can indicate correlation between different data. The degree of the phase delay of the reproduced output from the commanded input is considered by observing the correlation factor ρ. The degree of the errors between the time-series data is considered by observing the RMS ratio.

Table 2.9 shows the degree of the errors about the reproducibility of the shaking table for the various kinds of input motions. In this table, the reproducibility of the displacement and the acceleration are evaluated from the correlation factor ρ and the RMS ratio. The indicates for the displacements are calculated using the whole time-series data from the start to the end of the target input motion. The indicates for the accelerations are calculated using the data of 6 s interval which includes both 3 s before and after the timing of the maximum acceleration. As seen in Table 2.9, it is confirmed that both the correlation factor and the RMS ratio for both the displacement and the acceleration take values quite close to 1 for every case of the different input motions. Namely, it is regarded that the displacement commanded to the shaking table by the RTHS test system can reproduce accurate motion and that the control operation is highly reliable. Moreover, since the reproduced acceleration can be also confirmed to simulate accurately the target input, the simulated interaction motion on the shaking table for the RTHS test is also regarded to provide highly preciseness and reliability.

Table 2.9 Reproducibility of the shaking table for various type of input motions

Earthquake input motion	Correlation factor ρ	RMS ratio (Response/Command)
(a) Error ratio of displacements		
El Centro 1940 NS (90%)	0.998	0.984
Hachinohe 1968 NS (147%)	0.998	0.942
JMA Kobe 1995 NS (30%)	0.979	0.950
JR Takatori 1995 NS (40%)	0.979	0.973
Tomakomai 2003 NS (60%)	0.986	0.917
Earthquake input motion	Correlation factor ρ	RMS ratio (Response/Target)
(b) Error ratio of accelerations		
El Centro 1940 NS (90%)	0.966	0.955
Hachinohe 1968 NS (147%)	0.942	1.011
JMA Kobe 1995 NS (30%)	0.950	0.996
JR Takatori 1995 NS (40%)	0.973	0.997
Tomakomai 2003 NS (60%)	0.917	1.008

2.6 Conclusions

The RTHS tests for the mid-story isolated buildings are operated by using the shaking table of Kobe University. This report introduced the detail of the special function of the shaking table for using as the RTHS test system and explained about the test specimen of the building model and the semi-active control device provided for this RTHS test. The newly designed semi-active damper (the MR-RIM damper) equipped on the isolated layer and the upper structure part of the mid-story isolated building model are actually provided as the test specimens. The lower structure part below the isolation layer is simulated as a lumped mass system model in a computer. By combining both of the really-tested part and the simulated part for composing the mid-story isolated building model, this RTHS tests are conducted. Through verification for the RTHS test system performance, validity of this RTHS test system is assured and well reproducibility of this system is also confirmed.

Acknowledgements Part of this research was supported by JSPS KAKENHI Grant Number 23560672. The MR rotating inertia mass damper using for this study was manufactured with assistance of Sanwa Tekki Co. in Japan. The authors wish to acknowledge for these supports. The authors also acknowledge to Dr. Mai Ito and Mr. Shohei Yoshida for their cooperation to operate shaking table tests and data analyses.

References

Aoyama Y, Fujitani H, Mukai Y (2013) Proposal of semi-active control with variable rotational inertia force damper in base-isolated layer. Summaries of Technical Papers of Annual Meeting AIJ, Structure B-2, pp 401–402 (in Japanese)

Aoyama Y, Fujitani H, Ito M, Mukai Y, Shibata K, Sato Y (2014) Response reduction by passive and semi-active control of base-isolation system by using magneto-rheological rotary inertia

damper against near-fault pulse and long-period ground motions. In: Proceedings 6th world conference on structural control and monitoring (6WCSCM), Barcelona, Spain, pp 1223–1230
Architectural Institute of Japan (2007) Structural response and performance for long period seismic motions. AIJ, Japan (in Japanese)
Architectural Institute Japan (2013a) Extreme ground motions and seismic performance evaluation of building—how to prepare for mega subduction and inland earthquakes. AIJ, Japan (in Japanese)
Architectural Institute Japan (2013b) Design recommendations for seismically isolated buildings, 4th edn. AIJ, Japan (in Japanese)
Architectural Institute Japan (2016) Design recommendations for seismically isolated buildings (Written in English). AIJ, Digital Contents Distribution, Japan
Asai T, Chang C, Spencer BF Jr (2015) Real-time hybrid simulation of a smart base-isolated building. J Eng Mech ASCE 141(3):04014128
Christenson R, Lin Y, Emmons A, Bass B (2008) Large-scale experimental verification of semiactive control through real-time hybrid simulation. J Struct Eng ASCE 134(4):522–534
Fujii H, Hiwatashi T, Fujitani H, (2007) Shaking table test of semi-active control for base-isolation system by MR-damper applied with optimal regulator theory: effect of changing weighting matrix. J Struct Constr Eng AIJ 72(618):73–79 (in Japanese)
Isoda K, Hanzawa T, Tamura K, (2009) A study on response characteristics of a SDOF model with rotating inertia mass dampers. J Struct Constr Eng AIJ 74(642):1469–1476 (in Japanese)
Ito A, Kawasaki R, Fujitani H (2011) Effectiveness of real time hybrid test in semi-active controlled base isolation system. J Struct Constr Eng AIJ 76(663):891–897 (in Japanese)
Ito M, Yoshida S, Fujitani H, Sato Y (2015) Semi-active control of mid-story isolated building by MR rotary inertia mass damper: Part 1 and Part 2. Summaries of Technical Papers of Annual Meeting AIJ, Structure B-2, pp 623–626 (in Japanese)
Nagashima I, Shinozaki Y, Maseki R, Sanui Y, Kitagawa Y (2010) Sliding mode control of base-isolation system using semi-active hydraulic damper. J Struct Constr Eng AIJ 75 (649):511–519 (in Japanese)
Shao X, Reinhorn AM, Sivaselvan MV (2011) Real-time hybrid simulation using shaking tables and dynamic actuators. J Struct Eng ASCE 137(7):748–760
Taga K, Kamei I, Sumi A, Kondo K, Hayashi Y, Miyamoto Y, Inoue K (2011) Research on design input ground motion and design method of building for Uemachi fault earthquake. Summaries of Technical Papers of Annual Meeting AIJ, Structure B-1, pp 127–130 (in Japanese)
Yoshida K, Fujio T (2001) Bilinear optimal control theory and its application to semi-active vibration isolation control. J Trans JSME 67(656):992–998 (in Japanese)
Yoshida S, Fujitani H, Mukai Y, Ito M (2016) Real time hybrid tests of semi-active control using shaking table—proposal and verification of test method for mid-story isolated buildings. J Struct Constr Eng AIJ 81(729):1821–1829 (in Japanese)

Chapter 3
Horasan Mortar Bearings in Base Isolation with Centuries Experiences

Azer A. Kasimzade, Sertac Tuhta and Gencay Atmaca

Abstract The purpose of this study is to demonstrate the applicability of the Horasan mortar friction interface, which has been used for many long years in the past historical periods, as a seismic isolator in lightweight building structures with no overturning moment. To this end a four-storey hospital building has been selected as an example. Using the beyond-linear mathematical model simulation the results for Horasan mortar, stainless steel, gray-iron/mortar friction interfaces have been compared. For each interface the base acceleration is seen to be reduced in the ratio of 45, 38, 33% compared to the built-in building. For complete these favorable results-NSI device's material stress- and strain rang assessment have been modeled by the FEM in ANSYS, LS-DYNA environments and compute results to be compared to the related material ultimate values. Generally different devices are used to restore the superstructure to the initial position. In this study it has been proven that the earthquake has a restoring property and this phenomena have been used as the indirect restoring device. Full details of the application of Horasan mortar, which is economically favorable and has been used for many centuries in the history as a friction interface, has been given.

Keywords Earthquake response control · Durable natural seismic isolation (NSI) device · Nonlinear ANSYS · LS-DYNA finite element modeling Pure friction

A. A. Kasimzade (✉) · S. Tuhta
Department of Civil Engineering, Ondokuz Mayis University, Samsun, Turkey
e-mail: azer@omu.edu.tr

S. Tuhta
e-mail: stuhta@omu.edu.tr

G. Atmaca
Provincial Directorate of Disaster and Emergency, Samsun, Turkey
e-mail: gencayatmaca@hotmail.com

A. A. Kasimzade et al. (eds.), *Seismic Isolation, Structural Health Monitoring, and Performance Based Seismic Design in Earthquake Engineering*,
https://doi.org/10.1007/978-3-319-93157-9_3

3.1 Introduction

Seismic isolation separates the structure from the harmful motions of the ground by providing the flexibility and energy dissipation capability through the insertion of the isolated device, called the isolators, between the foundation and the building structure. Depending on the superstructure and the mass ratio of the base, isolation is classified as base isolation (a standard building with a higher mass ratio) or foundation isolation (a single-storey building with a lower mass ratio). The simplest sliding system is the pure–friction (P-F) system, which does not have any restoring force and a period and it is very effective for a wide range of frequency inputs (Mostaghel et al. 1983a, b).

In order to prevent and control large sliding and residual displacement in the P-F base isolator system, base isolation systems with a restoring force have been developed and investigated. These systems include: the resilient–friction system (Mostaghel and Khodaverdian 1987), the Electricite de France (Gueraud et al. 1985), the friction pendulum (Zayas et al. 1990), the sliding resilient friction (Su et al. 1989a, b), the elliptical rolling rods (Jangid and Londhe 1998) and others period-dependent base isolator systems.

The future P-F systems will most likely be developed by investigating various sliding interface materials. In the meantime, period-dependent sliding systems with restoring force can be developed using P-F and other damper devices to avoid the possibility of the vulnerability of base-isolated building to near-fault pulse ground motion. The history of the development of the P-F base isolation systems is given below.

Using the sinusoidal shake table test, the friction coefficient values of the some materials have been reported as follows; 0.25 for graphite powder, 0.34 for dry sand, and 0.41 for wet sand (Arya 1984; Qamaruddin et al. 1986). The friction coefficients for other sliding interface materials that have been proposed for masonry buildings are; 0.2 for viz. derlin, 0.6 for asphalt, and 0.7 for vinyl florin (Zongjin et al. 1989). Several bricks with and without sliding joint have also been tested under lateral loads with simulated dead load (Lou et al. 1992). The walls with the sliding joints were observed to slide at lateral force amplitude 50% of the force, at which the cracking began in the wall without the sliding joint. The friction coefficients obtained by the shaking table test for the suggested friction interface materials were; 0.23 for graphite/concrete interface at a peak acceleration of 0.2–0.3 g; and 0.4 for the same interface at a higher peak acceleration of 0.3–0.6 g. Tehrani and Hasani (1996) investigated an adobe building in Iran using dune sand (0.25), clay (0.16), and lightweight expanded clay (0.2–0.3) as the friction interface and concluded that dune sand and lightweight expanded clay can be good materials to create sliding layers in adobe buildings in Iran. A composite friction interface consisting of upper and lower reinforced concrete ring beams filled with two layers of asphalt felt interposed with graphite powder has also been used to test two six-storey brick masonry models on the shake table (Song et al. 1990). These models were excited with acceleration from 0.1 to 0.5 g and the sliding was initiated at 0.25 g. No cracking was detected in the two models even when the base acceleration was 0.5 g. In another study, the multi-level isolation in the form of discrete isolation Teflon and stainless steel at base level and continuous isolation in the form of sliding joint at upper floor levels for multi-storey masonry building were explored (Nikolic-Brzev and Arya 1996). Compared to the fixed base, in the isolated structure, an average reduction by 40% was observed in the maximum response accelerations and maximum base shear, respectively. In addition, various base isolation systems (including pure friction) have been used (Park et al. 2002; Fan and Ahmadi 1992; Fan et al. 1990a, b, 1991; Patil and Reddy 2012) or considered to have a significant potential for wider applications, and they have been systematically compared and discussed in terms of their aseismic performance on the multi-span continuous bridge. The authors concluded that an adequate friction element is effective in reducing the deck displacement and recommended the use of a friction coefficient smaller than 0.18. Simple slide isolation using a different type of stainless steel has also been researched (Nishimura et al. 2004a, b), where the experimental damping coefficient, static frictional coefficient, dynamic frictional coefficient and frictional coefficients were assigned as $c = 0.03 - 0.352$, $\mu_{max} = 0.17 - 0.22$, $\mu_{min} = 0.07 - 0.09$, $\mu = 0.05 - 0.1$, respectively. It was also shown that even if $\mu = 0.1 - 0.2$, it could be effective in the case of a big earthquake. In the same study, two different types of stainless steel were proposed as sliding bearings and both the accelerations and inter-storey drifts were reported to be reduced by 20–40%. The response reduction effect was analyzed using the multi-degree-of-freedom system. In this analysis, the static and the dynamic frictional coefficient were assigned to 0.2 and both the accelerations and inter-storey drifts were reduced by 40–80% in the case of four-or 11-degree of freedom, which shows that the response reduction effect was more than the expected under a significantly larger earthquake even when the friction coefficient was $\mu = 0.2$. The difference between lubricant and no-lubricant cases was hardly seen. Dolce et al. (2005) studied the frictional behavior of steel-PTFE (Polytetrafluoroethylene) interfaces for seismic isolation to evaluate the effect of the sliding velocity, contact pressure, air temperature and state of lubrication on the mechanical behavior of steel-PTFE sliding. In another study, the optimum design for a resilient sliding isolation system was used in the seismic protection equipment have been investigated by sinusoidal excitation on the shaking table (Iemura et al. 2007) and maximum values of friction coefficient were found to be 0.1 and 0.2, respectively. Nishimura et al. (2007) reported on the results of two previous studies, where they investigated a metal-touched type base isolator, for which they assigned the experimental damping coefficient, static frictional coefficient and dynamic frictional coefficient as $c = 0.096 - 0.153$, $\mu_{max} = 0.17 - 0.21$, and $\mu_{min} = 0.11 - 0.16$, respectively, and used a simulation to evaluate the response reduction effect of a multi-degree-freedom system model. The responses were reduced in the range of 50–80%. McCormick et al. (2009) reported that the friction developed between a steel base plate and a mortar base added shear resistance to the building system during a seismic event. The authors evaluated the possible sliding behavior of a steel base plate and a mortar base on a shake table (under various horizontal input accelerations and at various magnitude and frequencies). The results showed a constant friction coefficient during sliding with an average value of approximately $\mu = 0.78$. A P-F isolation system using coarse dry sand as a friction material ($\mu = 0.36$) was proposed for low-rise buildings in economically developing countries (Ahmad et al. 2009). It was observed that more than 70% of the energy was dissipated through this pure friction isolation technique. Experimental studies were performed with the shaking table test to investigate a frictional base isolation system using geo-textiles and smooth marble at the plinth level of a brick masonry building (Nanda et al. 2010). A 65% reduction was observed in the absolute response acceleration at the roof level when compared with the response of the fixed base structure at the cost of 0.025 m peak sliding displacement, which is well within the plinth projection of 0.075 m. A three-dimensional shaking table test was investigated (Ikawa et al. 2011) using the sliding isolation devices composed of surface-treated steel plates when $\mu = 0.01$, $\mu = 0.02$, and it was reported that in terms of the damage to the building, the decrease in the inter-storey displacement is more important. This was shown by Mukaibo et al. (2011), who investigated a traditional Japan timber building under a strong earthquake action and found that the friction coefficient between the base and the foundation stone could be dependent on the sliding direction and the friction coefficient may vary when the sliding occurs. For three types of specimens, the friction coefficient was obtained as $\mu_{average} = 0.38$. In another study (Nanda et al. 2012), four different sliding interfaces, green marble/high density poly ethylene (HDPE), green marble/green marble, green marble/geo-synthetic, and green marble/rubber layers were used in a P-F system and the average friction coefficients were found to be $\mu = 0.075, 0.085, 0.105$, and 0.17, respectively. The results showed that most of these sliding interfaces were effective in reducing the spectral accelerations up to 50% and the sliding displacement was restricted within 0.075 m.

The P-F base isolation composed with other models was also investigated in terms of their effectiveness in improving the seismic resistance superstructure. Hong and Kim (2004) evaluated the performance of a multi-storey structure with a resilient-friction base isolation system on the excitation for the stationary random process. The responses that were obtained using the equivalent linearization technique were compared with those calculated from the time history analysis. A discontinuous nonlinearity P-F system was developed and evaluated using linear laminated rubber bearings, viscous dampers, and nonlinear friction elements (Suy et al. 2007). In another research (Jalali et al. 2011), steel-Teflon flat sliding bearings were used to support the gravity loads while allowing large horizontal displacements, and simply connected Shape Memory Allow (SMA) truss elements, to provide the necessary horizontal stiffness as well as proper restoring capability. The system was referred to as the Smart Restorable Sliding Isolation System (SRSBIS). Gajan and Saravanathiiban (2011) evaluated the structures with energy dissipating devices and found that approximately 70–90% of the seismic input energy is dissipated by these devices while foundation rocking dissipates around 30–90% of the total seismic input energy in the foundation soil (depending on the static factor of safety). The results also indicated that if the system was properly designed (with reliable capacity and tolerable settlements), the adverse effect of foundation rocking could be minimized due to the favorable features of foundation rocking and hence they can be used as efficient and economical seismic energy dissipation mechanisms in buildings and bridges. Self-lubricating materials were proposed to be used in sliding isolation systems; such as the curved surface sliding isolator system, or pendulum bearing system, for the protection of buildings and structures (Quaglini et al. 2012; Becker and Mahin 2011). In another study (Gur and Mishra 2013), a P-F system supplemented with shape memory allow-restraining was presented on a framework of multi-objective optimization. A reduction in the residual bearing displacement was observed.

The base isolation technology can be used for both new structural design and seismic retrofit. A more detailed historical progression of the isolation systems are presented in the studies by Ismail et al. (2010) and Islam et al. (2011). Various types of isolators were also schematically classified (Naeim and Kelly 1999), where the design force approach and design displacement approach were set up for the design of isolation systems as defined by the ATC-Applied Technology Council (ATC 1996). Since the first implementation of seismic isolation (Foothill Community Law and Justice Center, Rancho Cucamonga, CA, United States 1985) successful energy dissipaters have been developed all around the world to control the response of an isolated structure by limiting both the displacement and the forces. At the same time, relevant regulations have been issued; such as the FEMA (2009) and the International Code Council (2018) computer software based on the finite element has been developed for the design.

3.2 Problem Identification

The overview of the literature shows that, a P-F isolation system needs to be developed after studying and exploring various sliding interface materials and improving the system by incorporating the restoring mechanism. The other important point is that this friction interface materials must be researched at least during the repetition time of major earthquakes. As seen from the references there are no or very few such research studies and the time for research have not been able to be sufficient for the applications.

The results of the studies comparing the repeat times of strong earthquakes also indicate that the majority of the materials used in the experiments were not sufficiently investigated, particularly in terms of their behavior over an extended period. In another word, in references, friction interface material was not researched at least during for the repetition time of major earthquakes. For example, Nagashima (1998) investigated Teflon only for ten years and reported that the friction coefficient had increased by 20% after ten years. In other research studies where the restoring mechanism was included in the P-F system, the resulting combined isolation system were period-dependent. Therefore, there is a need to search (a) for alternative interface materials of which behavior was researched and have gained experience in time that have the potential to be used efficiently as friction interface materials and (b) present and prove the restorable particularity of the earthquake action, which can partly be used in the P-F base isolation systems.

Being experienced in time Horasan mortar was used as friction interface material which are provided above requirements. Horasan mortar was widely used in historical buildings across the world, known as "cocciopesto" in the Roman Empire, as "surkhi" in India, as "Korassa" in Greece and as "Homra" in Arabia. The historical use of the Horasan mortar was investigated in many studies and its characteristics were compared to the material produced under the laboratory conditions (Çakmak et al. 1995; Çamlibel 1998; Valluzzi et al. 2005; Binal 2008).

The aim of the present study is to explore (a) the feasibility of using Horasan mortar as the sliding interface material on the presented single friction interface for natural seismic isolation or protection (NSI or NSP, Kasimzade 2009) device (Fig. 3.1); and (b) to explore the restorable particularity of the earthquake action.

The NSI with Horasan mortar as the interface material was devised when the "Walled Obelisk" monument was modeled for investigation in terms of its structural safety (Kasimzade 2009, 2012; Kasimzade et al. 2009, 2011, 2014). The results of the research showed that the bottom part of the monument acts as a frictional responsive seismic isolator. This bottom part of the monument consists of a three-step marble stone and a massive marble stone (enablement) that form the four friction/sliding surfaces with Horasan mortar as the interface. So, the bottom part of the monument acts as the NSI device with three steps and four sliding interfaces (Fig. 3.2).

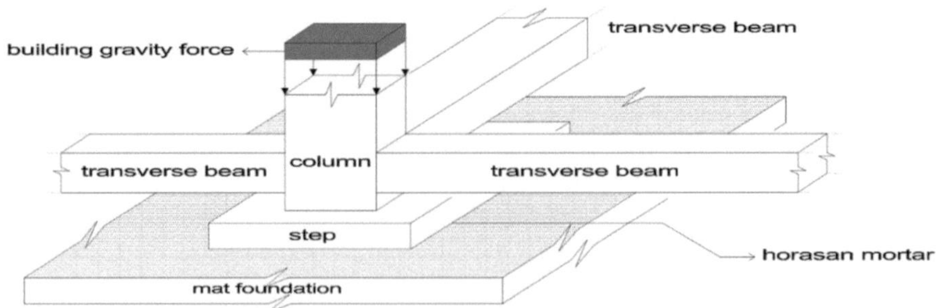

Fig. 3.1 Illustration of the NSI devices with single step and single friction interface of Horasan mortar

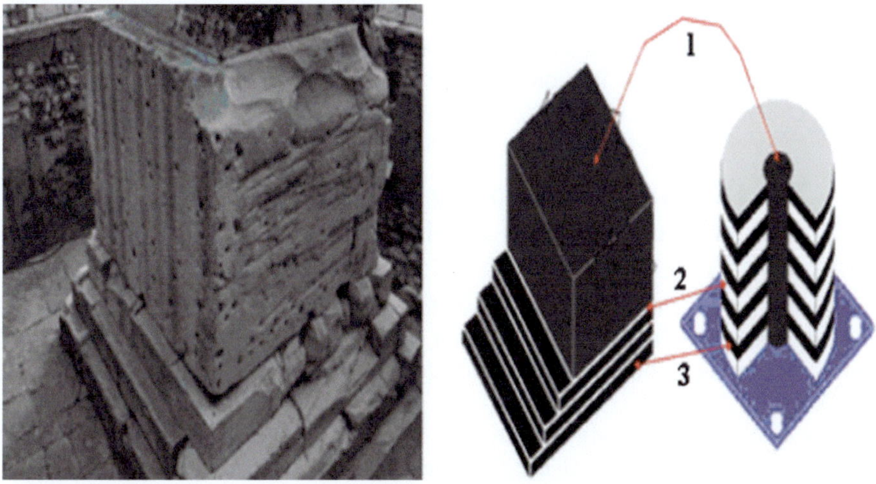

Fig. 3.2 Comparison of the related peculiarities of the Walled Obelisk monument's (Istanbul/Turkey) foundation "Natural Seismic Isolator" (on left) and elastomeric seismic isolation system with lead (LCRB) core (on right) (Kasimzade et al. 2011)

Fig. 3.3 Tomb of cyrus

Tomb of Cyrus (Fig. 3.3) (in Pasargadae, a city in ancient Persia, now Iran) is said to be the oldest and first base-isolated structure in the world where used similar isolation technology and friction interface materials, were used.

3.3 The Discovery of Restoring Mechanism Phenomenon of Earthquake and Its Application

The nature of the earthquake excitation force has been reported to undertake the role of a restoring mechanism, in other words, it maintains the absolute rigid structure around equilibrium (Kasimzade 2009; Kasimzade et al. 2009, 2011, 2013, 2014).

This asserts that, excitation force must be undertaking the role of the restoring mechanism. In another word keeping the structure around equilibrium, may be proved only if, applied to the structure excitation force energy-the total supplied energy equals to the sum of the stored energy (E_{stored}) and the dissipated energy ($E_{\text{dissipated}}$) and total energy during over time must be equal to zero. In another word the following relation must be satisfied (Eq. 3.1):

$$E_{\text{stored}} + E_{\text{dissipated}} = \int_{0}^{t_{\ddot{u}g}} m_{\text{total}} \ddot{u}_g(t) u(t) \mathrm{d}t = 0 \tag{3.1}$$

$$m_{\text{total}} = \text{const} \quad \text{and} \quad m_{\text{total}} \neq 0 \tag{3.2}$$

and taking into consideration that (Eq. 3.2) the above relation becomes as (Eq. 3.3)

$$E_{\text{stored}} + E_{\text{dissipated}} = m_{\text{total}} \int_{0}^{t_{\ddot{u}g}} \ddot{u}_g(t) u(t) \mathrm{d}t = 0 \tag{3.3}$$

Applying integration by part, chosen to be (Eq. 3.4–3.5)

$$u = u(t), \quad \ddot{u}_g(t)\mathrm{d}t = \mathrm{d}v \tag{3.4}$$

$$v = \int_{0}^{t_{\ddot{u}g}} \ddot{u}_g(t)\mathrm{d}t \tag{3.5}$$

Equation (3.5) and from here it is obtained that (Eq. 3.6)

$$\int_0^{t_{\ddot{u}_g}} \ddot{u}_g(t)u(t)\mathrm{d}t = uv - \int_0^{t_{\ddot{u}_g}} v\mathrm{d}u, \tag{3.6}$$

where m_{total} is the total mass of the structure; $\ddot{u}_g(t), t_{\ddot{u}_g}$ is the earthquake acceleration and duration time respectively; $u(t)$ is NSI system device displacement.

As seen from the relations (Eqs. 3.1 or 3.3) for it proof, it is necessary to prove that total earthquake acceleration area over excitation time (Eq. 3.5) must be equal to zero (and supporting it, relation (Eq. 3.6) and consequently (Eq. 3.1) or (Eq. 3.3) becomes zero). Real earthquake's acceleration-time relation has many discontinuity points, because relation (Eq. 3.6) was investigated numerically (trapezoidal numerical integration) for the various acceleration records from PEER (PEER 2011), with different characteristics. Results are presented in Table 3.1.

As seen from Table 3.1 target assertion (Eq. 3.5 equal to zero) was reached with error range which may be acceptable. This very little deviation from the target assertion may be explicable with numerical integration method error.

This earthquake action phenomenon presented by Eqs. 3.1, 3.3 and 3.5 was found out, proved and was used for mathematic modeling monument's seismic isolation system by the author (Kasimzade 2009; Kasimzade et al. 2009, 2011, 2012, 2014) for the first time. It may enable the design of a seismic isolation system simpler and very cheaper as far as maximum possible (Kasimzade et al. 2013).

The results were obtained with the assumption of absolutely rigidity of superstructure. In this case there is a P-F isolation sliding interface between the superstructure and foundation. The foundation maintains its equilibrium during an earthquake, but superstructure slides. In other words, the earthquake partly keeps the superstructure around equilibrium while another part is controlled by the friction coefficient of the isolation interface. As a result, the duration of the earthquake excitation has a residual displacement on the superstructure (Fig. 3.4).

Hence it has been found out that the NSI device has no restoring mechanism and no natural period. This particularity indicates that excitation force must be undertaking the role of the restoring mechanism, in another word, keeping the structure around equilibrium. Based on this phenomenon and the NSI device found out, for

Table 3.1 Characteristics of various earthquake records and appropriate ground acceleration area obtained by trapezoidal numerical integration and the integration error from target assertion

Earthquake name, date, time/ characteristics	Duzce, Turkey, 12.11.1999 18:57:00	Erzincan Turkey, 27.12. 1939 01:57:00	Kocaeli Turkey, 19.07. 1999 03:02:00	Cape Mendocino California USA 25.04.1992 11:06:00	Chichi Taiwan Nantou County 21.09.1999 01:47:12	Tabas Iran 16.09.1978 19:18:00
Acceleration area & error in %	−0.0035 0.35	0.00083 0.083	−0.00074 0.074	−0.00095 0.095	−0.000045 0.0045	0.00034 0.034

Fig. 3.4 Illustration of the residual displacement at the bottom of the "Walled Obelisk" monument in Istanbul in December 2008. The massive stone surface was shifted above the level of 1.24 m, which is not entirely the original position (Kasimzade et al. 2011)

application to day's structures (Fig. 3.1). The new isolation bearing presented in this paper is referred to as no restoring mechanism seismic base isolation—NSI system and the first implementation is illustrated in Figs. 3.5a–d.

3.4 Frictional Model for the Pure Friction (P-F) Seismic Isolation System

The Coulomb friction model defines an equivalent shear stress (τ), at which sliding on the surface begins as a friction of the contact pressure (Kaldal 2013) (Fig. 3.6) (Eq. 3.7)

$$\tau = \mu P + C_{\text{cohesion}}, \tag{3.7}$$

where C_{cohesion} specifies the cohesion sliding resistance and μ is the friction coefficient is expressed in (Eq. 3.7).

Considering velocity dependency, the frictional coefficient is expressed as (Kato et al. 1974; Constantinou et al. 1990) (Fig. 3.7) (Eq. 3.8)

$$\mu\cdot(v) = \mu_{\min} + (\mu_{\max} - \mu_{\min})e^{-d|v|}, \tag{3.8}$$

where d is the parameter for velocity dependency; v is the sliding velocity; $\mu_{\min}$, $\mu_{\max}$ are the minimum and maximum friction coefficients respectively (Enokida and Nagae 2017; Symans et al. 2000).

For mortar contact, $\mu = \mu_{\max}$, i.e., dynamic effects are ignored (LS-DYNA Keyword User's Manual R10.0 Vol 1 2017)

Fig. 3.5 First natural seismic isolated (NSI) private building in Turkey/Samsun (Incesu Mah. Sirincak Sok. No: 2/1) **a** Preparation of the friction interface-Level 1 **b** Preparation of the friction interface-Level 2 **c** General view of building—SN direction **d** General view of building—NS direction

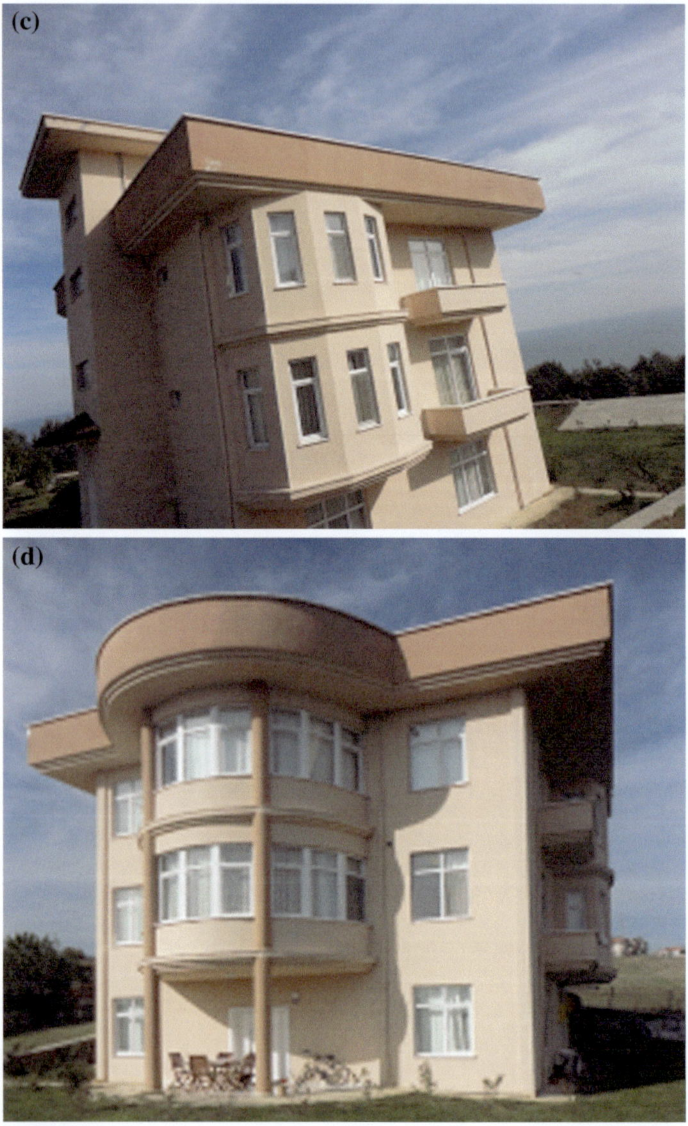

Fig. 3.5 (continued)

Here μ_{max}, μ_{min} are the minimum and maximum friction coefficients, d is the parameter for velocity dependency and v is the sliding velocity, respectively. For definition of the mentioned coefficients numerical values for selected friction interface materials, the equation of motion experimental equipment on shaking table is expressed as follows (Fig. 3.8).

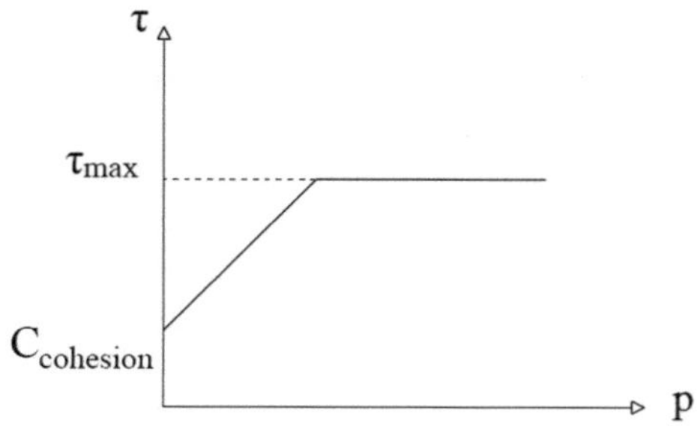

Fig. 3.6 Relation between pressure and shear stress

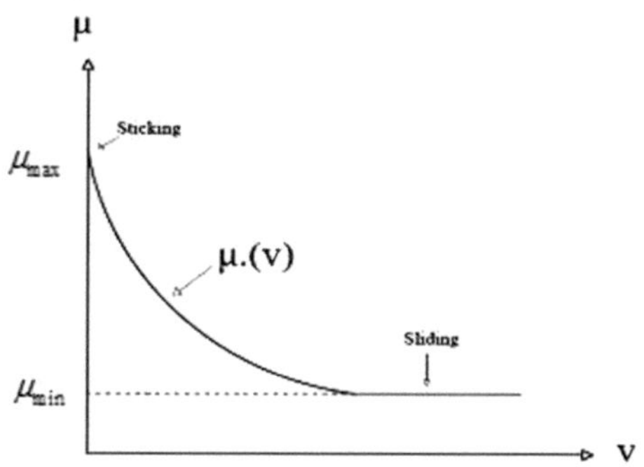

Fig. 3.7 Illustration of the velocity and friction coefficient

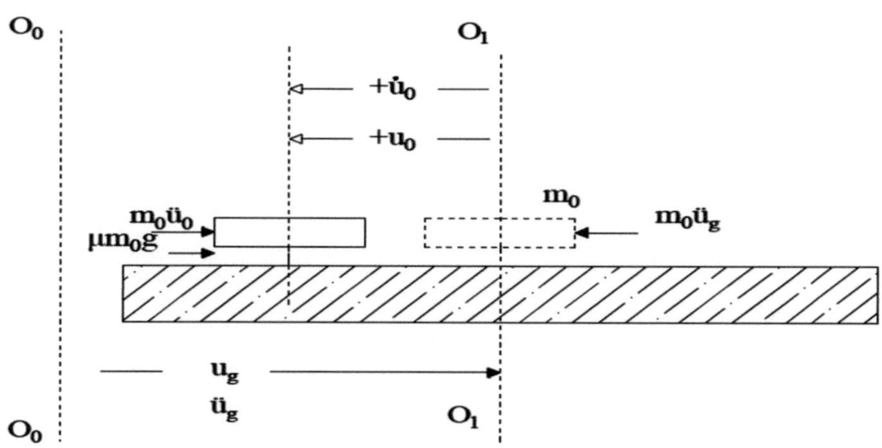

Fig. 3.8 Experimental equipment on shaking table

Assuming that the underground part of the structure (or the shaking table) is moved u_g meter from initial position $O_0 - O_0$ to position $O_1 - O_1$ with an acceleration $\ddot{u}_g$ in the same direction and superstructure with mass m_0 under adopted force $m\ddot{u}_g$, slides u meter relative to the position $O_1 - O_1$, and superstructure takes the acceleration $\ddot{u}$ and velocity $\dot{u}$ respectively. By arranging the projection of forces on the positive moving direction of the superstructure (m_0), its equation of motion becomes as (Eqs. 3.9 and 3.10)

$$\sum F_{+u} = 0; \quad m_0\ddot{u}_g - m_0\ddot{u}_0 - \mu g m_0 = 0 \tag{3.9}$$

$$\ddot{u}_0 + g\mu(+\dot{u}_0) = \ddot{u}_g \tag{3.10}$$

Equation of motion superstructure (Eq. 3.10) valid for the positive velocity of the superstructure. By generalizing it for both directions it becomes (Eq. 3.11);

$$\ddot{u}_0 + g\mu \text{ sgn } (\dot{u}_0) = \ddot{u}_g \tag{3.11}$$

where $u_0, \dot{u}_0, \ddot{u}_0$ are the relative displacement, velocity, acceleration of the superstructure relatively to the underground part of the structure; $(\ddot{u}_0 - \ddot{u}_g)$ is the absolute acceleration of the superstructure (m_0).

The observed response $u_{\bullet 0}(t_i) = u_{\bullet 0i}$ for any time t_i during shaking table excitation duration will be defined experimentally.

The adopted target function (Eq. 3.12)

$$F_t = \sum (u_{\bullet 0i} - u_{0i})^2 \tag{3.12}$$

(F_t) will be close to zero, when $\mu_{\max}$, $\mu_{\min}$, d approach appropriate values.

By the minimization (the slowdown Newton method) of the target function for Horasan mortar, parameters are obtained as:

$$\mu_{\max} = 0.37, \mu_{\min} = 0.26, d = 11$$

Here, Duzce Earthquake is used with duration 55.9 s and PGA = 8.0588 m/s^2.

For the gray-iron/mortar and stainless steel above parameters was obtained as (Iemura et al. 2007) respectively:

$$\mu_{\max} = 0.28, \ \mu_{\min} = 0.214, \ d = 11, \ \mu_{\max} = 0.2, \ \mu_{\min} = 0.14, \ d = 11$$

Here was used horizontal earthquake excitations JMA Kobe- NS (PGA = 8.18 m/s^2) and JMA Kobe- EW (PGA = 6.17 m/s^2) respectively.

Due to the relation given by Eq. 3.8, friction coefficient diagrams for the above frictional interfaces are presented on Fig. 3.9a–c respectively.

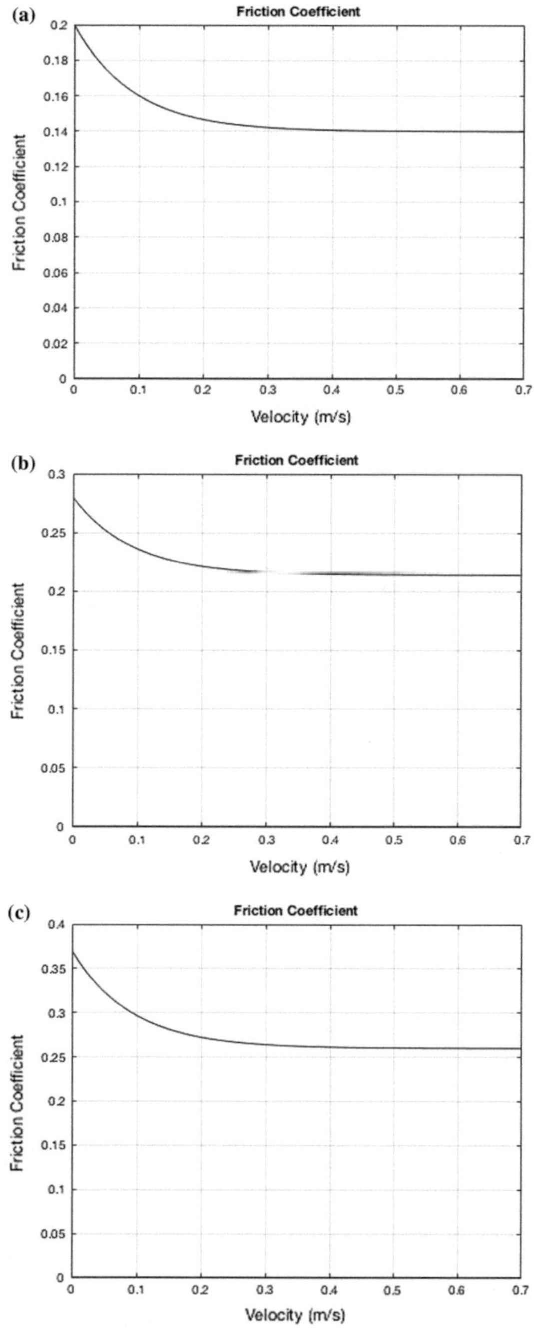

Fig. 3.9 Friction coefficient diagrams by the relation Eq. 3.8 for the frictional interfaces **a** Friction coefficient behavior for the stainless steel friction interface **b** Friction coefficient behavior for the gray/iron mortar friction interface **c** Friction coefficient behavior for the Horasan mortar friction interface

3.5 Four-Storey Building's Mathematical Model with Frictional Bearing

Based on the relation presented by Eqs. 3.13–3.15,

$$F_{\mu} = \mu N = \mu M g \tag{3.13}$$

$$M = m_0 + m_{ss} \tag{3.14}$$

$$m_{ss} = \sum_{i=1,n} m_i \tag{3.15}$$

the sliding phase defined by relation (Eq. 3.16)

$$\left| \frac{m_0}{M} (\ddot{u}_g - u_0) + \frac{c_1 \dot{u}_1 + k_1 u_1}{M} \right| \geq \mu g \tag{3.16}$$

Equation of motion can be exhibited by the set of system differential equations as (Eqs. 3.17a and 3.17b)

$$m_0 \ddot{u}_0 + \mu M g \text{ sgn } (\dot{u}_0) - c_1 \dot{u}_1 - k_1 u_1 = m_0 \ddot{u}_g \tag{3.17a}$$

$$[m]\{\ddot{u}\} + [c]\{\dot{u}\} + [k]\{u\} = \left[\frac{\mu M g}{m_0} \text{ sgn } (\dot{u}_0) - \frac{c_1 \dot{u}_1 + k_1 u_1}{m_0} \right] [m]\{1\} \tag{3.17b}$$

Otherwise, if the condition (Eq. 3.16) is not met, equation of motion because as (Eq. 3.18),

$$[m]\{\ddot{u}\} + [c]\{\dot{u}\} + [k]\{u\} = +\ddot{u}_g [m]\{1\}, \tag{3.18}$$

where sgn is defined as (Eq. 3.19)

$$\begin{array}{l} \text{sgn} = \frac{1}{\mu g} \left[\frac{m_0}{M} (\ddot{u}_g - \ddot{u}_0) + \frac{c_1 \dot{u}_1 + k_1 u_1}{M} \right], \text{ if} |\dot{u}_0| = 0 \\ \text{else} \quad \text{if} \quad u_0 > 0, \text{ sgn} = 1, \text{ else} \quad \text{sgn} = -1, \end{array} \tag{3.19}$$

Parameters included in Eqs. 3.13–3.19 illustrated in Fig. 3.10 and explained.

Here $[m]$, $[c]$, $[k]$ are mass, damping, and stiffness matrices respectively, u is displacement, $\dot{u}$ is velocity, u_0 is base displacement, $\ddot{u}_g$ is earthquake acceleration.

Set of equations (Eqs. 3.17a–3.17b) are highly nonlinear. It is created by FEM and Runge–Kutta method is used with Matlab tools (MATLAB R2017b Documentation 2017).

In the presented seismic base isolation application, the superstructure's mass distribution is designed such that the occurrence of overturning moments is prevented. Reinforced concrete transfer beams (they are red color in Figs. 3.11 and 3.12) with the floor slabs on it (Slab types are hardy slabs. All beams are hidden

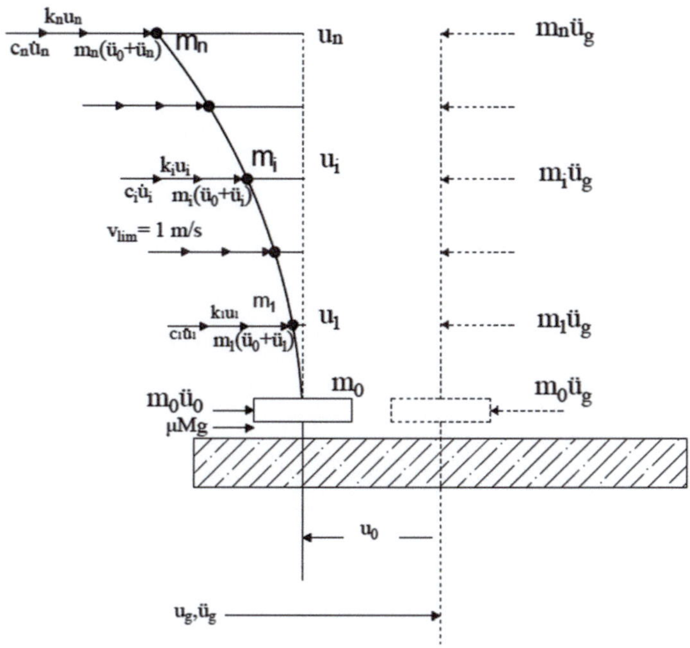

Fig. 3.10 Illustration of the four story building's mathematical model with frictional bearing

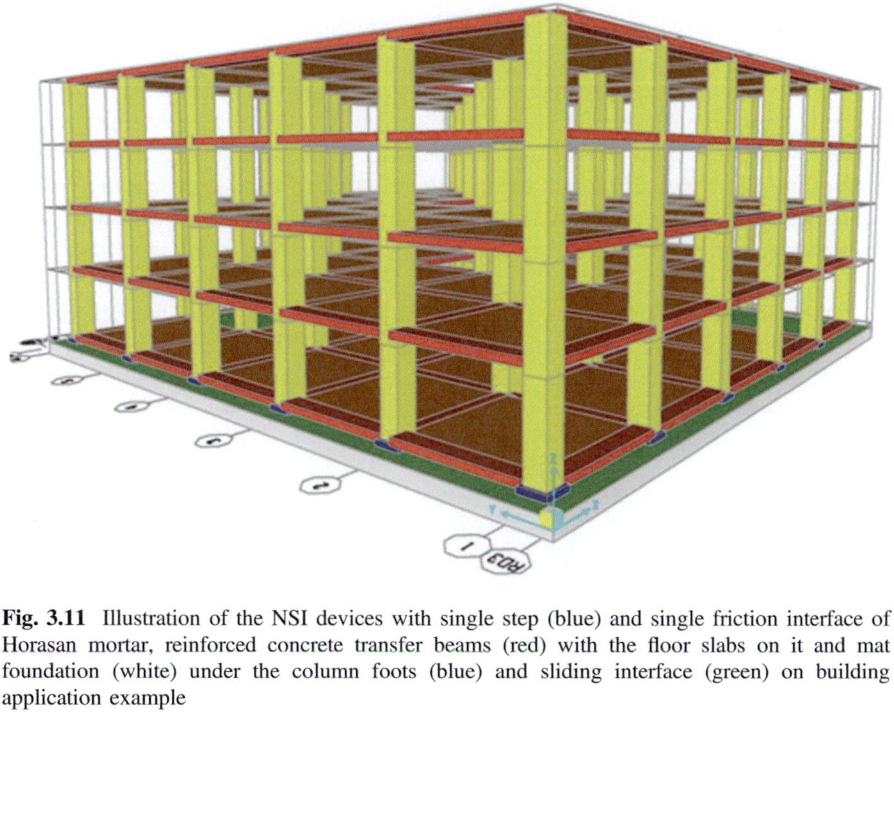

Fig. 3.11 Illustration of the NSI devices with single step (blue) and single friction interface of Horasan mortar, reinforced concrete transfer beams (red) with the floor slabs on it and mat foundation (white) under the column foots (blue) and sliding interface (green) on building application example

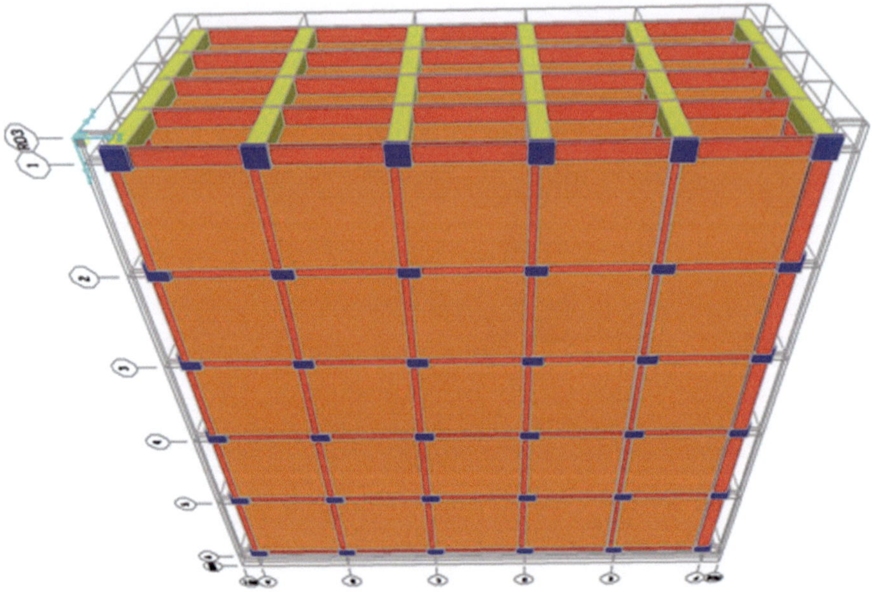

Fig. 3.12 Bottom view illustration of the NSI devices with single step (blue) on building application example

beams. It is not shown in Fig. 3.1) are designed to ensure the synchronous working condition of the column foots (they are blue color in Figs. 3.11 and 3.12) on the sliding interface (they are green color in Fig. 3.11) under earthquake excitation and building application example illustration shown in Figs. 3.11 and 3.12 superstructure separated from the mat foundation (it is white color in Fig. 3.11) only by sliding interface (they are green color in Fig. 3.11). In another word, the ground floor slabs (not shown in Fig. 3.1) with counter transfer beams is designed so that, under horizontal earthquake excitation they are in their own plane may keep rigid body behavior.

For application a four-storey hospital building is used, which previously was designed under Duzce earthquake with $PGA = 8.0599\ \text{m/s}^2$ and duration time $t_d = 59.99$ s. Extructed from here are the mass $[m]$, damping $[c]$ and stiffness $[k]$ matrices presented as follow (Eqs. 3.20–3.22)

$$[m] = 1e^6 \begin{bmatrix} 2.025 & 0 & 0 & 0 \\ 0 & 2.0033 & 0 & 0 \\ 0 & 0 & 2.0033 & 0 \\ 0 & 0 & 0 & 1.6755 \end{bmatrix} (\text{kg}) \tag{3.20}$$

$$[k] = 1e^{10}\begin{bmatrix} 3.7486 & -2.1587 & 0 & 0 \\ -2.1587 & 4.3173 & -2.1587 & 0 \\ 0 & -2.1587 & 4.3173 & -2.1587 \\ 0 & 0 & -2.1587 & 2.1587 \end{bmatrix}(\text{N/m}) \tag{3.21}$$

$$[c] = 1e^{7}\begin{bmatrix} 2.6101 & -0.8576 & -0.1669 & -0.0749 \\ -0.8576 & 2.6571 & -0.9001 & -0.2060 \\ -0.1669 & -0.9001 & 2.5759 & -0.9917 \\ -0.0749 & -0.2060 & -0.9917 & 1.6596 \end{bmatrix}(\text{Nsn/m}) \tag{3.22}$$

The response quantities of interest are the top and base levels of the examined hospital building and are presented in Table 3.2 and Figs. 3.14, 3.15 and 3.16 respectively from the Duzce earthquake (Fig. 3.13).

As seen the structure base acceleration is decreased by 45.6437, 38.3919, 33.0212%; top floor acceleration is decreased by 64.9959, 61.4724, 58.4987%; top floor drift is decreased by 58.1769, 54.0705, 48.1211% regarding to fixed based structure for stainless steel, gray-iron mortar and Horasan mortar respectively. Pure frictional systems maximum displacement was 0.068, 0.0469, 0.0276 m for stainless steel, gray-iron mortar and Horasan mortar respectively.

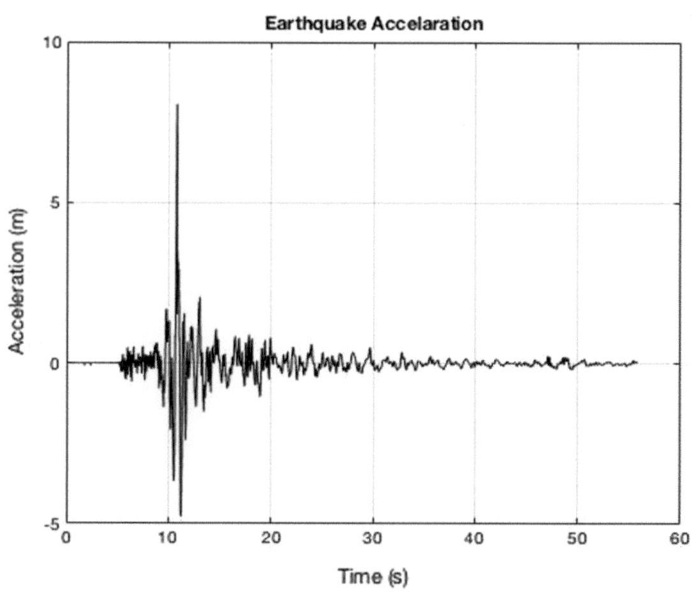

Fig. 3.13 Horizontal acceleration component of the Duzce (12/11/1999) earthquake

Table 3.2 Response of the pure frictional isolated structure with variable frictional interfaces from the Duzce earthquake

Maximum response (relative)	Sliding base interface			Fixed base
	Stainless steel (μ_{min} = 0.14, μ_{max} = 0.2, d = 11)	Gray-iron/mortar (μ_{min} = 0.214, μ_{max} = 0.28, d = 11)	Horasan mortar (μ_{min} = 0.26, μ_{max} = 0.37, d = 11)	
u_b (m) (residual)	0.068 0.068	0.0469 0.0469	0.0276 0.0276	0
a_b (m/s^2)	4.3805	4.9649	5.3977	8.06
100 $(a_g - a_b)/a_g$ (reduction)	45.6437	38.3919	33.0212	0
u_4 (m)	0.0029	0.0035	0.0043	0.0105
$\Delta u_4 = u_4 - u_3$ (inter story drift)	3.8376e-4	4.2144e-4	4.7603e-4	9.1758e-4
100 $(\Delta u_{4FB} - \Delta u_4)/$ Δ u_{4FB} (reduction)	58.1769	54.0705	48.1211	
a_4 (m/s^2)	4.6721	5.1424	5.5393	13.3473
100 $(a_{4FB} - a_4)/a_{FB}$ (reduction)	64.9959	61.4724	58.4987	

Fig. 3.14 Structure response with Stainless steel interface from the Duzce earthquake **a** Pure friction isolated building top floor displacement **b** Pure friction isolated building top floor velocity **c** Pure friction isolated building top floor acceleration **d** Pure friction isolated building bearing displacement **e** Pure friction isolated building bearing velocity **f** Pure friction isolated building bearing acceleration

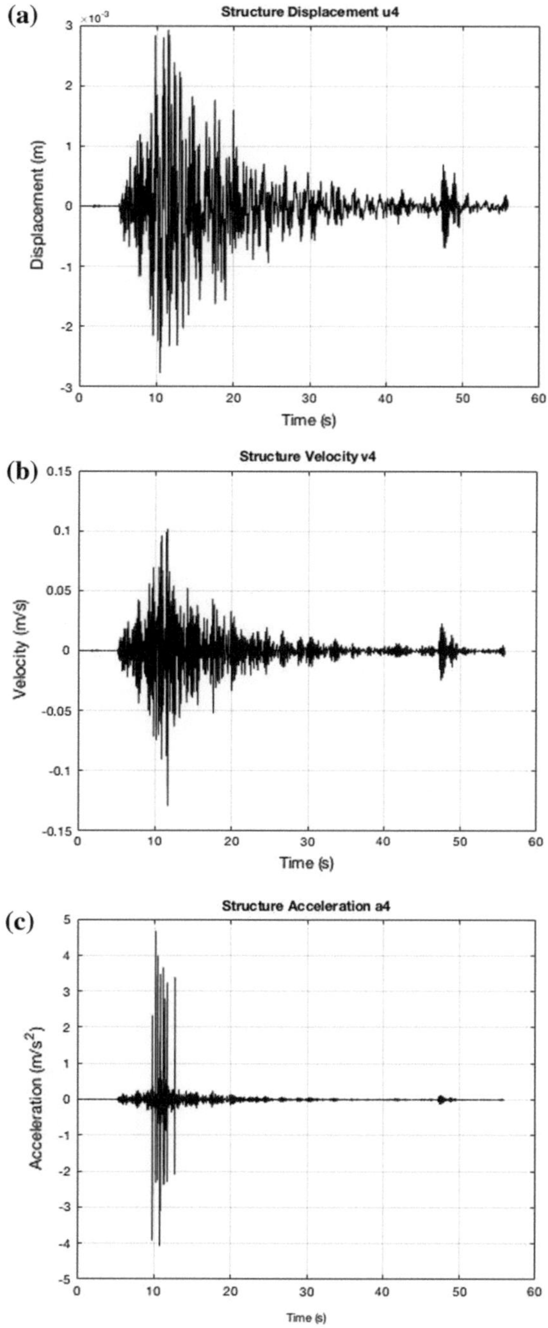

Fig. 3.14 (continued)

(d)

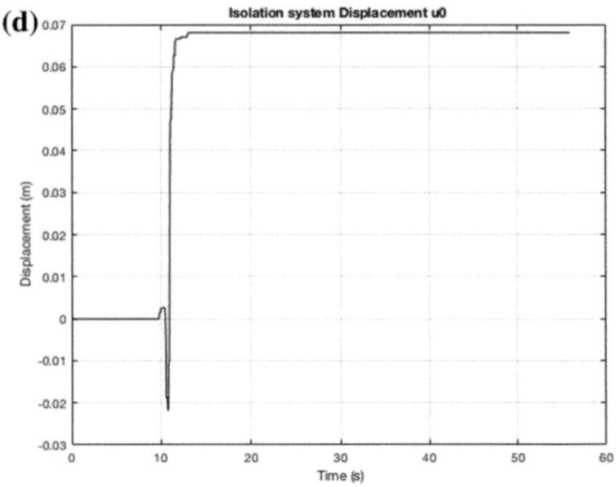

(e)

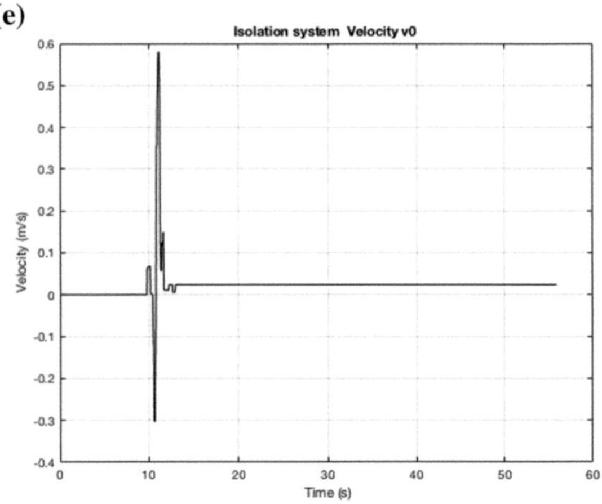

(f)

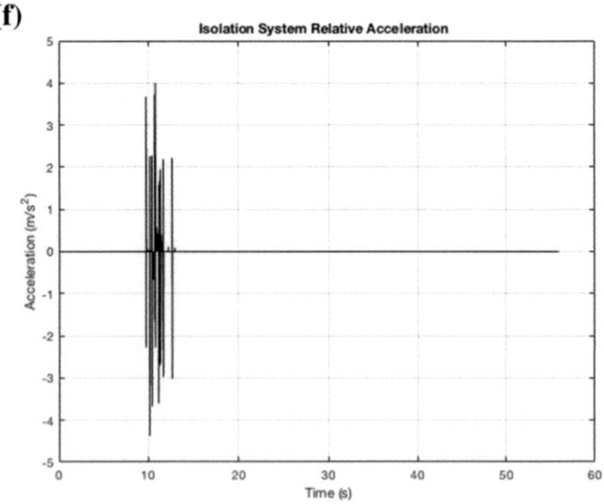

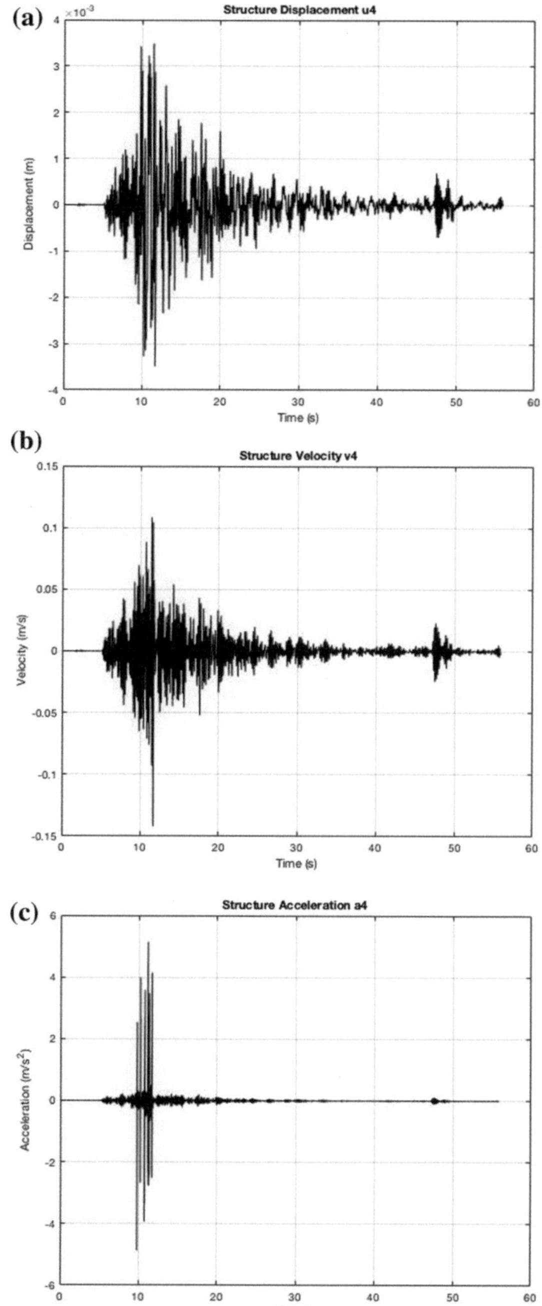

Fig. 3.15 Structure response with gray-iron/mortar from the Duzce earthquake **a** Pure friction isolated building top floor displacement **b** Pure friction isolated building top floor velocity **c** Pure friction isolated building top floor acceleration **d** Pure friction isolated building bearing displacement **e** Pure friction isolated building bearing velocity **f** Pure friction isolated building bearing acceleration

Fig. 3.15 (continued)

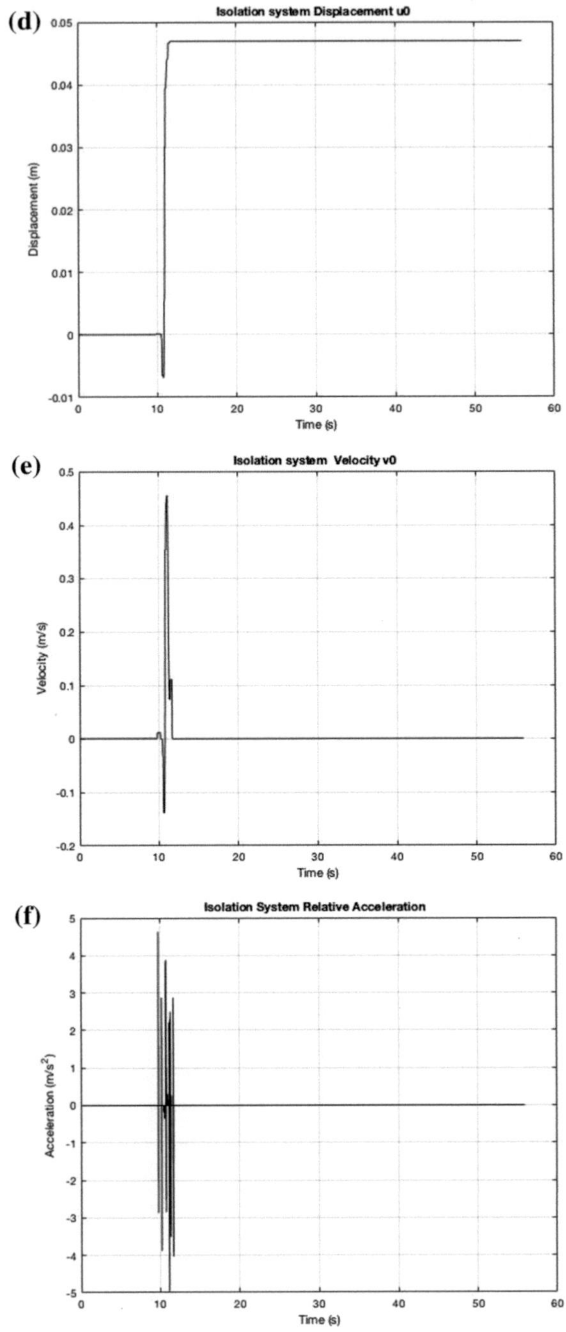

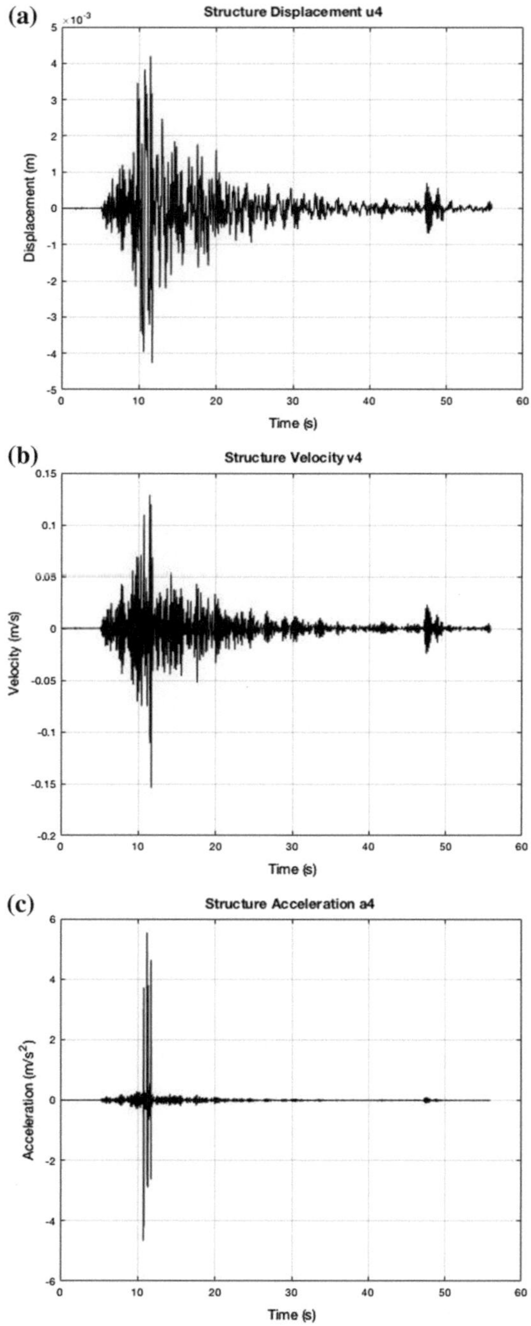

Fig. 3.16 Structure response for Horasan mortar from the Duzce earthquake **a** Pure friction isolated building top floor displacement **b** Pure friction isolated building top floor velocity **c** Pure friction isolated building top floor acceleration **d** Pure friction isolated building bearing displacement **e** Pure friction isolated building bearing velocity **f** Pure friction isolated building bearing acceleration

Fig. 3.16 (continued)

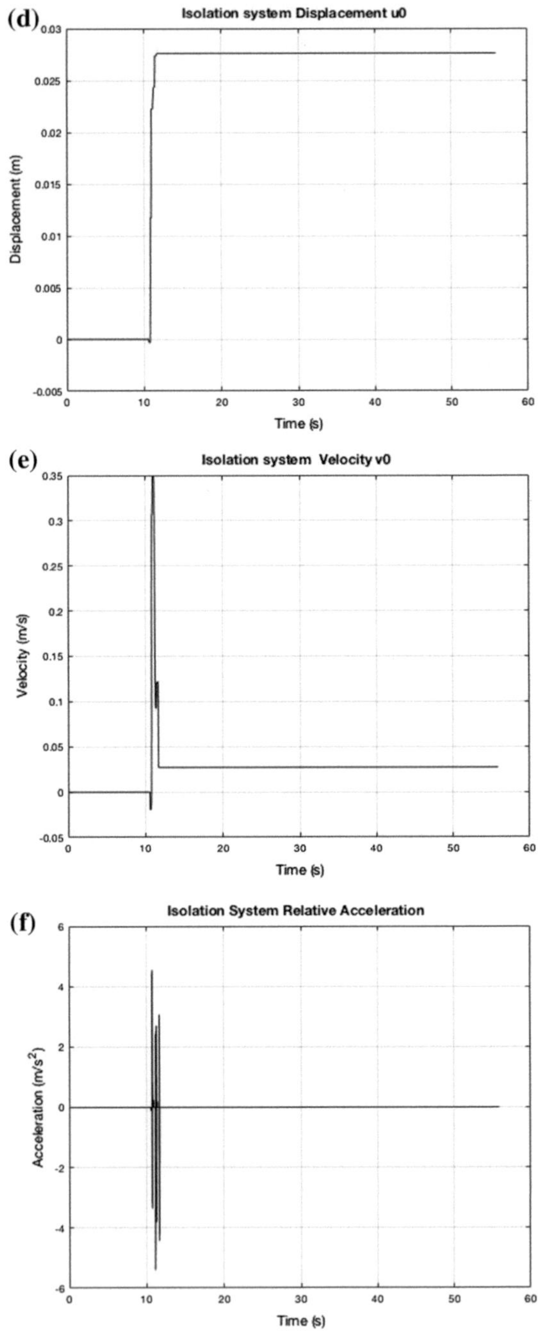

3.6 Assessment of the NSI Device's Material Stress–Strain Range

For assessment of the NSI device's material stress-and strain range, the NSI device finite element modeling has been provided by ANSYS software under the vertical 5×10^6 N (approximately equivalent to the force on a single column foot's in a four-storey reinforced concrete building) force and horizontal acceleration of the earthquake and analysis result have been compared to the related ultimate values presented in Table 3.3. For this force one step reinforced plates with dimensions of $1.4 \times 1.4 \times 0.6$ m and 1% reinforcing percentages of the steel bars is for each of the *x*, *y*, *z* coordinate direction. The thickness of the Horasan mortar between the step and mat foundation is 0.02 m. The parameters of the materials presented NSI device model (Fig. 3.1) are presented in Table 3.3.

In the presented example, steel bars have been used for reinforcing. Other possible reinforcing materials include fiber, which provides more protection against corrosion. It is also possible to replace the reinforced concrete with a more durable material. According to the Von Mises yield criterion the yielding of materials begins when the second deviator stress invariant reaches a critical value, otherwise, the system remains in linear case. As shown in Table 3.3, the second deviator stress invariant (the ultimate tensile stress value) for Horasan mortar is low. Therefore, under the strong lateral motion of the earthquake, Horasan mortar is cracked and will exhibit the characteristics of granular material. This was taken into consideration in the estimation of bound of the friction coefficient. The known bound of the friction coefficient by fuzzification or stochastic methods (in Kasimzade 2002; Kasimzade and Tuhta 2004) can be defined the bound of response. Inversely, the known response bound can be defined (Kasimzade and Tuhta 2012) the bound of friction coefficient. Plastic properties of the mortar in the construction phase as well as after earthquakes horizontal deformation causes largely maintain its shape. Granular material is devoid of these features alone.

Three major parts of the plasticity model and the nonlinear material properties of Horasan presented here are: flow rule, yield criterion, and hardening rule. The bilinear stress–strain curve is used to study the Horasan mortar as nonlinear contact

Table 3.3 Physical and mechanical parameters of the concrete, reinforced concrete and Horasan mortar used in the seismic NSI device

Material	Module of elasticity (N/m^2)	Poisson ratio	Density (kg/m^3)	Compressive strength (N/m^2)	Tensile strength (N/m^2)	Shear strength (N/m^2)
Concrete (C30)	0.32×10^{11}	0.2	0.22×10^4	0.3×10^8	0.19×10^7	0.27×10^8
Reinforced concrete	0.38×10^{11}	0.25	0.24×10^4	0.76×10^8	0.76×10^7	0.304×10^8
Sliding interface	3.89×10^9	0.167	13.342×10^2	0.49×10^7	1.71×10^6	0.275×10^7

interface material behavior, which is also characterized by rate independent (independent of time) plasticity. The yield criterion determines the stress level at which yielding of a material will occur, Von Mises criterion is used. In associative flow rule, which is a type of flow rule pertaining to bilinear isotropic hardening behavior, the plastic potential gradient is normal to the yield surface. The hardening rule describes the changing of the yield surface with progressive yielding, and the two fundamental hardening types, isotropic (usually assumed for small strain deformation) and kinematic are existent. Here the kinematic hardening plasticity is used. In this investigation, plasticity and the nonlinear material properties of Horasan mortar has been modeled and implemented by ANSYS software (ANSYS Online Manual's Release 5.5 2017).

For the contact element, the conventional Coulomb frictional model was used (Eq. 3.8). The structure with presented aseismic NSI device (Fig. 3.1), superstructure's load transfer to the mat foundation has been realized by the one step reinforced plate which was placed under each column (as column foot). Horasan mortar as the friction interface material has been used between the reinforced plate (step) and mat foundation.

For modeling contact and sliding by the above relations between Horasan mortar —one step reinforced plate and Horasan mortar–mat foundation interfaces "contact" (CONTA174) and "target" (TARGE170) finite elements are used. This element is located on the surfaces of three-dimensional solid elements (one step reinforced plate and reinforced mat foundation has been modeled by SOLID65 element) with midside nodes.

Penalty (penalize the nodes which penetrate) method is used as contact algorithm. All contact formulations require the solver to identify the elements where the surface penetration has occurred. Surface integration (Gauss) points for contact detection method are used. Surface integration point method allows for additional points to detect penetration between surfaces. Surface to surface contact is truly assembled, which has been used in presented implementation of contact problem. Contact stiffness has been updated for every step of iteration. Updating the penalty stiffness allow the solver to calculate the penalty stiffness smartly. In order to assess the nonlinear performance (Belytschko et al. 2001), the stress–strain state of 3D model and analysis of the contact problem has been modeled and implemented by ANSYS software.

The reinforced steps and the mortar between them are modeled using the reinforced solid and structural solid elements. Horasan mortar, between the step and the mat foundation, form one sliding surfaces and the contact elements. The modeling system has 8927 elements in total including 1561 joint elements. Contact problems are highly nonlinear and required significant computer resources. The finite element model of the system has been excited by the Duzce earthquake (12/11/1999) and the results have been obtained through the nonlinear transient analysis.

Some of the results of the analysis are presented below (Table 3.4).

Analysis results have been illustrated in Figs. 3.17, 3.18, 3.19, 3.20, 3.21, 3.22 and 3.23. Units in the manuscript are presented in the international system of units SI (length meters, speed meters/second, acceleration m/sec^2, stresses Newton/m^2).

Table 3.4 The absolute displacement and acceleration of the single-step NSI device from Duzce earthquake with an excitation time of t_{end} = 55.88 s

Dimension of the friction interface under the every column's foot (*m*)	Absolute displacement- *m* (reduction regarding)			Absolute acceleration—m/sec^2 (reduction regarding $\|\ddot{u}_g\|_{max}$= 8.058780 m/s^2, in %)	
	Interval	Average	Displacement state for t_{end}	Interval	Average
1.4 × 1.4	(−0.0464384) −(0.00497443)	0.0025706	−0.035062	−4.17922 −3.30435 (48.14–59.00)%	3.741785 (53.57) %

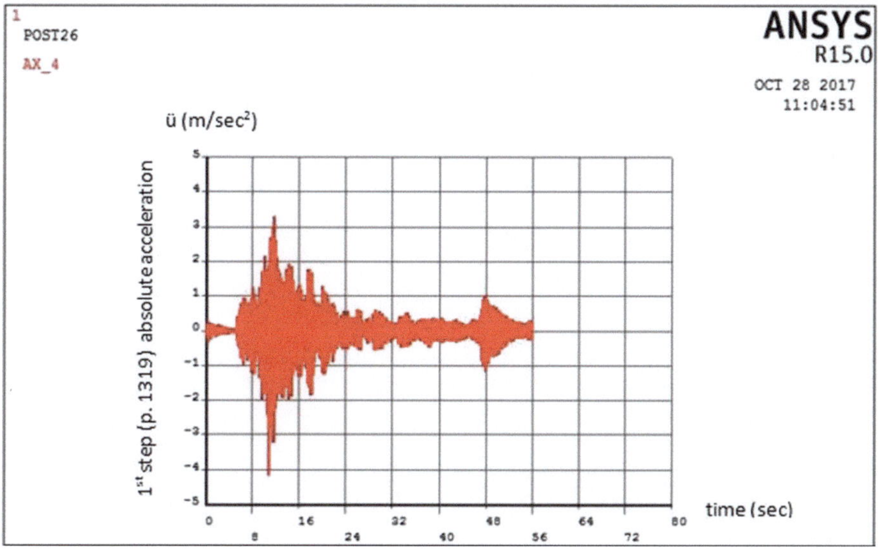

Fig. 3.17 The absolute acceleration of the step

As can be seen, the reinforced plate (step) obtained von Mises stress ranges between (+0.107 × 10^7) and (+0.321 × 10^7) N/m^2, normal (in *Z* direction) stress ranges between (−0.34 × 10^7) and (−323884) N/m^2, shear (in *XY* plane) stress ranges between (−282901) and (+294814) N/m^2 are not exceeded the material ultimate compressive strength (0.76 × 10^8) N/m^2, ultimate tensile strength (0.76 × 10^7) N/m^2 and ultimate shear strength (0.304 × 10^8) N/m^2, respectively. The NSI device in the earthquake action direction obtained a value of 4.97–46.44 mm total average displacement (Fig. 3.23) and maximum acceleration

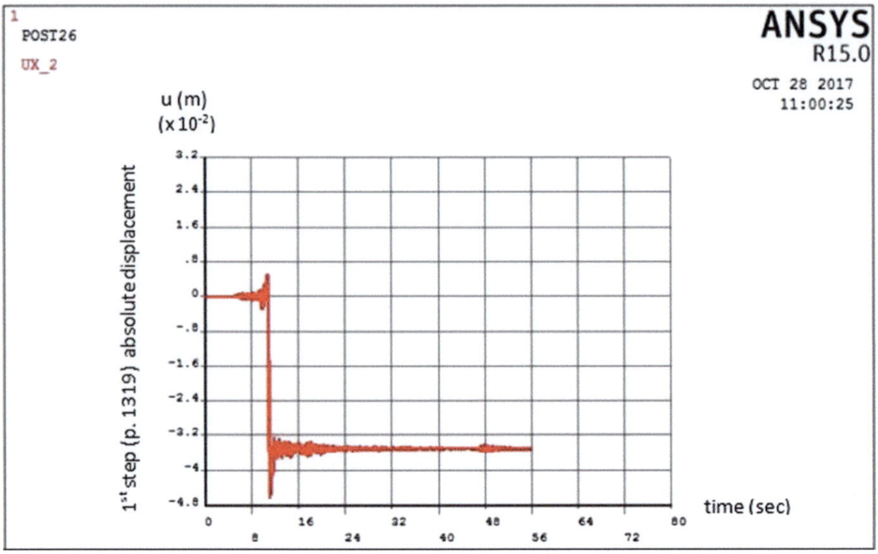

Fig. 3.18 The absolute displacement of the step

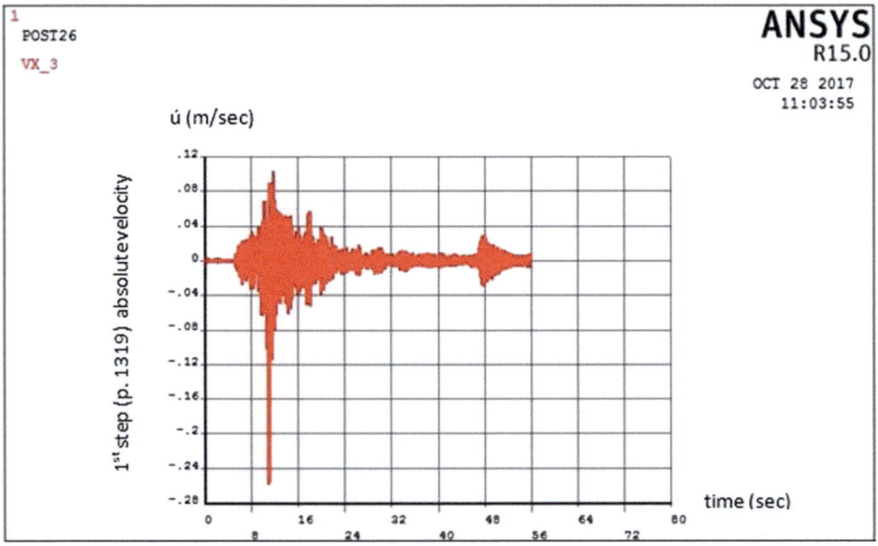

Fig. 3.19 The absolute velocity of the step

responses are reduced by an average of 48–59% (Table 3.4 and Fig. 3.17). The residual displacement is 35.06 mm (Fig. 3.18). The NSI device's absolute velocity and time relation is presented in Fig. 3.19.

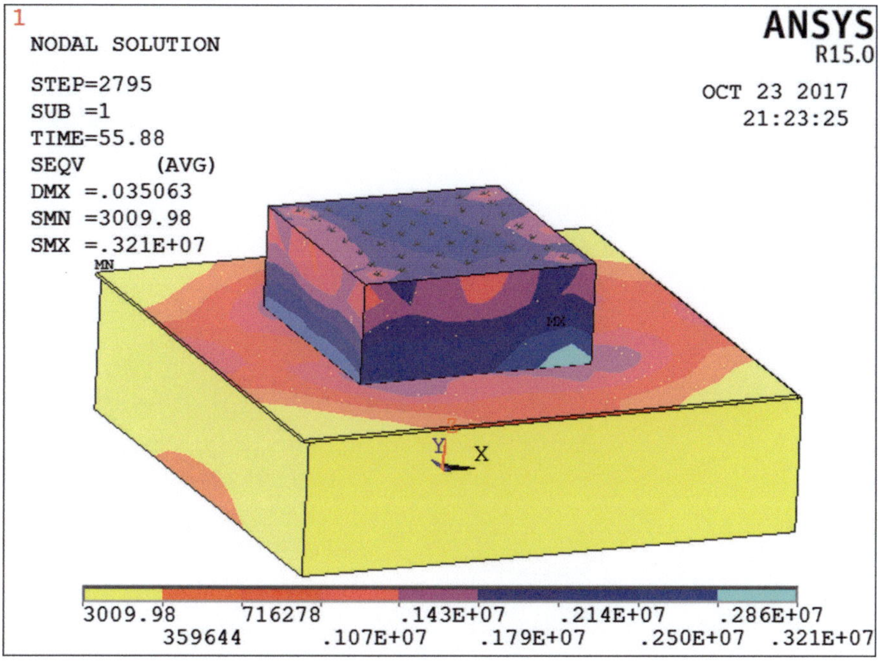

Fig. 3.20 Von Mises stress of the NSI device

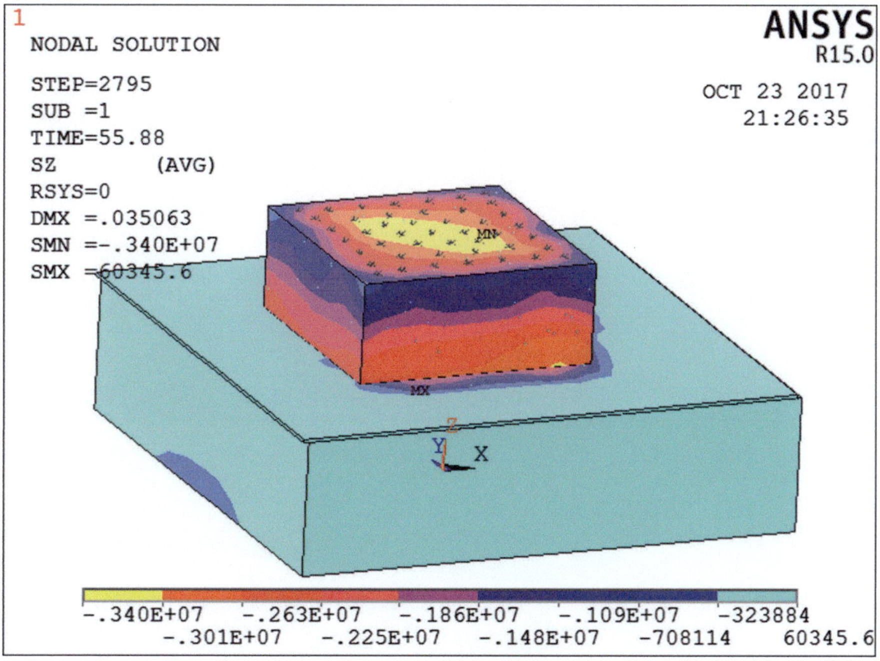

Fig. 3.21 Normal (in *Z* direction) stresses of the NSI device

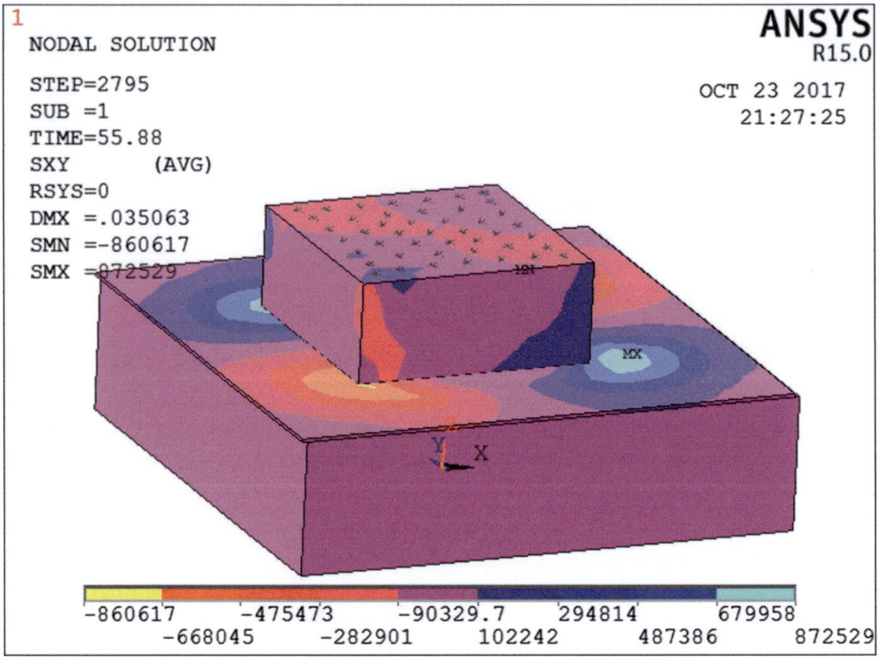

Fig. 3.22 Shear (in *XY* plane) stresses of the NSI device

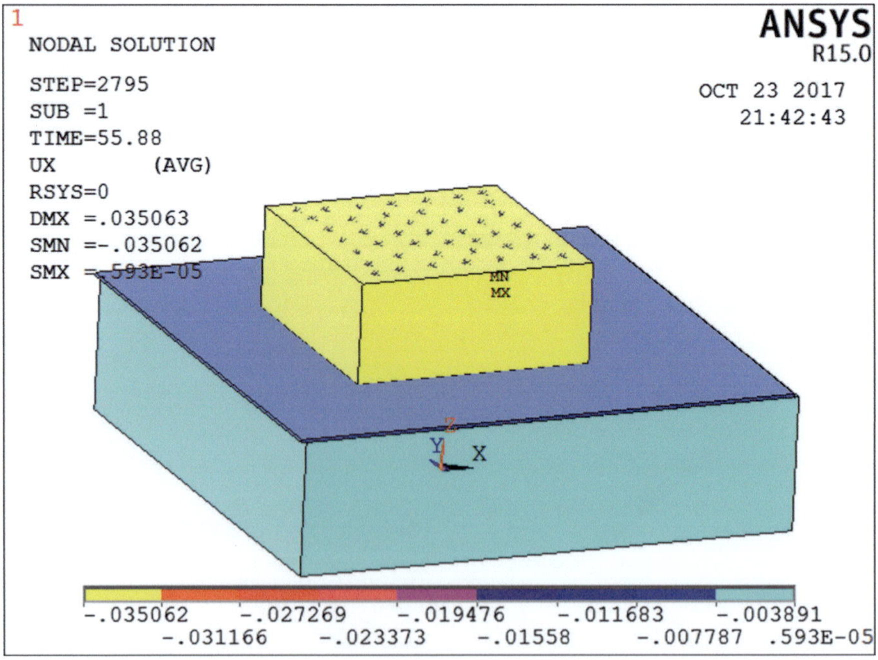

Fig. 3.23 Lateral (in global *X* direction) displacement of the NSI device

As can be seen from the Ls-Dyna software modeling results for the four story building (Figs. 3.11 and 3.12) with the NSI devices with Horasan mortar sliding base interface under Duzce/Turkey, 1999 earthquake (Appendix, Figs. 3.24–3.36): 1) Horasan mortar Von Mises stresses do not exceed safety stress limit; 2) Stresses obtained according the inside forces acting to the structural elements do not exceed safety stress limits also; 3)Averaged absolute maximum of the base (bearing-step) level acceleration was decreased 35% (Fig. 3.31); 4) Averaged absolute maximum of the base level velocity is 0.51 m/s (Fig. 3.32) and this velocity is much lower than the velocity limit (0.7^{-1} m/s) that can damage sensitive equipment involved inside of the building; 5) Averaged absolute maximum of the base level displacements (Fig. 3.33) indicate that the superstructure does not exceed the boundaries of the mat foundation area. The same structure's base and top levels responses (Appendix, Figs. 3.34, 3.35 and 3.36) using lightweight concrete (superstructure total mass equal to 4.08 e^{+6} kg), with Horasan mortar sliding base interface minimum, maximum decay parameters 0.17, 0.37, 11 respectively under same Duzce/Turkey, 1999 earthquake also indicate decreasing (44%) of the superstructure accelerations (Fig. 3.34) and averaged absolute maximum of the velocity (Fig. 3.35) and displacement(Fig. 3.36) of the base and top levels of the superstructure are in the reliable acceptable limits respectively.

3.7 Conclusions

Assessment of the NSI device's materials (reinforced concrete, Horasan mortar as friction interface) stress–strain range shows that stress ranges of the appropriated materials of the aseismic control device are around ultimate values which were presented in Table 3.4.

In whole isolated with NSI devices four-storey hospital building's base acceleration is decreased by 45.6437, 38.3919, 33.0212%; top floor acceleration is decreased by 64.9959, 61.4754, 58.4987%; top floor drift is decreased by 58.1769, 54.0705, 48.1211% regarding to fixed based structure for stainless steel, gray-iron mortar and Horasan mortar respectively. Building NSI device maximum displacement is 0.068, 0.0469, 0.0276 m for stainless steel, gray-iron mortar and Horasan mortar respectively.

Using lightweight concrete with Horasan mortar sliding base interface, its increases the performance of the mentioned system even further (for example the acceleration reduction became 44% for presented same structure, Appendex 1).

The research reveals that the NSI device with a sliding interface provided by Horasan mortar is feasible for aseismic base isolation applications.

The particularity of the earthquake action to act as a restoring force has also been presented. In other words, the earthquake action can keep the superstructure around equilibrium while another part is controlled by the friction coefficient of the isolation interface. From this point of view, the device incorporates isolation, energy dissipation, and the restoring force mechanism in a single unit. Also, the durability of this device has been shown to extend over ten centuries requiring virtually no maintenance.

The simplicity of the model and the long-term experience with the sliding interface material involved make the NSI device attractive for application in the isolation and retrofit of building structures.

Appendix

Figs. (3.24–3.36) Ls-Dyna software modeling response results of the four story building with the NSI devices with Horasan mortar sliding base interface under Duzce/Turkey, 1999 earthquake for time 10.83 s: Fig. (3.24) Horasan Mortar's Von Mises stress; Fig. (3.25) Superstructure beams axial forces; Fig. (3.26) Superstructure beams tor-sional forces; Fig. (3.27) Superstructure beams bending moments in section S direction; Fig. (3.28) Superstructure beams bending moments in section T direction; Fig. (3.29) Superstructure beams shear forces in section S direction; Fig. (3.30) Superstructure beams shear forces in section T direction; (Fig. 3.31) Averaged base level (green) acceleration; (Fig. 3.32) Averaged base level (green) velocity; (Fig. 3.33) Averaged base level (green) displacement; (Fig. 3.34) Averaged base level (pinkish) and toplevel (yellow) acceler-ations; (Fig. 3.35) Averaged base level (pinkish) and top-level (yellow) velocities; (Fig. 3.36) Averaged base level (pinkish) and top level (yellow) displacement.

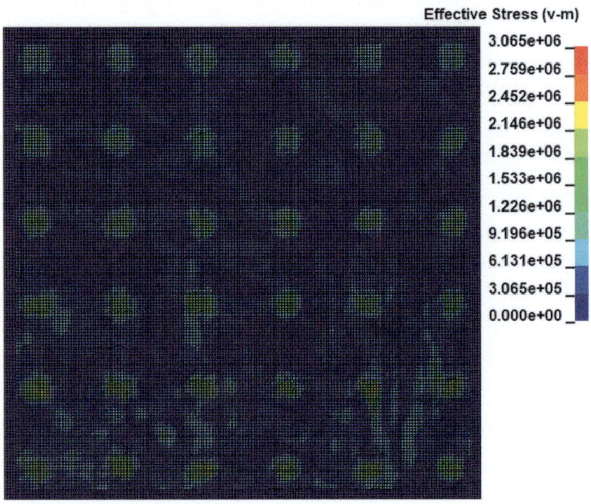

Fig. 3.24 Horasan Mortar's Von Mises stress

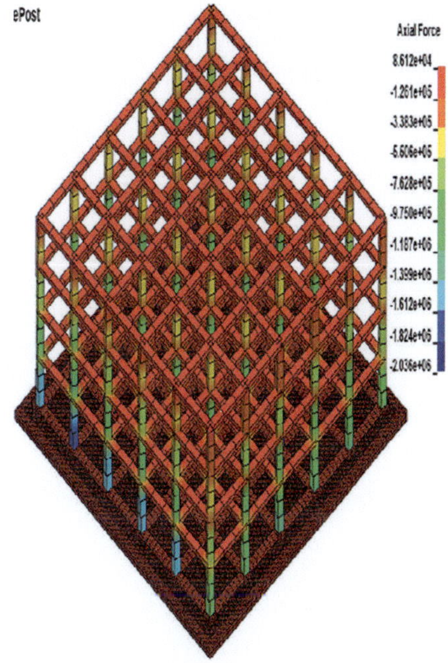

Fig. 3.25 Superstructure beams axial forces

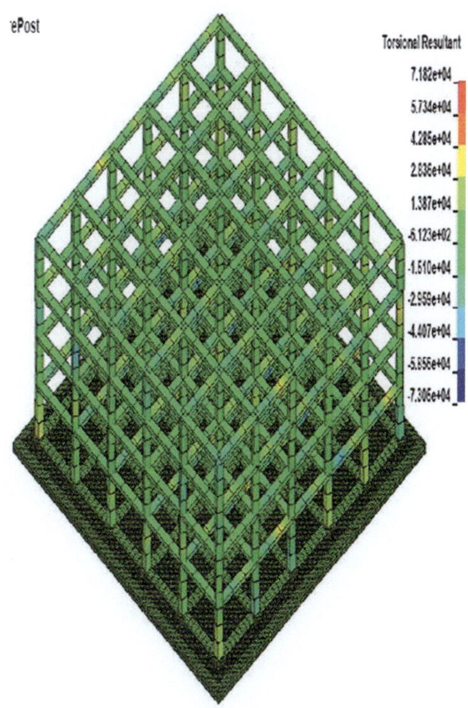

Fig. 3.26 Superstructure beams torsional forces

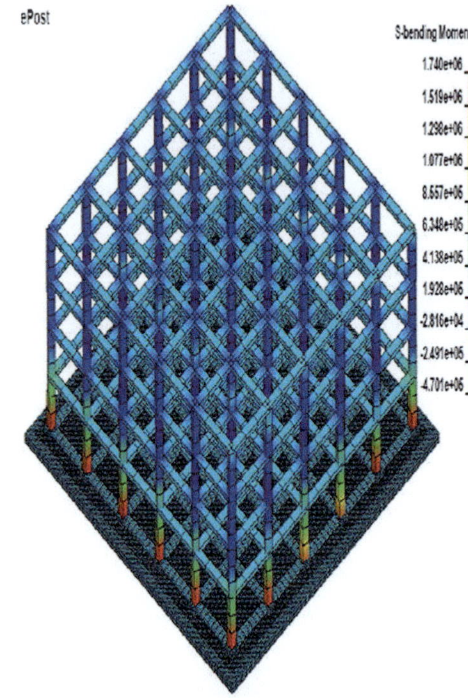

Fig. 3.27 Superstructure beams bending moments in section S direction

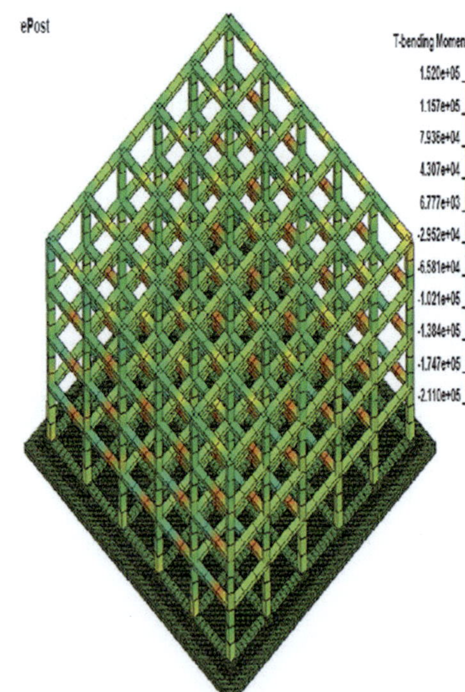

Fig. 3.28 Superstructure beams bending moments in section T direction

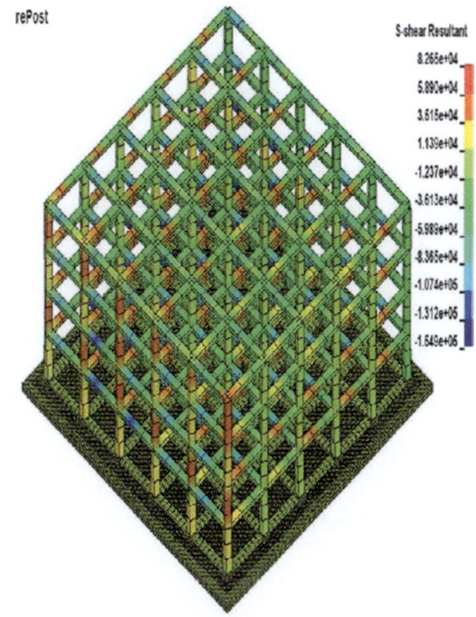

Fig. 3.29 Superstructure beams shear forces in section S direction

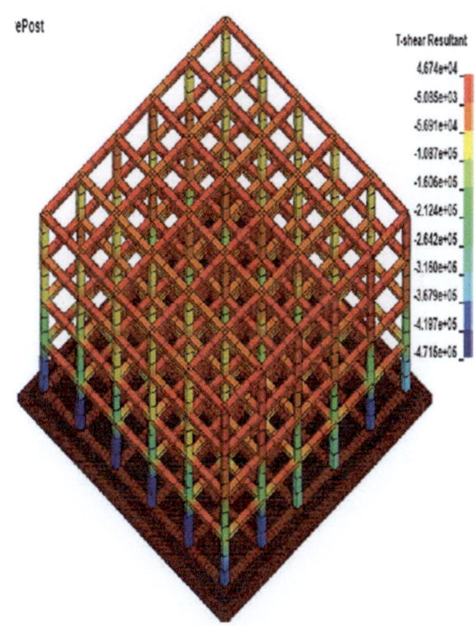

Fig. 3.30 Superstructure beams shear forces in section T direction

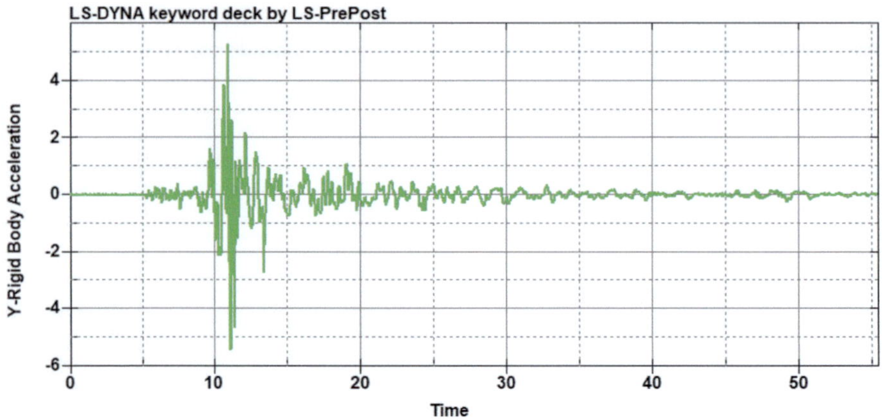

Fig. 3.31 Averaged base level (green) acceleration

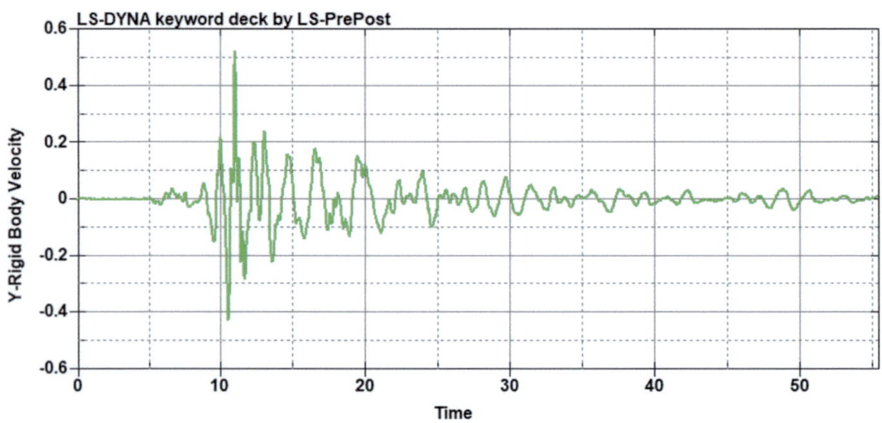

Fig. 3.32 Averaged base level (green) velocity

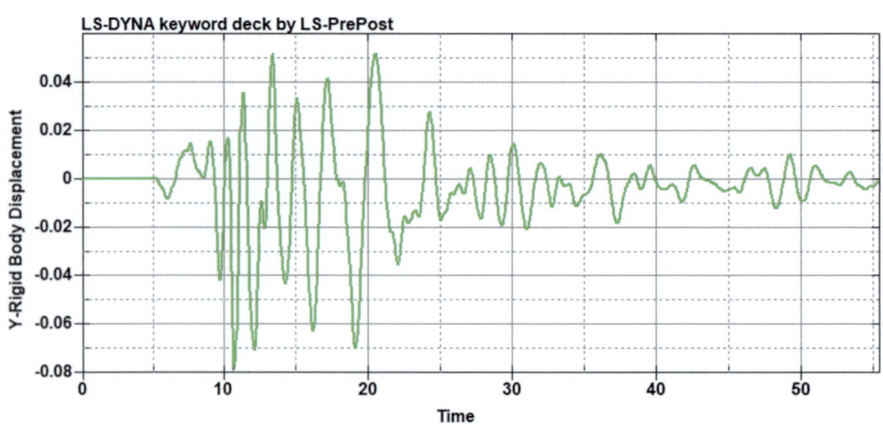

Fig. 3.33 Averaged base level (green) displacement

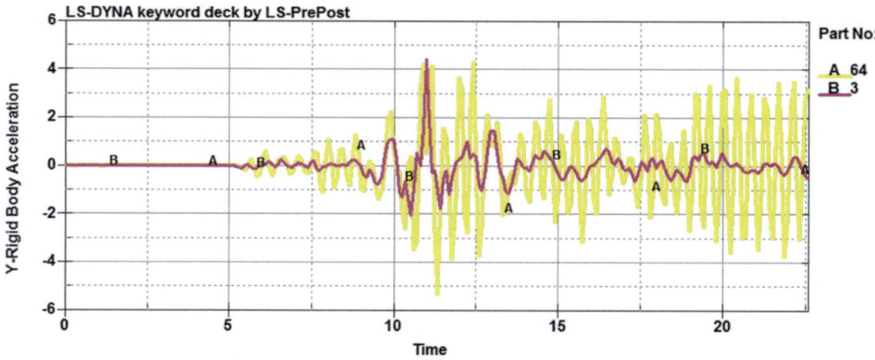

Fig. 3.34 Averaged base level (pinkish) and top-level (yellow) accelerations

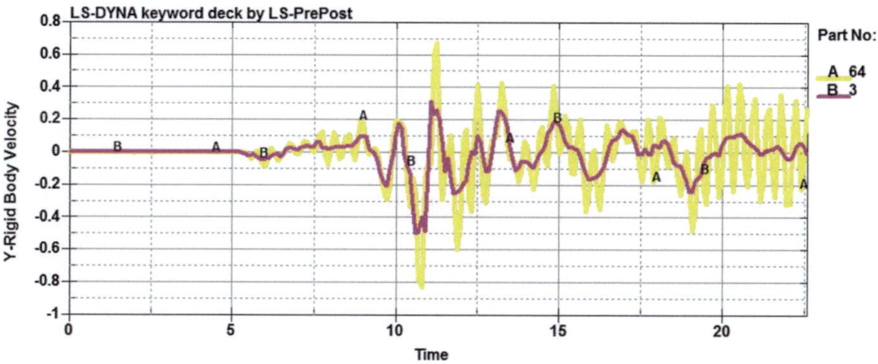

Fig. 3.35 Averaged base level (pinkish) and top-level (yellow) velocities

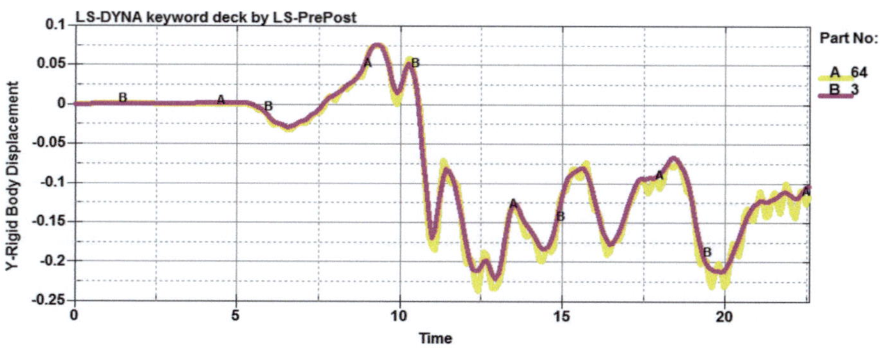

Fig. 3.36 Averaged base level (pinkish) and top level (yellow) displacements

References

Ahmad S, Ghani F, Adil MR (2009) Seismic friction base isolation performance using demolished waste in masonry housing. Constr Build Mater 23:146–152

ANSYS Online Manual's Release 5.5 (2017) Bratislava. http://www.ansys.stuba.sk. Accessed 2 Sept 2017

Arya AS (1984) Sliding concept for mitigation of earthquake disaster to masonry buildings. In: Proceedings of 8th world conference on earthquake engineering, San Francisco, USA, July 1984

ATC (1996) Seismic evaluation and retrofit of concrete buildings. Report ATC-40, Applied Technology Council, Redwood City

Becker TC, Mahin SA (2011) Experimental and analytical study of the bi-directional behavior of the triple friction pendulum isolator. Earthquake Eng Struct Dynam 41:355–373

Belytschko T, Liu WK, Moran B (2001) Nonlinear finite elements for continua and structures. Wiley, Chichester

Binal A (2008) The properties of Khorasan mortar and brick used in stone and brick masonry buildings in Kula (Manisa-Turkey). In: 1st historical mortars conference (HMC 08), Lisbon, Portugal, 24–26 Sept 2008

Constantinou M, Mokha A, Reinhorn A (1990) Teflon bearings in base isolation II: modeling. J Struct Eng ASCE 116(2):455–474

Çakmak AŞ, Moropoulou A, Mullen CL (1995) Interdisciplinary study of dynamic behaviour and earthquake response of Hagia Sophia. Soil Dyn Earthq Eng 14(2):125–133

Çamlıbel N (1998) Sinan Mimarlığında Yapı Strüktürünün Analitik İncelenmesi. Yıldız Teknik Üniversitesi Basım-Yayın Merkezi, Istanbul

Dolce M, Cardone D, Croatto F (2005) Frictional behaviour of steel-PTFE interfaces for seismic isolation. Bull Earthq Eng 3:75–99

Enokida R, Nagae T (2017) Seismic damage reduction of a structural system based on nontraditional sliding interfaces with graphite lubrication. J Earthq Eng 1–21. https://doi.org/10.1080/13632469.2016.1264326

Fan FG, Ahmadi G (1992) Seismic responses of secondary systems in base-isolated structures. Eng Struct 14(1):35–48. https://doi.org/10.1016/0141-0296(92)90006-C

Fan FG, Ahmadi G, Tadjbakhsh IG (1990a) Multi-story base-isolated buildings under a harmonic ground motion—Part I: a comparison of performances of various systems. Nucl Eng Des 123 (1):1–16. https://doi.org/10.1016/0029-5493(90)90266-Z

Fan FG, Ahmadi G, Tadjbakhsh IG (1990b) Multi-story base-isolated buildings under a harmonic ground motion—Part II: sensitivity analysis. Nucl Eng Des 123(1):17–26. https://doi.org/10.1016/0029-5493(90)90267-2

Fan FG, Ahmadi G, Mostaghel N, Tadjbakhsh IG (1991) Performance analysis of aseismic base isolation systems for a multi-story building. Soil Dyn Earthq Eng 10(3):152–171. https://doi.org/10.1016/0267-7261(91)90029-Y

FEMA (2009) NEHRP (National Earthquake Hazards Reduction Program) recommended sesismic provisions for new buildings and other structures. https://www.fema.gov/media-library-data/20130726-1730-25045-1580/femap_750.pdf. Accessed 1 Nov 2017

Gajan S, Saravanathiiban DS (2011) Modeling of energy dissipation in structural devices and foundation soil during seismic loading. Soil Dyn Earthq Eng 31:1106–1122

Gueraud R, Noel-Leroux JP, Livolant M, Michalopoulos AP (1985) Seismic isolation using sliding-elastomer bearing pads. Nucl Eng Des 84:363–377

Gur S, Mishra SK (2013) Multi-objective stochastic-structural-optimization of shape-memory-alloy assisted pure-friction bearing for isolating building against random earthquakes. Soil Dyn Earthq Eng 54:1–16. https://doi.org/10.1016/j.soildyn.2013.07.013

Hong WK, Kim HC (2004) Performance of a multi-story structure with a resilient-friction base isolation system. Comput Struct 82(27):2271–2283. https://doi.org/10.1016/j.compstruc.2004.06.002

Iemura H, Taghikhany T, Jain SK (2007) Optimum design of resilient sliding isolation system for seismic protection of equipments. Bull Earthq Eng 5:85–103

Ikawa N, Yamashita Y, Tachibana E (2011) Verification of the response reduction of base isolation system using sliding bearing. In: International symposium on disaster simulation & structural safety in the next generation (DS'11), Kobe, Japan, September 2011

International Code Council (2018) International building code. https://www.iccsafe.org/codes-tech-support/codes/2018-i-codes/ibc/. Accessed 5 Jan 2018

Islam ABMS, Jameel M, Jumaat MZ (2011) Seismic isolation in buildings to be a practical reality: behaviour of structure and installation technique. J Eng Technol Res 3(4):99–117

Ismail M, Rodellar J, Ikhouane F (2010) An innovative isolation device for aseismic design. Eng Struct 32:1168–1183

Jalali A, Cardone D, Narjabadifam P (2011) Smart restorable base isolation system. Bull Earthq Eng 9:657–673

Jangid RS, Londhe YB (1998) Effectiveness of elliptical rolling rods for base-isolation. J Struct Eng, ASCE 124:469–472

Kaldal GS (2013) Collapse analysis of the casing in high temperature geothermal wells. In: Proceedings, 38th workshop on geothermal reservoir engineering Stanford University, Stanford, CA, SGP-TR-198, 11–13 Feb 2013

Kasimzade AA (2002) Bounds of the structural response imprecisely-defined systems under earthquake action. PJAS 2(10):969–974

Kasimzade AA, Tuhta S (2004) Estimation of sensitivity and reliNSIlity base isolation buildings under earthquake action. In: International symposium on network and center-based for smart structure technologies and earthquake engineering, Osaka, Japan, July 2004

Kasimzade AA (2009a) Mathematical model of the natural seismic protection: modelling, implementation, application. In: The 2nd international multi-conference on engineering and technological innovation (IMETI 2009), July, Orlando, USA 2009

Kasimzade AA, Tuhta S, Bal S (2009b) Natural seismic protection peculiarities of historical walled obelisk structure in Istanbul: modelling and numerical approach. In: Smart structures and materials (SMART 2009), July, Porto, Portugal 2009

Kasimzade AA, Bal S, Bayar T (2011) Structural safety secret of "walled obelisk" monument from the tenth century and its simulated model for applications. In: International symposium on disaster simulation & structural safety in the next generation (DS'11), Kobe, Japan, September 2011

Kasimzade AA, Tuhta S (2012) Stochastic parametrical system identification approach for validation of finite elements models. TWMS J Pure Appl Math 3(1):11–31

Kasimzade AA (2012) Earthquake resistant structure design and new seismic base isolation system. In: Earthquake risk of Samsun province and necessary precautions, Samsun, Turkey, May 2012

Kasimzade AA, Tuhta S, Atmaca G (2013) The discovery of restoring mechanism phenomenon for earthquake and applications in creating new seismic base isolation system. J Theor Appl Mech 3(3):17–30

Kasimzade AA, Tuhta S, Atmaca G (2014) Modelling, computing and application particularities of no restoring mechanism aseismic control device. IERI Procedia 7(1):77–83

Kato S, Yamaguchi K, Matsubayashi T (1974) Stick-slip motion of machine tool sideway. J Eng Ind 96(2):557–566

Lou Y, Wang M, Su J (1992) A research of sliding shock absorbing multi-storey brick building. In: Proceedings of 10th world conference on earthquake engineering, Barcelona, Spain, July 1992

LS-DYNA Keyword User's Manual R10.0 Vol 1 (2017) Livermore Software Technology Corporation, Livermore, CA. http://www.dynasupport.com/manuals/ls-dyna-manuals/ls-dyna-manual-r10.0-vol-i. Accessed 25 Sept 2017

MATLAB R2017b Documentation (2017) The MathWorks Inc., Natick, MA. https://www.mathworks.com/help/matlab/index.html. Accessed 1 Dec 2017

McCormick J, Nagae T, Ikenaga M, Zhang PC, Katsuo M, Nakashima M (2009) Investigation of the sliding behaviour between steel and mortar for seismic applications in structures. Earthquake Eng Struct Dynam 38:1401–1419
Mostaghel N, Hejazi M, Tanbakuchi J (1983) Response of sliding structures to harmonic support motion. Earthquake Eng Struct Dynam 11:355–366
Mostaghel N, Tanbakuchi J (1983) Response of sliding structures to earthquake support motion. Earthquake Eng Struct Dynam 11:729–748
Mostaghel N, Khodaverdian M (1987) Dynamics of resilient-friction base isolator (R-FBI). Earthquake Eng Struct Dynam 15:379–390
Mukaibo K, Kawakami T, Ariyama M, Suzuki Y (2011) Experimental and analytical study on sliding behaviour of traditional timber building under strong earthquakes. In: International symposium on disaster simulation & structural safety in the next generation (DS'11), Kobe, Japan, September 2011
Naeim F, Kelly JM (1999) Design of seismic isolated structures: from theory to practice. Wiley, New York
Nagashima I (1998) Experimental evaluation of aging effects on properties of a building with sliding base isolation system. In: Proc. 10JEES, p 2659
Nanda RP, Agarwal P, Shrikhande M (2010) Frictional base isolation by geotextiles for brick masonry buildings. Geosynthetic Int 17(1):48–55
Nanda RP, Agarwal P, Shrikhande M (2012) Suitable friction sliding materials for base isolation of masonry buildings. Shock Vibr 19(6):1327–1339. https://doi.org/10.3233/SAV-2012-0675
Nikolic-Brzev S, Arya, AS (1996) Seismic isolation of masonry buildings-an experimental study. In: Proceedings of 11th world conference on earthquake engineering, Acapulco, Mexico, June 1996
Nishimura H, Ono S, Tachibana E (2004a) Identification of frictional coefficients of stainless steel sliding base isolators. In: Proceedings of 13th world conference on earthquake engineering, Vancouver, Canada, August 2004, p 380
Nishimura H, Ono S, Xing L, Tachibana E (2004b) Dynamic characteristics and response reduction of the sliding type of base-isolation. In: Proceedings of international symposium on network and center-based research for smart structures, technologies, and earthquake engineering (SE'04), Osaka, Japan, July 2004
Nishimura H, Abe S, Tachibana E (2007) Dynamic characteristics and response reduction of the metal -touched type of base-isolator. Struct Control Health Monit 14(4):537–555
Park KS, Jung HJ, Lee IW (2002) A comparative study on aseismic performances of base isolation systems for multi-span continuous bridge. J Eng Struct 24:1001–1013
Patil S, Reddy G (2012) State of art review—Base isolation systems for structures. Int J Emerg Technol Adv Eng 2(7):438–453. Retrieved from http://www.ijetae.com/Volume2Issue7.html
PEER (2011), Pacific Earthquake Research Center: NGA Database, University of California, Berkeley, CA, USA, http://peer.berkeley.edu/nga & http://peer.berkeley.edu/svbin/Detail?id=P1536
Qamaruddin M, Arya AS, Chandra B (1986) Seismic response of masonry buildings with sliding substructure. J Struct Eng (ASCE) 3:558–572
Quaglini V, Dubini P, Poggi C (2012) Experimental assessment of sliding materials for seismic isolation systems. Bull Earthq Eng 10:717–740
Song B, Yin C, Zhang X, Tao S (1990) Experimental study and seismic response analysis of multi-storeyed brick buildings with friction base isolation. In: Proceedings of 5th North American Masonry conference, Illinois, USA, June 1990
Strong Ground Motion Database of Turkey. AFAD, Ankara, Turkey, 2017. http://kyhdata.deprem.gov.tr/2K/genAcc.php?dst=TU9EVUxFX05BTUU9ZXZ0RmlsZSZNT0RVTEVfVEFTSz1zaG93Jk1PRFVMRV9TVUJUQVNLPUFMTCZNT0RVTEVfVEFSR0VUPW9sZCZUQVJHRVQ9MTk5OTExMTIxNjU3MjJfMTQwMSZUQVJHRVRfU0VSSUFMPTQyMTc%3D. Accessed 5 Aug 2017
Su L, Ahmadi G, Tadjbakhsh IG (1989a) Comparative study of base isolation systems. J Eng Mech (ASCE) 115(9):1976–1992

Su L, Ahmadi G, Tadjbakhsh IG (1989b) A comparative study of performances of various base isolation systems, part I: shear beam structures. Earthquake Eng Struct Dynam 18(1):11–32. https://doi.org/10.1002/eqe.4290180104

Suy HMR, Fey RHB, Galanti FMB, Nijmeijer H (2007) Nonlinear dynamic analysis of a structure with a friction-based seismic base isolation system. Nonlinear Dyn 50(3):523–538. https://doi.org/10.1007/s11071-006-9182-1

Symans MD, Madden GJ, Wongprasert N (2000) experimental study of an adaptive base isolation system for buildings, 12th world conference on earthquake engineering, Auckland, New Zealand, Sunday 30 Jan—Friday 4 Feb 2000, p 1965

Tehrani FM, Hasani A (1996) Behaviour of Iranian low rise buildings on sliding base to earthquake excitation. In: Proceedings of 11th world conference on earthquake engineering, Acapulco, Mexico, June 1996

Valluzzi MR, Binda L, Modena C (2005) Mechanical behaviour of historic masonry structures strengthened by bed joints structural repointing. Constr Build Mater 19:63–73

Zayas VA, Low SS, Mahin SA (1990) A simple pendulum technique for achieving seismic isolation. Earthquake Spectra 6:317–333

Zongjin L, Rossow EC, Shah SP (1989) Sinusoidal forced vibration of sliding masonry system. J Struct Eng (ASCE) 115(7):1741–1755

Part II
New Developments on Structural Health Monitoring and Earthquake Early Warning Systems for Performance Assessment of Structures

Chapter 4
BC Earthquake Early Warning System, a Program for Seismic Structural Health Monitoring of Infrastructure

Carlos E. Ventura, Yavuz Kaya and Alireza Taale

Abstract The University of British Columbia (UBC) and the British Columbia Ministry of Transportation and Infrastructure in Canada, have been instrumenting key structures in recent years to provide confirmation of their seismic capacity, assist in focusing retrofit efforts, detect damage from any cause, and provide rapid damage assessment of those structures following a seismic event. The program now includes more than 15 monitored bridges and is expanding to include public schools and other buildings. The newest addition to the bridge monitoring program is the incorporation of an earthquake early warning system developed at UBC. This paper describes how the EEW system will be used as part of the existing seismic structural health monitoring program in British Columbia.

Keywords Earthquake early warning system · Structural health monitoring system
Smart infrastructure · Performance-based earthquake engineering
Automated decision support systems · Damage detection

4.1 Introduction

The British Columbia Ministry of Transportation and Infrastructure (MoT) has been instrumenting structures in collaboration with the Earthquake Engineering Research Facility (EERF) at The University of British Columbia (UBC) since the late 1990s. The west coast of British Columbia (BC) lies in Canada's highest seismic zone, and as a result, the primary, initial purpose of the instrumentation systems was to

C. E. Ventura (✉) · Y. Kaya · A. Taale
Department of Civil Engineering, The University of British Columbia,
6250 Applied Science Ln, Vancouver, BC V6T 1Z4, Canada
e-mail: ventura@civil.ubc.ca

Y. Kaya
e-mail: kayaya@mail.ubc.com

A. Taale
e-mail: taale@ece.ubc.ca

A. A. Kasimzade et al. (eds.), *Seismic Isolation, Structural Health Monitoring, and Performance Based Seismic Design in Earthquake Engineering*,
https://doi.org/10.1007/978-3-319-93157-9_4

capture the ground motion input and effect on the structure in the event of an earthquake. More recently, the instrumentation has been expanded to incorporate Structural Health Monitoring (SHM) capabilities.

In addition to the structural monitoring, the Geologic Survey of Canada (GSC), through the Pacific Geosciences Centre (PGC), maintains the Provincial Strong Motion Network (SMN) comprised of over 130 ground monitoring stations. The bridge instrumentation and SMN have been integrated into a system called the BC Smart Infrastructure Monitoring System (BCSIMS). The system integrates data from the instrumented structures and strong motion network, organizes and processes the information in an efficient manner, to deliver that information to the appropriate parties.

A parallel collaborative project between the EERF, the MoT, the Roman Catholic Archdiocese of Vancouver (RCAV), and the BC Ministry of Education (MoE) was started in 2013 to develop and implement an Earthquake Early Warning System (EEWS) for urban areas in the province.

This paper describes the rationale and approach that has been followed to integrate these two systems and provides an update on how the systems are being integrated.

4.2 Description of BCSIMS

The (FDD) ambient modal identification is an extension of the Basic Frequency Domain (BFD) technique or called the Peak-Picking technique. This method uses the fact that modes can be estimated from the spectral densities calculated, in the case of a white noise input, and a lightly damped structure. It is a non parametric technique that determines the modal parameters directly from signal processing. The FDD technique estimates the modes using a Singular Value Decomposition (SVD) of each of the measurement data sets. This decomposition corresponds to a Single Degree of Freedom (SDOF) identification of the measured system for each singular value.

All of the data and results are available via the web interface as shown at the bottom right of the figure. The http://www.bcsims.ca website is a gateway for user interaction and operational management. The primary display consists of a shakemap which provides information on the severity of shaking across the region; Fig. 4.1 shows the website homepage. The circles represent the strong motion network stations, and the squares are structural stations.

The page is displayed as an interactive map allows zooming in/out and focus on a particular station. Additional metadata for the structures such as location information and live links to webcams are also provided. Lists of recent events and recent seismic activities are provided from which the user can access published information for the corresponding events and activities (Fig. 4.2).

The user has the option to display several different shake maps such as Instrumental Intensity (1999), Katayama Spectra Intensity (1998), and Japan

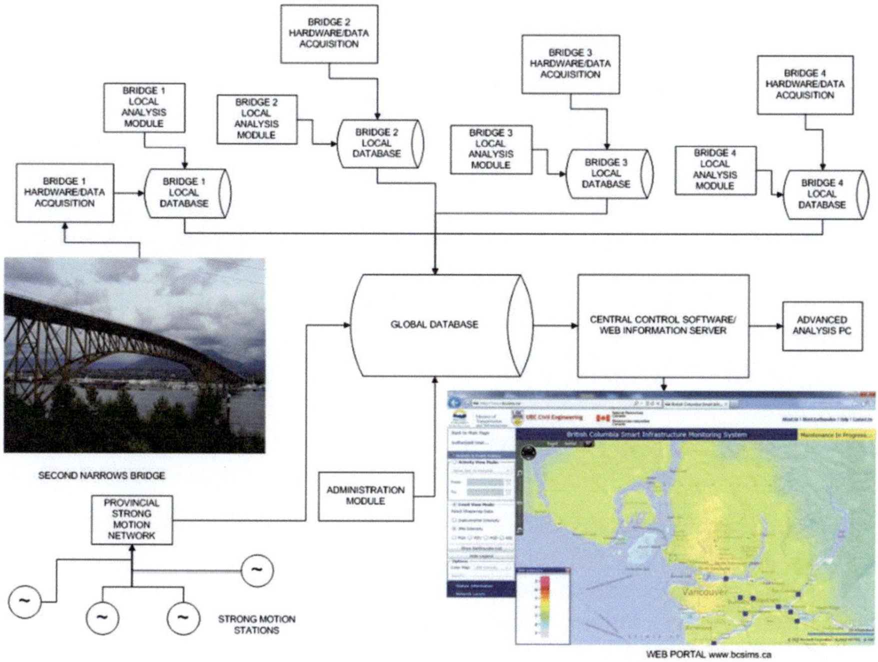

Fig. 4.1 BCSIMS system architecture

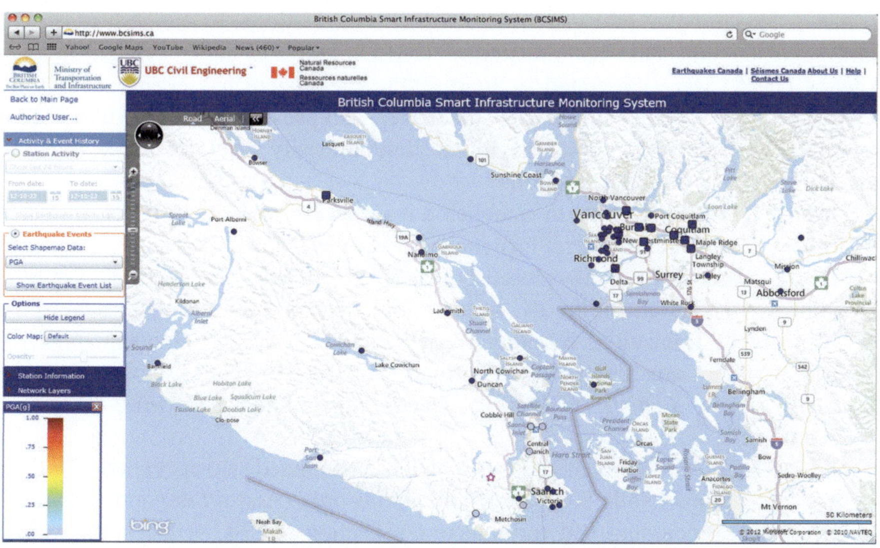

Fig. 4.2 Screenshot of the BCSIMS homepage

Meteorological Agency Intensity (Karim and Yamazaki 2002). Other options include PGA, PGV, and PGD values. The epicenter for is event is shown as a star with size being proportional to event magnitude.

In addition to the homepage and shakemaps, links from each bridge icon in the map directs to the Structures Information Pages (SIP). The SIP provides an overview of the status of the structure and more detailed results of the various structural assessments carried out by the system. The information available depends on the structure, what is monitored and how it is analyzed. The system is setup to have many capabilities of analysis including statistics, modal analysis, damage detection, FE analysis and FE model updating. Details of the system can be found in Ventura and Kaya (2012).

4.3 Description of BC-EEWN

Earthquake Early Warning (EEW) is achieved by monitoring ground vibrations to detect the arrival of a P-wave from an earthquake and provide a warning before the more damaging waves arrive. The EEW output can be an audio tone and message that can be passed to a Public Address system for distribution throughout the site of interest. The amount of warning time may be only a few seconds for a nearby earthquake to as many as 40 s or more for a distant event.

In the past decade, progress has been made towards implementation of earthquake early warning in Japan, Taiwan, Mexico, Southern California, Italy, and Romania. In particular, the systems developed at the National Research Institute for Earth Science and Disaster Prevention (NIED) and the Japan Meteorological Agency (JMA) were integrated in June 2005. The system has been successfully activated during several earthquakes since 2007, and provided accurate information regarding the source location, magnitude, and intensity at about 3.8 s after the arrival of P-wave at nearby stations.

There is an ongoing effort to establish a regional EEW network in the province of British Columbia and several Provincial Government institutions are involved in this effort. The institutions presently involved include the Roman Catholic Archdiocese of Vancouver, the BC Ministry of Transportation & Infrastructure, the BC Ministry of Education, Emergency Management BC, BC Housing, NRCan, Ocean Networks Canada (NEPTUNE project) and the EERF at UBC. The name of the network is ***British Columbia Earthquake Early Warning Network*** (BC-EEWN).

The BC-EEWN consists of sites with highly customized sensors called "TETRA sensors" and alarms (Node Site), sites with alarms only (Alarm Site) and the central server for data processing and earthquake notification at UBC (UBC site). All the sites are interconnected via the Internet as shown in the schematic diagram in Fig. 4.3.

It is expected that by the end of 2016 there will be a large network of TETRAS distributed throughout the province that will be part of the BC-EEWN. And,

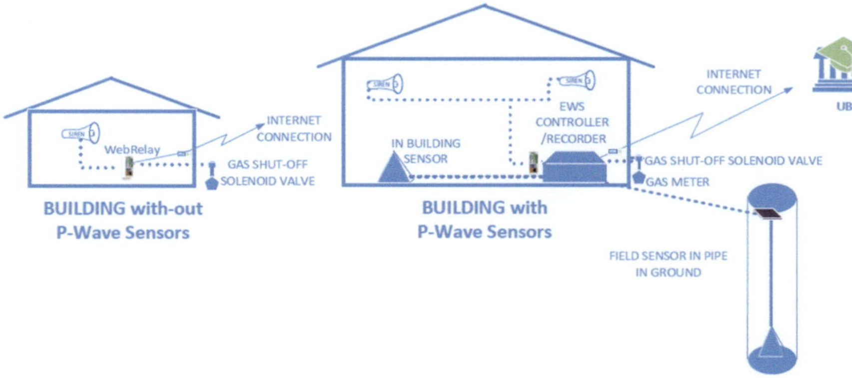

Fig. 4.3 BC EEWS network components

eventually, the BC-EEWN will be integrated with the EEWS being developed by the US Geological Survey for California, Oregon and Washington.

The sites that have sensors in British Columbia each have a controller and a pair of Tetras. Ideally the two sensors are installed in free-field sites, about 30 m feet apart, but if this is not possible, at least one sensor is installed in a free-field site about 30 m away from any major construction and the other sensor will be installed inside the building where the controller will be installed. These EEW systems will operate in standalone mode.

For any warning system, reliability is always important and it is desirable to have redundancy built into the system to make it more robust. As part of this project, we are exploring the feasibility of using several early warning algorithms to increase the speed (efficiency) and reliability (efficacy) of early warning. In these algorithms, the information from the initial part (up to a few seconds) of P-wave is used to estimate the magnitude and the strength of the impending ground motion at the same site. A network of sensors will improve earthquake early warning. A network of sensors can quickly identify the earthquake's epicenter, improve predictions of the earthquake's magnitude and reduce the incidence of false alarms. All the sites where the EEWS has been installed are connected to the network in order to improve the reliability of the warning system. Once a warning is confirmed, notifications are provided to all the sites that subscribe to the network in BC.

At the present time the BC-EEWN includes more than 30 Node sites and over 50 Alarm sites. The system is displayed on a webpage showing all of the sites, which are online, in alarm and other information, as shown in Fig. 4.4. The system has been in operation since the fall of 2014, and has shown to be capable of detecting far away earthquakes of magnitude 4.5 and higher. However, the system has not been fully tested since no significant earthquake has affected BC in recent months. Figure 4.5 shows a recently developed version of a webpage that shows in real time the seismic activity detected at each Node (measuring station) in the Vancouver region.

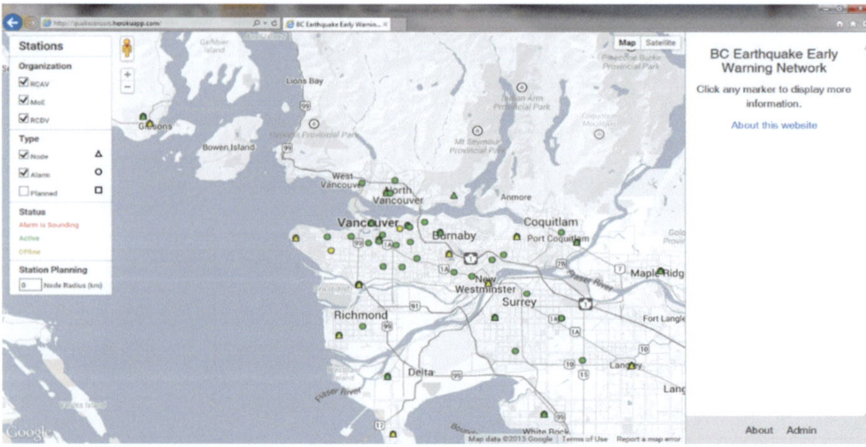

Fig. 4.4 BC earthquake early warning system webpage

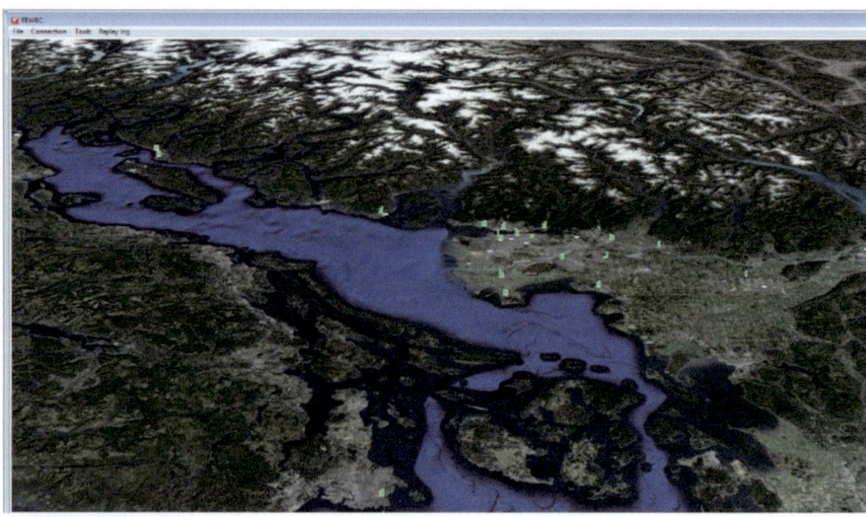

Fig. 4.5 Overview of Node sites (green color bars) in the Vancouver region. Each node site includes two sensors

4.4 Integrating the EEW System and the SHM Program

Efforts to integrate an EEW system with a seismic SHM program have been recently discussed by Rainieri et al. (2010) and Wu and Beck (2012). Earlier work discussing the interaction of SHM and EEW systems for effective structural control of buildings can be found in Kanda et al. (1994) and Occhiuzzi et al. (2004).

Rainieri and his colleagues present the results of a research study aimed at the development of an SHM architecture that integrates the information and data available from EEW systems. They proposed hardware solutions for continuous monitoring of the dynamic response of buildings that allows to host information and data provided by EWW applications and pointed out key aspects for the effective integration of both systems.

These included the need for proactive seismic risk mitigation programs incorporating proper seismic design, retrofit of existing buildings and the installation of an EEW network; the need for the EEW system to be able to calculate in real-time important seismic parameters, and to effectively transmit a warning signal to the interested receivers in order to take adequate countermeasures; and effective emergency management program that accounts for all actions to be taken immediately after the main event including the assessment of the safety of critical buildings.

Wu and Beck (2012) proposed a synergistic framework to utilize information from both EEW and SHM systems to provide more reliable loss estimation and safety alerts. In this approach, EEW information can be used independently to provide pre-event loss estimation and decision support based on the Performance Based Earthquake Engineering (PBEE) methodology developed at the Pacific Earthquake Engineering Center (PEER). By making use of the EEW information to reduce the inherent (i.e., Epistemic) uncertainty in the probabilistic SHM system to predict damage. The posterior information from the SHM system is used as feedback information to update the EEW loss estimation and provide more reliable future safety alerts. The approach proposed by Wu and Beck is illustrated in Fig. 4.6.

The study concludes that the combination of SHM and EEW systems can provide a tool for safety enhancement of existing critical buildings. The hardware architecture of the monitoring system can incorporate alarms and triggers of countermeasures and at the same time evaluate the health of the structure.

A modified version of the proposed approach outlined by Wu and Beck is presently being incorporated as part of the integration of the BC-EEWN and

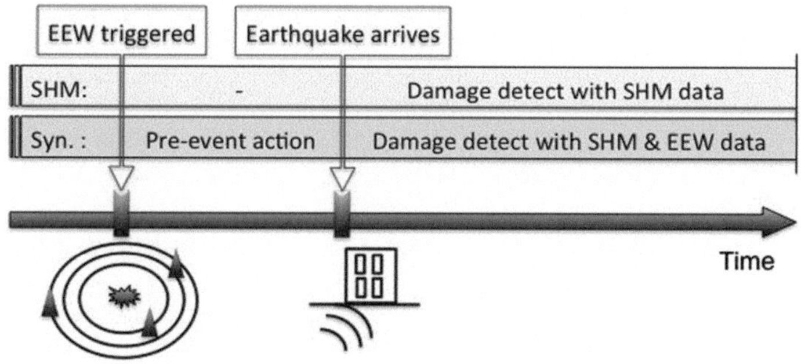

Fig. 4.6 Framework that combines information from EEW and SHM systems [Adapted from Wu and Beck (2012)]

BCSIMS. Instead of implementing the PBEE methodology developed by PEER, the various damage estimation and localization techniques already incorporated in BCSIMS are being used. Future work will incorporate the adaptation of the PBEE (i.e., fragility curves and seismic hazard estimation) to bridges and other structures in Western Canada.

4.4.1 Major Challenges

There are a handful competing demands within the context of Earthquake Early Warning Platforms (EEWP). One common viewpoint is to characterize effectiveness versus efficiency. These two quantities may be applied to both hardware and software (Table 4.1).

Literally, it is desired to maximize both efficiency and efficacy, however, it is likely impractical. Therefore, it may be more realistic to optimize a platform to practically meet a specific set of requirements. On one hand, in case of a regional EEWP, it is apparent that efficiency becomes more important than efficacy in terms of software and software. On the other hand, should an on-site EEWP be desired then the effectiveness of the system becomes central. In both cases, regional or

Table 4.1 Representation of efficacy and efficiency for H/W and S/W for an earthquake early warning platform

	Efficacy	Efficiency
Hardware	Sensor network	Latency of network
	Precision	Cost of equipment
	Accuracy	Maintenance of platform
	Signal to noise ratio	Redundancy
	Dynamic range	Processing capacity
	Resolution	Installation
		Commissioning
Software	Detection algorithm	Processing time
	Estimation algorithm (M, R, PGA, PGV, PGD)	Implementation of algorithms
	Warning time estimation algorithm	Integration of algorithms
	Decision-making algorithm	Robustness of algorithms
	Site characterization (Geotechnical assessment)	Maintenance
	Structural characterization (SHM)	Accessibility to software platform
		Data storage capacity
		Graphical User Interface
		Data communication procedure
		Programming language

on-site the driving factor by which efficacy can be compromised is *uncertainty*. Better efficacy demands less uncertainty. Lastly, the major driving factor for efficiency is *cost* in general. More efficient platform would most likely require more expensive hardware and software.

Typical range of EEW lead time is very short, from just a few seconds up to a minute or more in western Canada; this can be a huge challenge for applications taking advantage of EEW. Therefore, a robust automated decision process about whether to initiate a mitigation action is essential.

The effect of lead time (in particular a short lead time) has a critical effect on a decision process which is required to be performed in real time. Consequently, it is desirable to conduct detailed investigation of any actions than can be implemented to account for this short lead time in initiating a decision process. Concepts such as the earthquake probability-based automated decision-making (ePAD) proposed by Wu et al. (2013) are being considered for this integration of systems to maximize the benefits of EEW given a short lead time. The ePAD method uses concepts of decision function, decision contour, and surrogate model to perform fast computations and to allow comparison between various decision criteria.

Once the EEW system issues a warning, the three main steps in a mitigation application are performing the probabilistic ground motion prediction for the site, the response prediction for the target structure/system, and the loss/damage prediction. Because the last two steps do not involve information from the EEW, they can be done ahead of time.

Since some of the damage detection tools incorporated in the SHM and BCSIMS program rely on a priori information about the status of the structure, the implementation of the ePAD concept is of great interest for the integration of the EEW and SHM systems.

4.4.2 Proposing Framework

The central idea of integrating BCSIMS and BC-EEWN is to incorporate suitable event characterization algorithms into a Server which makes an informed decision based on current dynamic properties of the structure at hand (e.g., Hydropower Plants, Bridges, Hospitals) and activates a predetermined set of procedures (i.e., actions) when estimated seismic event compromises the serviceability of the structure. The *Server* hosts a handful of algorithms by which seismic P-Wave is detected and analyzed in order to evaluate the risk, based on structural health monitoring information stored in *Database* and it subsequently packages and transmits customized instructions for clients A, B, C and so on by means of an *Automatic Response Unit*. The physical location of the Server is recommended by the client (Fig. 4.7).

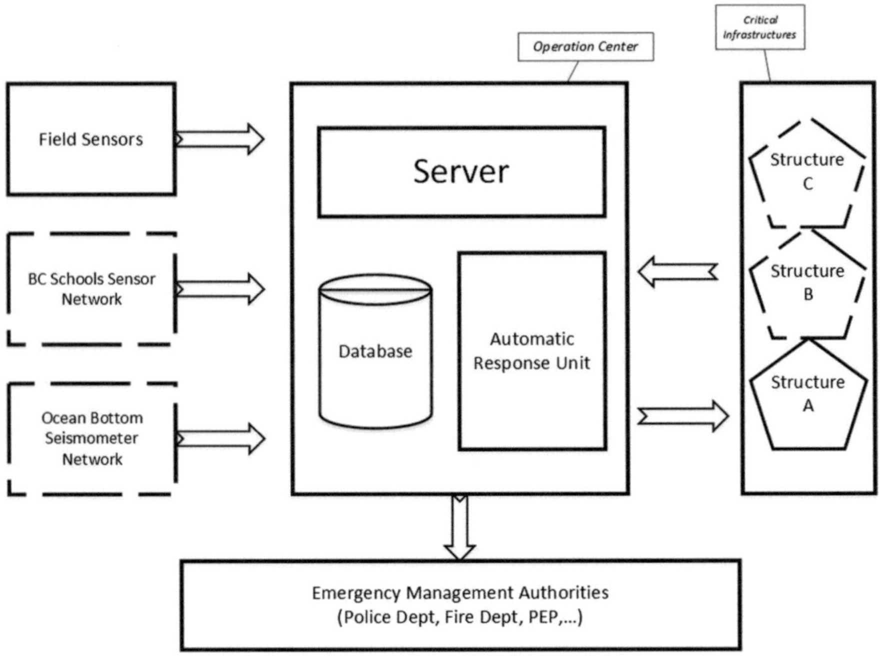

Fig. 4.7 Visualization of proposed EEWP; dashed lines demonstrate potential future developments; pentagons represent Structural Health Monitoring (SHM) deployments

The proposed platform will comply with current protocols employed by the British Columbia Smart Infrastructure Monitoring System (BCSIMS) as well as British Columbia Earthquake Early Warning Network (BCEEWN) which indeed provides the capacity for future developments.

4.4.3 Client-Oriented Automated Deciding Concept (Cadence)

There have been (or are currently underway) major EEW deployments in countries such as Japan, US, Canada, Mexico, Turkey, Italy, Romania, Taiwan, and China. Explicit similarities are evidently present among all deployments. For example, the LTA/STA algorithm has nearly become a standard procedure in P-wave detection process, although there are examples in which this procedure has been enhanced.

All these deployments are, however, implicitly different in terms of the adopted methodologies which may be based on seismology, probabilistic concepts or even crowd sourcing techniques. None of the reported methods (including but not limited to the abovementioned ones) has claimed to be the best practice so far. Nonetheless, all deployments have practically met a set of requirements defined by their clients.

In general, the complexity of algorithms (e.g., Detection, Estimation, etc.), which is hosted by the server, decides the quality of hardware (e.g., sensor type, resolution, SNR, processing capacity, communication bandwidth and so on). The primary focus of the alternative algorithm is to optimize efficiency and efficacy at a reasonable cost.

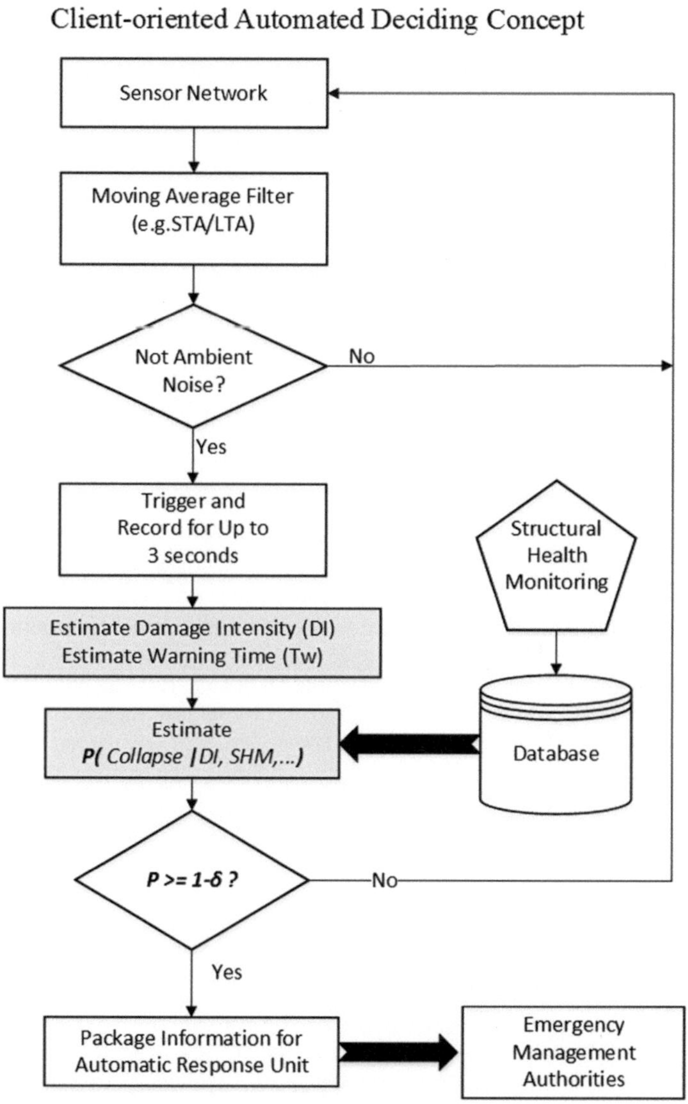

Fig. 4.8 Flow of information in *Cadence* algorithm hosted by the server; δ is a very small number (e.g. 0.1, 0.01) that may be specified by the client/user

Cadence block diagram (Fig. 4.8) showcases the flow of information by which the proposed decision-making process is realized. The general characteristics of the concept can be summarized in the followings:

- Automation: To make an informed and reliable decision without human intervention
- Efficient and Effective: To incorporate both regional and site specific information for estimating the possibility of failure (i.e., structural capacity)
- Cost efficiency: To minimize the cost of installation and configuration of the system
- Network Connectivity: To deliver the earthquake information, warning or any notification, the system must be easy to connect network.

4.5 Conclusions

The University of British Columbia, in collaboration with several Provincial Government Institutions have undertaken development of both a Structural Health Monitoring network and an Earthquake Early Warning network over the last 10 years. The next major step in their development is in integrating the two systems to take advantage of the benefits of each.

A synergy can be achieved by combining an EEW system with installed SHM systems to enhance pre-event prognosis and post-event diagnosis of structural health. For pre-event prognosis, the EEW system information can be used to make predictions of the anticipated performance or potential damage to a structure. These predictions can support decision-making regarding the activation of appropriate risk mitigation systems suitable for the type of structure that is being monitored.

Acknowledgements The authors would like to thank Ms. Sharlie Huffman of the MOT, and Dr. Martin Turek, Mr. Kent Johansen and Mr. Felix Yao of UBC, for their support in the development and installation of the networks described in this paper.

References

Kanda K, Kobori T, Ikeda Y, Koshida H (1994) The development of a pre-arrival transmission system for earthquake information applied to seismic response controlled structures. In: Proceedings 1st world conference on structural control, California, USA

Karim KR, Yamazaki F (2002) Correlation of JMA instrumental seismic intensity with strong motion parameters. Earthq Eng Struct Dyn 31:1191–1212

Katayama T, Sato N, Saito K (1998) SI-sensor for the identification of destructive ground motion. In: Proceeding ninth world conference of earthquake engineering, Tokyo-Kyoto, vol VII, pp 667–672

Occhiuzzi A, Grasso VF, Manfredi G (2004) Early warning systems from a structural control perspective. In: Proceedings 3rd European conference on structural control, Vienna, Austria

Rainieri C, Fabbrocino G, Cosenza E (2010) Integrated seismic early warning and structural health monitoring of critical civil infrastructures in seismically prone areas. Structural Health Monitoring, SAGE Publications 10(3):291–308

Ventura C, Kaya Y (2012) Seismic structural health monitoring of bridges in British Columbia, CANADA. In: Proceeding the 15 World conference of earthquake engineering, Lisbon, Portugal

Wu S, Beck JL (2012) Synergistic combination of systems for structural health monitoring and earthquake early warning for structural health prognosis and diagnosis. In: Proceedings of SPIE, vol 8348, 83481Z, p 10

Wu S, Beck JL, Heaton TH (2013) ePAD: Earthquake probability-based automated decision-making framework for earthquake early warning. Comput Aided Civil Infrastruct Eng 28:737–752

Chapter 5
Structural Health Monitoring: Lessons Learned

Eser Çaktı and Erdal Şafak

Abstract Department of Earthquake Engineering of Kandilli Observatory and Earthquake Research Institute, Boğaziçi University (DEE-KOERI) has designed and been operating a significant number of structural monitoring networks in Istanbul. They are installed in a large number of historical structures (i.e., mosques, minarets, and museums), lifeline structures across the Bosphorus (i.e., suspension bridges and tunnels) and several tall buildings including the Sapphire tower, currently the tallest building in Europe. The structural monitoring networks record the dynamic motions of the structures continuously, and the data are transmitted in real time to the monitoring center at the DEE-KOERI. The majority of the systems use accelerometers for monitoring. Some structures are also instrumented with tiltmeters and GPS sensors. In-house real-time modal analysis software is used to process and analyze the data. The software includes data processing, spectral identification, and animation modules. The results are displayed in real time, showing the time variations of modal properties and the structure's configuration. This chapter provides an overview of these monitoring systems in Istanbul. Moreover, it presents major findings related to the dynamic response properties of monitored structures particularly focusing on structural response to long-distance, long-period earthquakes; on the sensitivity of dynamic modal parameters to variations in atmospheric conditions; on structural response characteristics due to explosions; and on damping in tall buildings.

Keywords Structural health monitoring · Tall buildings · Damping
Surface waves · Large-distant earthquakes

E. Çaktı (✉) · E. Şafak
Department of Earthquake Engineering, Kandilli Observatory
and Earthquake Research Institute, Boğaziçi University, Istanbul, Turkey
e-mail: eser.cakti@boun.edu.tr

A. A. Kasimzade et al. (eds.), *Seismic Isolation, Structural Health Monitoring, and Performance Based Seismic Design in Earthquake Engineering*,
https://doi.org/10.1007/978-3-319-93157-9_5

5.1 Introduction

Structural Health Monitoring (SHM) networks are designed and installed in structures to monitor their dynamic motions continuously in order to track any changes in their structural integrity and detect damage. From the earthquake engineering perspective, the majority of the systems use accelerometers for monitoring. Some structures are also instrumented with tiltmeters and GPS sensors. The continuous recording of dynamic motions requires that the data are processed and analyzed continuously and the results are displayed in real time.

The analyses of data from SHM systems with 24/7 data feed have led to significant improvements in our understanding of structural behavior, particularly for flexible structures and masonry buildings. Such analyses have also underlined the importance of factors other than earthquakes in structural response, as continuous monitoring and analysis enable a fresh look at long-term effects such as variation in atmospheric conditions, unforeseen events such as explosions; or transient phenomena such as storms or strong winds. Acknowledging the fact that the term SHM should be used in a broader sense including monitoring of vibrations, strains, displacements, stresses, and local atmospheric conditions, herein we focus on systems for vibration monitoring. This contribution gives some examples of SHM systems that are being operated in Istanbul, and the lessons learned from the data.

5.2 SHM Networks in Istanbul

The Department of Earthquake Engineering of Kandilli Observatory and Earthquake Research Institute, Boğaziçi University (DEE-KOERI) has designed and been operating a significant number of SHM networks in Istanbul. They are installed in a large number of historical structures (i.e., mosques, minarets, museums), suspension bridges on the Bosphorus, the Marmaray rail tube tunnel and several tall buildings. The monitoring systems installed in historical structures in Istanbul are unique, as they form the largest group of monumental structures in the world monitored in one city. In-house real-time modal analysis software is used to process and analyze the data. The software includes data processing, spectral identification, and animation modules. The results are displayed in real time, showing the time variations of modal properties and the structure's configuration.

5.2.1 SHM Networks in Historical Structures

The first installations of SHM systems in Istanbul have started with historical structures back in 1991. Today they include the 35-channel Hagia Sophia Museum;

Fig. 5.1 Hagia Sophia Museum SHM network

47-channel Fatih Mosque; 27-channel Süleymaniye Mosque; 30-channel Sultanahmet Mosque; and the 35-channel Mihrimah Sultan Mosque systems. Each system consists of three-component accelerometers placed in structurally critical locations. They record the vibrations of the structures continuously and transmit the data in real time to the data center at the DEE-KOERI. Three of the systems (Hagia Sophia Museum, Fatih Mosque, and Mihrimah Sultan Mosque) also include four two-channel tiltmeters. Among the monitored historical structures, the sixth-century Hagia Sophia Museum is the oldest, and the seventeenth-century Sultanahmet Mosque (i.e., the Blue Mosque) is the youngest. The Hagia Sophia Museum and the existing sensor layout are shown in Fig. 5.1.

Additionally, we monitor two minarets in Istanbul. Because of their slenderness, minarets are vulnerable, not only to earthquake-induced vibrations, but also the wind-induced vibrations due to vortex shedding. The first minaret instrumented is one of the four masonry minarets of the Hagia Sophia Museum (Fig. 5.2). It was added to the complex in the sixteenth century. The second minaret is one of the four reinforced concrete minarets of the Maltepe Mosque, which is currently the tallest modern minaret in Istanbul. It was constructed in 1970. The Maltepe Mosque with its minarets is shown in Fig. 5.2.

5.2.2 SHM Networks in Lifelines

The three critical lifeline structures that are installed with SHM systems are the two suspension bridges on the Bosphorus, i.e., the 15 July Martyrs Bridge and the Fatih Sultan Mehmet Bridge, and the Marmaray rail tube tunnel under the Bosphorus.

The SHM system on the 15 July Martyrs Bridge (i.e., the First Bosphorus Bridge) was installed in 2007 by the Turkish Highway Department during a major maintenance work done on the bridge. It is composed of 258 channels of real-time data from 168 sensors, which includes accelerometers, GPS sensors, tiltmeters, laser-based displacement meters, load sensors, thermos couples, and strain gauges. The SHM system on the Fatih Sultan Mehmet Bridge (i.e., the Second Bosphorus

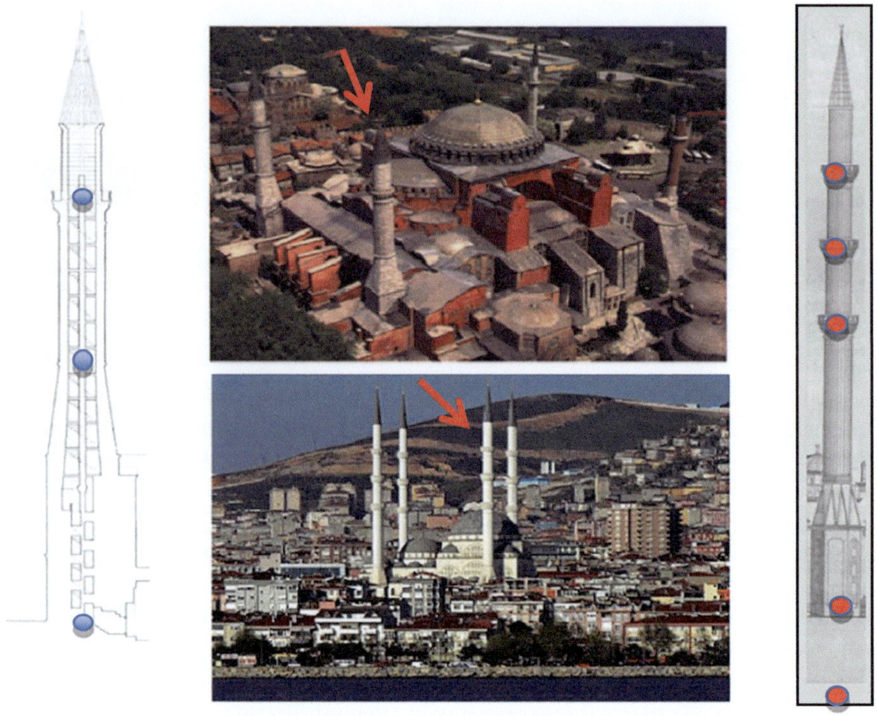

Fig. 5.2 Hagia Sophia in Istanbul (center top) and its instrumented minaret (left). The Maltepe Mosque (center bottom) and its instrumented minaret (right)

Bridge) was installed by DEE-KOERI in 2008 as part of a research project funded by EU. It is composed of 44 channels of accelerometers placed inside the deck, on the towers, and the center of suspension cables. The bridge and the instrument layout are shown schematically in Fig. 5.3.

The recently opened Marmaray rail tube tunnel runs at the bottom of the Bosphorus at a depth of 60 m (Fig. 5.4). It is composed of two adjoining tunnels and connects the European and Asian lines of Istanbul metro. The tube is composed of eleven 100 m reinforced concrete segments that were lowered to the sea bottom and joined together. Each segment is instrumented with four 3-channel accelerometers. The data are collected continuously and transmitted in real time to the data center at DEE-KOERI.

5.2.3 SHM Networks in Buildings

The SHM systems in buildings include the 36-channel Sapphire Tower, and the 15-channel each Kanyon Building, Isbank Tower and Polat Tower networks.

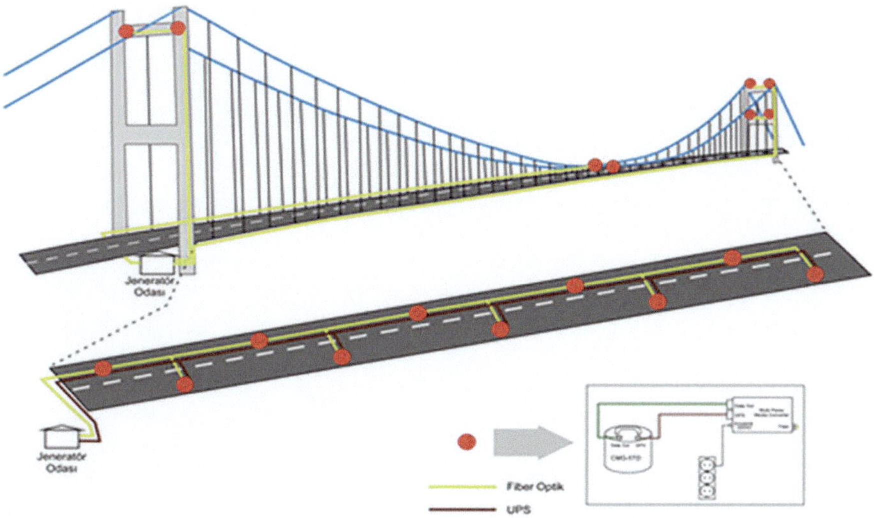

Fig. 5.3 FSM Bridge—Structural Health Monitoring system (44 channels)

Figure 5.5 shows the 62-story, 238 m high Sapphire Building, which is currently the tallest building in Turkey, and the sensor layout of the monitoring system.

5.3 Real-Time Analysis of SHM Data

Data from the structural monitoring networks are transmitted in real time to the monitoring center at DEE-KOERI. The in-house real-time modal identification software, KOERI_MIDS, is used to process and analyze the data. The software includes data processing, spectral identification, and animation modules. The results are displayed in real time, showing the time variations of modal properties and the structure's configuration. Figure 5.6 gives a typical screen plot of KOERI_MIDS. There are four main windows. The window in upper right shows the selected (recorded or processed) waveforms. The window in upper left shows the spectral content of the selected waveforms. The window in lower left shows the time variations of the identified structural parameters from the data, such as modal frequencies and damping ratios. The window in lower right shows the real-time animations of the total or modal responses.

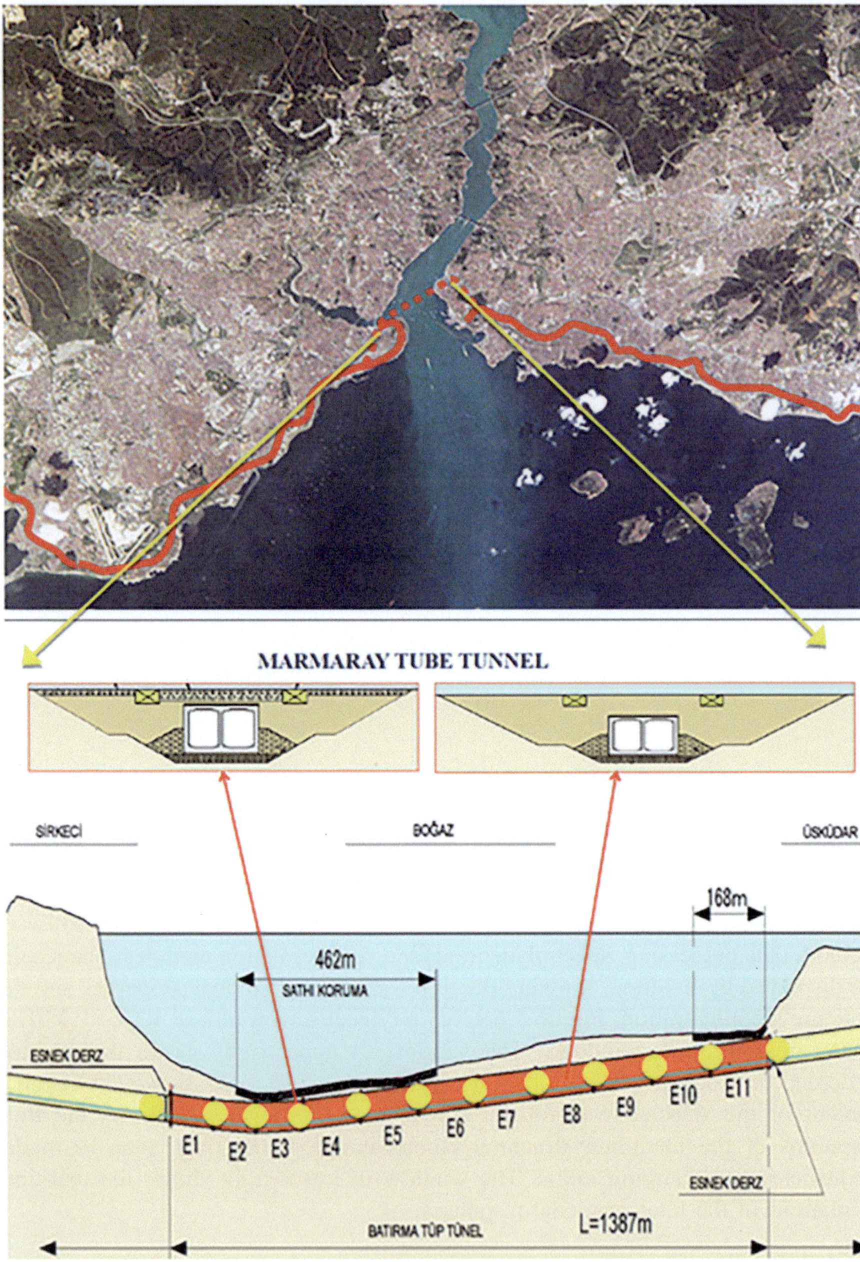

Fig. 5.4 The Marmaray Tube Tunnel across the Bosphorus (top, image from Wikipedia); the instrument layout in the tunnel (bottom), each section has accelerometers installed on both sides and tiltmeters (Tunc 2016)

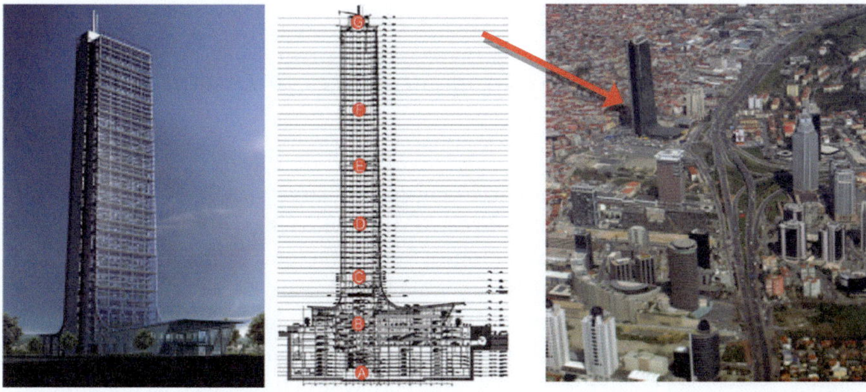

Fig. 5.5 62-story Sapphire Tower and the sensor layout

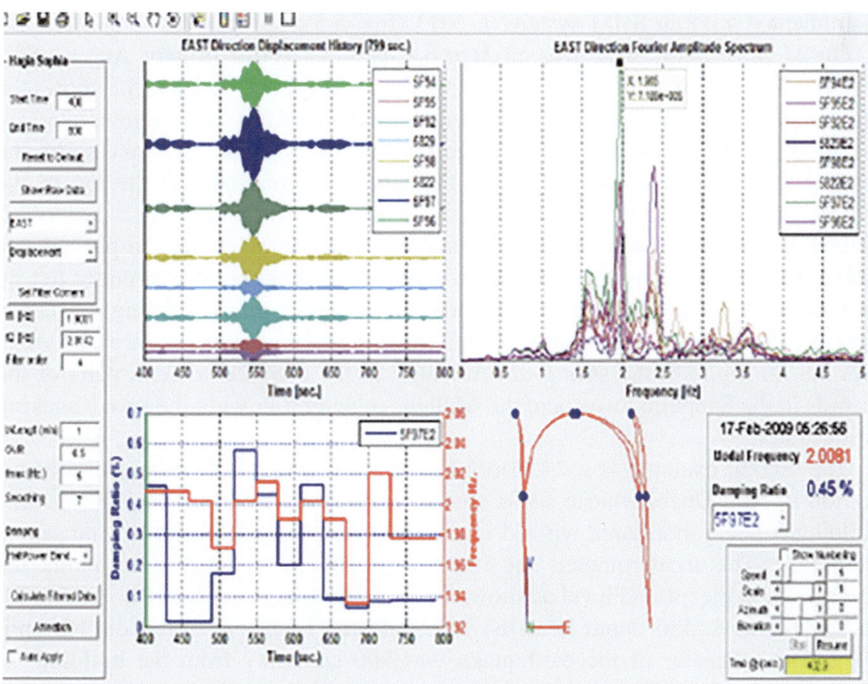

Fig. 5.6 Typical screenshot of the real-time modal identification software, KOERI_MIDS. Image shows the software running for Hagia Sophia

5.4 Lessons Learned from SHM Data

5.4.1 Response of Long-Period Structures to Distant Large Earthquakes

In recent years, seismic response of long-period structures (e.g., tall buildings, long-span bridges, pipelines, base-isolated structures) has become an important research subject. The seismic excitation of such structures is mainly controlled by long-period surface waves, which can travel very long distances without much attenuation. Consequently, even if they are located in low-seismicity regions, long-period structures can be vulnerable to large earthquakes occurring in far-away locations. SHM data clearly show that long-period structures, such as tall buildings, are susceptible to long-duration shaking from distant large earthquakes. Such waves can cause resonant vibrations in tall buildings, even when the earthquake is hundreds of kilometers away. We will give two examples of this phenomenon.

The first example is the response of the Sapphire tower, a 238 m tall reinforced concrete building with a narrow rectangular cross section. The building was instrumented with an SHM system in 2011 (Fig. 5.5).

On 24 May 2014, there was an $M = 6.5$ earthquake in Northern Aegean Sea, which is 300 km away from the building as shown in Fig. 5.7. Figure 5.8 shows the displacements calculated from the recorded accelerations by double integration in the longer and shorter directions of the building during this earthquake. As the figure shows, in the shorter direction, the vibration amplitudes at the top of the building are very large and last more than 10 min after the earthquake has stopped. Although the accelerations and displacements are not large enough to present any danger structurally, the long duration of building's vibrations was strongly felt by the occupants. Maltepe minaret responded in a similar manner during the earthquake. Accelerations and displacements at its uppermost balcony are shown in Fig. 5.8. In Fig. 5.9, the strong directionality in the displacements in parts of the records in the Sapphire tower and the Maltepe minaret following the strong shaking is illustrated.

The second example is a 74-story, 310 m-tall building with ellipse-like cross-section in Abu Dhabi, whose SHM system is designed by the DEE-KOERI. The building is being monitored with 31 channels of accelerometers plus a wind sensor at the top. The accelerometers are installed at 10 different levels, including the basement and the ground level as shown in Fig. 5.10 (Safak et al. 2014). A $M = 7.8$ earthquake at 82 km depth near the Iran–Pakistan border occurred on 16 April 2013. The epicenter of the earthquake was 900 km away from the building, as shown in Fig. 5.11. The shaking from the earthquake was felt over a wide area, including the Middle East, causing panic among people in tall buildings in the region.

Recorded top-story accelerations and corresponding displacements in the building from this earthquake are shown in Fig. 5.12. The figure clearly indicates the effects of long-period surface waves in the shaking. The acceleration plots show

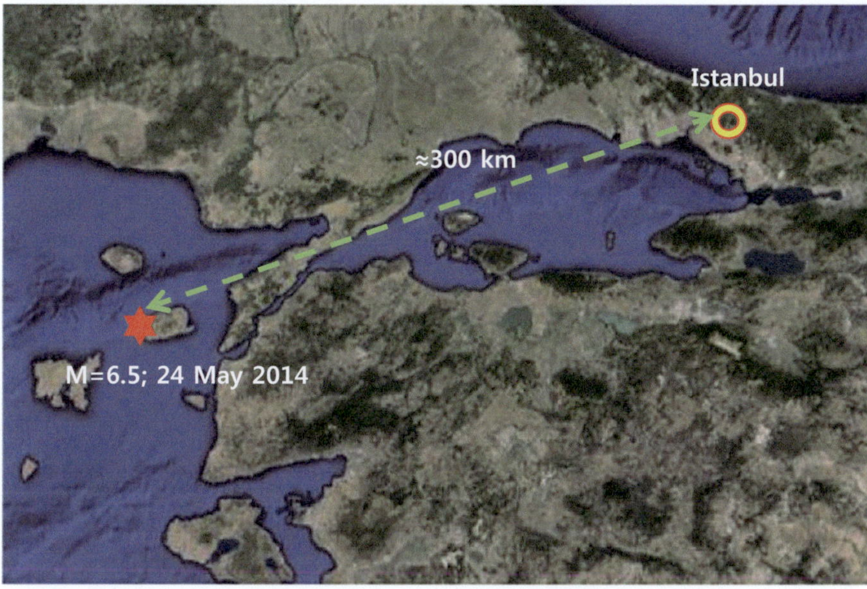

Fig. 5.7 Location of M = 6.5, 24 May 2014 Northern Aegean earthquake and its distance from Istanbul

strong shaking for about one minute. The strong shaking for displacements lasts much longer, about 5 min in the long direction and more than 15 min in the short direction. Although both the acceleration and displacement amplitudes were not critical in terms of structural safety, such a long duration of shaking was the main reason for the panic among people in this tall building. Similar panics were observed in all tall buildings throughout the Middle East during this earthquake.

The Marmaray rail tube tunnel provides another example representative of the response of large-scale engineering structures to distant earthquakes. During the 2014 Northern Aegean Sea earthquake the SHM system in the tunnel registered long-duration accelerations, which when converted to displacements yielded peak values that were controlled by long-period components and significant considering the size of the earthquake and the distance of the tunnel from the epicentre (Fig. 5.13). The level of long period displacements are large, but from the point of structural safety of the tunnel they are at the low-risk side.

5.4.2 Damping in Tall Buildings and Masonry Structures

In addition to long-period resonant excitations from surface waves, another reason for the long duration of vibrations in tall buildings is the very low levels of damping.

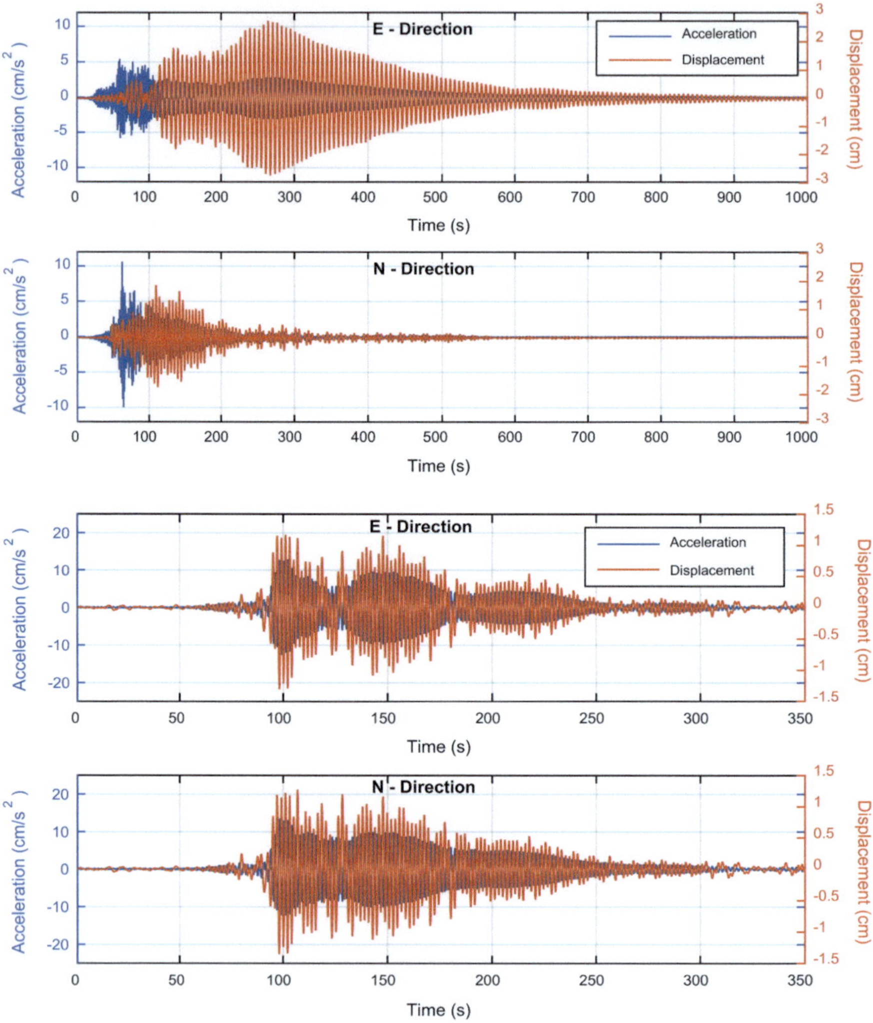

Fig. 5.8 Acceleration and displacement responses of the Sapphire building (top) and the Maltepe minaret (bottom) filtered between 0.1 and 2.0 Hz. Direction east, represented with E in the top figure, corresponds to the short direction of the Sapphire Tower. Direction north is parallel to its long direction. The figure shows data from the 24 May 2014 Northern Aegean earthquake (M6.5) recorded at the uppermost stations in the two structures

Figure 5.14 shows the modal damping values for the two horizontal and the torsional modes of the 62-story Sapphire building calculated from the recorded vibrations during the 2014 Northern Aegean earthquake. The damping values are calculated from the decay of the free vibration modal displacements of the building after the earthquake stopped. As the figure shows, the viscous damping ratios are all less than 1%.

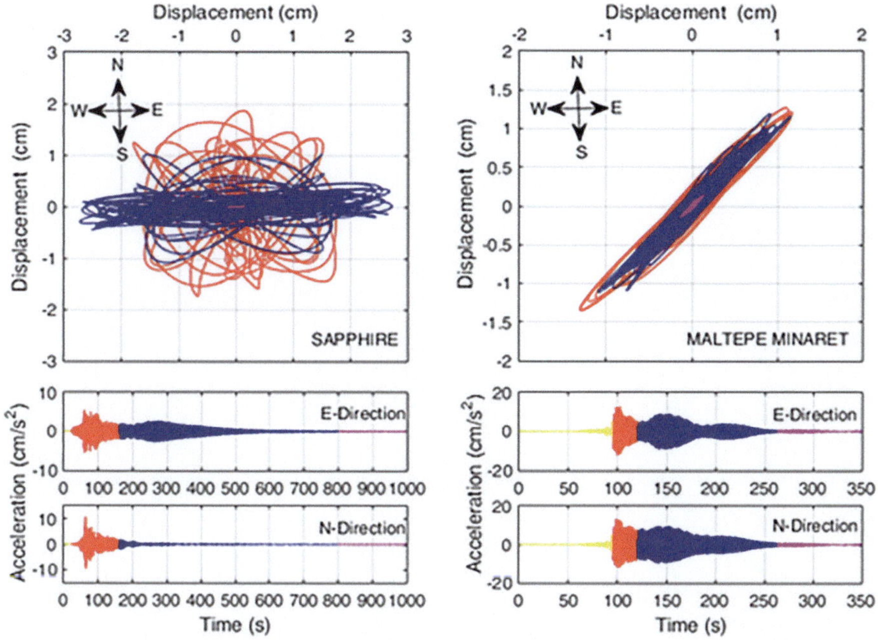

Fig. 5.9 Particle motions at the top level co-shown with accelerations recorded at the same level, all filtered between 0.1 and 2.0 Hz; Sapphire tower on the left, Maltepe minaret on the right

Similar results were observed by others as well. A study by Satake et al. (2003) on a large number of records from tall buildings confirms that there is a definite reduction in damping ratios as the building height increases (Fig. 5.15). On the other hand, it has been observed that damping in a structure is temperature dependent. Long-term, continuous observation of historical structures over several years suggests that damping in a masonry structure tends to decrease as the temperature increases. This phenomenon will be elaborated on in the following subsection.

5.4.3 Influence of Environmental Conditions on Structural Response

Structures are under the continuous influence of many external factors such as ambient vibrations, earthquakes, atmospheric conditions, and other human activities. One of the most important external factors is the change in the atmospheric conditions. Daily and seasonal variations in the temperature, for example, induce continuous cyclic variation in the dynamic response parameters such as modal frequencies and modal damping.

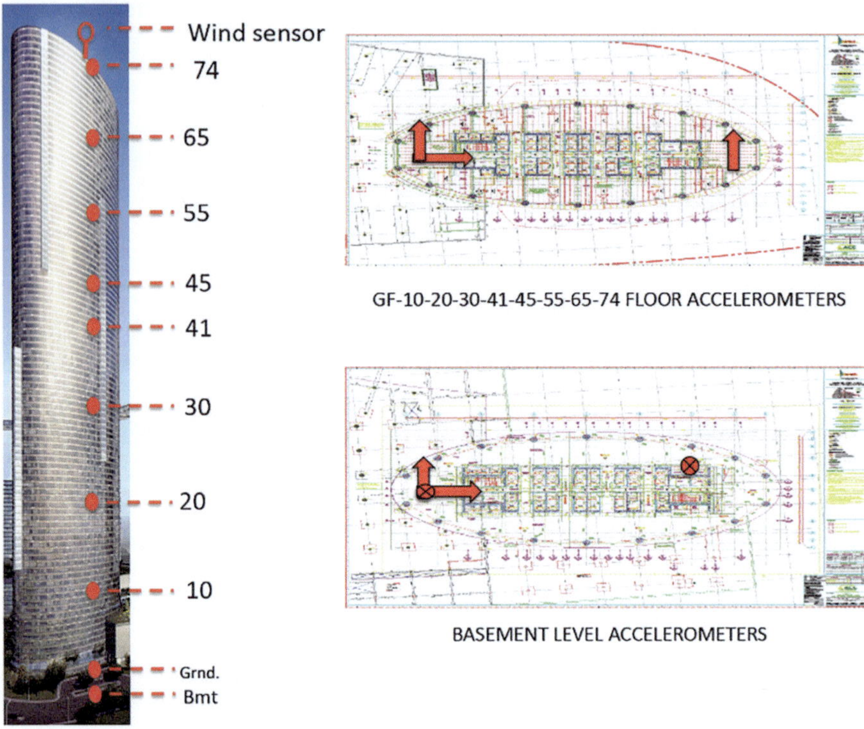

Fig. 5.10 Sensor locations in the Abu Dhabi building

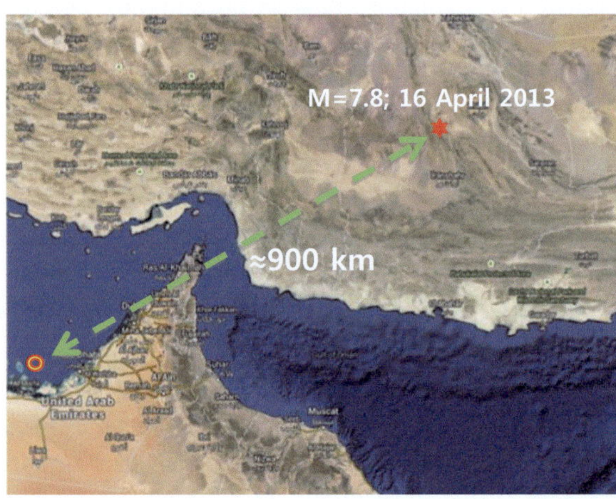

Fig. 5.11 $M = 7.8$, 16 April 2013 Iran–Pakistan border earthquake and its distance from Abu Dhabi

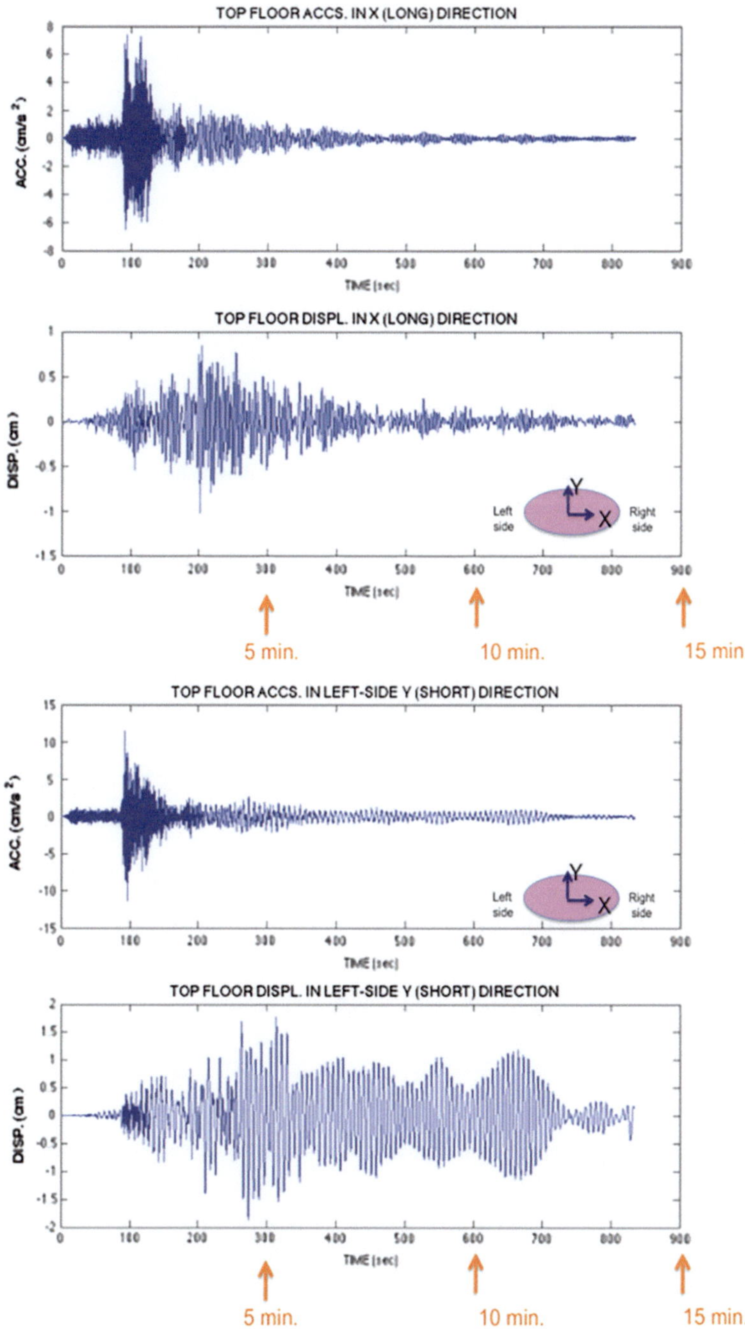

Fig. 5.12 Top floor accelerations and displacements in the Abu Dhabi building during the $M = 7.8$, 16 April 2013 Iran–Pakistan border earthquake

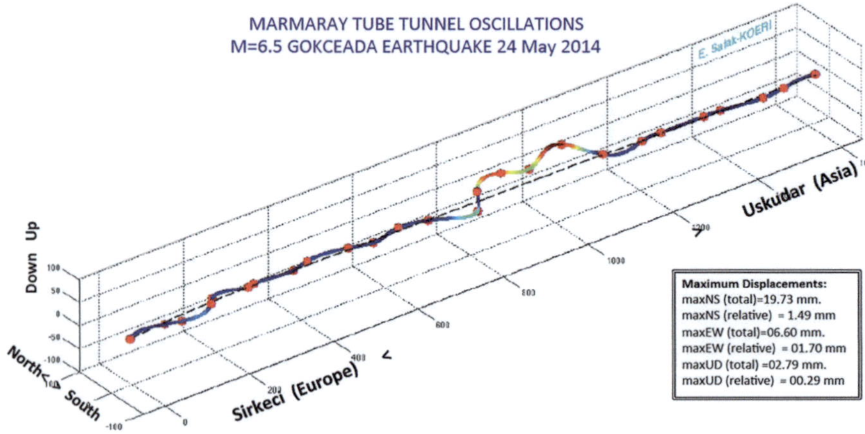

Fig. 5.13 Displacements across the Marmaray tube tunnel during the 24 May 2014 Northern Aegean Sea earthquake

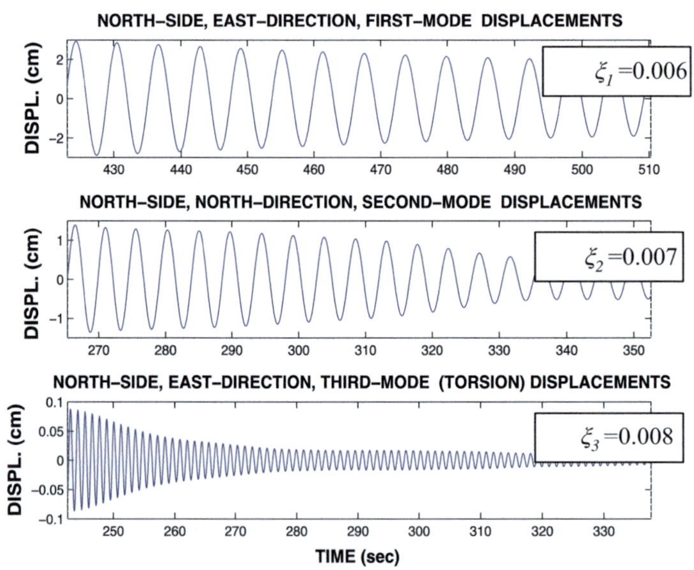

Fig. 5.14 Damping ratios for the first three modes of the Istanbul building calculated from the records

Cakti and Dar (2015) have shown that Hagia Sophia's first modal frequencies in two orthogonal directions increase with the rise of temperature in the transition from winter to summer and decrease with the temperature towards winter months (Fig. 5.16). The increase in frequency is 9.8% for the first mode and 7.4% for the second mode. It is found that modal frequencies are not only sensitive to long-term temperature variations across the seasons, but also to temperature changes within a month and even within one

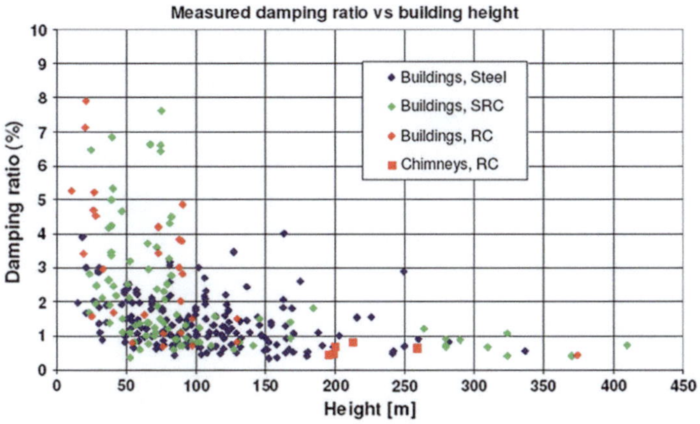

Fig. 5.15 Variation of measured damping with building height (after Satake et al. 2003)

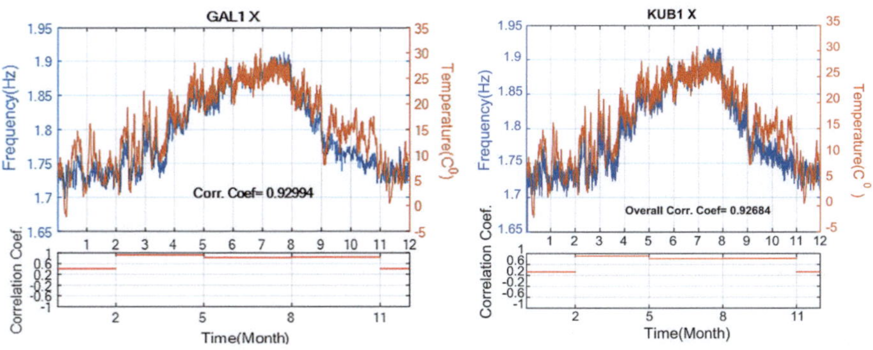

Fig. 5.16 Variation in the first modal frequency of Hagia Sophia in Istanbul at the gallery level (left) and at the dome level (right) over a one-year period, co-shown with temperature change in the same year

day. The rise in modal frequency starts at about 10 °C and has a more or less linear trend (Fig. 5.17). In lower temperatures, no variation in frequency was observed. It was found that modal damping is temperature dependent as well. The damping coefficients of Hagia Sophia drop over the period of time from winter to summer and then start to increase again in autumn and winter months (Fig. 5.18).

Although subtle, it appears that wind also has an influence on the vibration properties of the structure. It was observed that with increased wind speeds, the modal frequencies tend to drop. The observation is clear at times when the temperature change does not dominate the variation in frequency.

These findings indicate that the algorithms to estimate damage from frequency variations in buildings need to consider changes in atmospheric conditions and the structure's response to them. They also suggest that frequency should not be a single parameter to consider in damage detection, since frequency variations up to

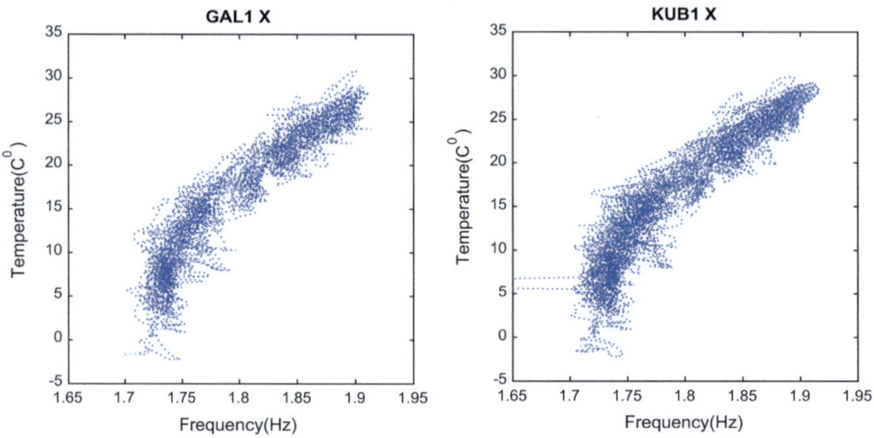

Fig. 5.17 Variation of the first modal frequency with temperature in Hagia Sophia's gallery level (left) and dome level (right)

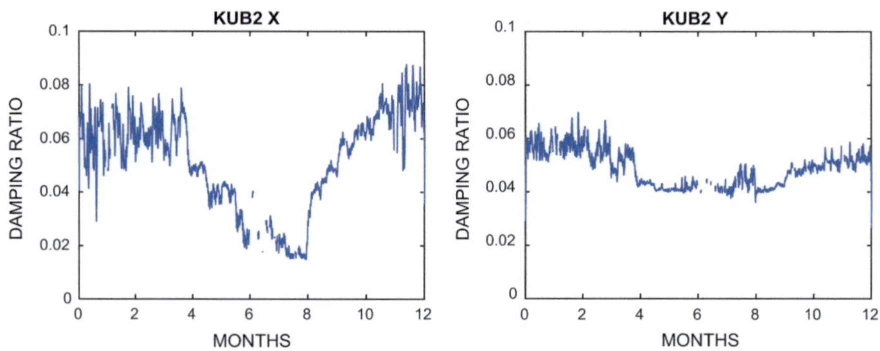

Fig. 5.18 Variation of modal damping in Hagia Sophia in Istanbul at the dome level over a one-year period

10% can be associated with temperature. Each structure responds in a different way to such factors. Therefore it is essential to monitor each building for a sufficiently long time, before coming up with a reliable and suitable algorithm that is carefully tuned to the conditions of the structure.

5.4.4 Amplitude and Duration Dependence of Modal Frequency

It is known that modal parameters, i.e., frequency and damping, depend on the level of shaking. Moreover, they are also dependent on the duration of strong shaking.

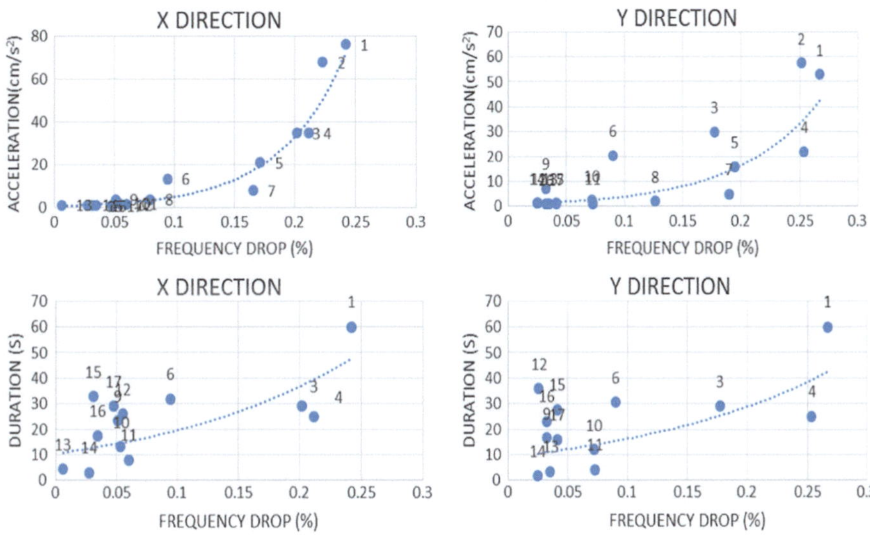

Fig. 5.19 Frequency drop in Hagia Sophia with respect to acceleration (top) and duration (bottom)

With increased levels of vibration amplitudes during earthquakes, the modal frequencies tend to drop. They recover their pre-event values following the earthquake. In Hagia Sophia, as an example, the frequency appears to drop up to 25% with respect to the pre-event modal frequency, estimated based on the study of earthquakes recorded over fifteen years in the structure (Fig. 5.19, top).

Duration of strong shaking gets longer with magnitude. However, there are many instances when the duration of two records is substantially different from each other, even though their peak amplitudes are comparable. Probable reasons for this could be distance, site conditions and structural properties. Data from Hagia Sophia suggest that the drop in modal frequencies during events of longer duration is larger than the events with shorter duration and similar peak amplitudes (Fig. 5.19, bottom).

5.4.5 Structural Response During Explosions

The blast as a result of the bombing attack that took place in the Sultanahmet Square in Istanbul, Turkey on 12 January 2016 was recorded by the strong motion networks installed in two nearby historical monuments, i.e., the Sultanahmet Mosque and the Hagia Sophia Museum. The explosion took place at about 100 m away from the Sultanahmet mosque, near the Egyptian Obelisk in the Hippodrome (Fig. 5.20). These very rare records by the two systems, i.e., Sultanahmet and Hagia Sophia, show the onset of the explosion that lasted about 1.5 s, followed by the free

Fig. 5.20 The location of the explosion on 12 January 2016 in the Hippodrome in Sultanahmet

vibration of the two buildings. Damping could be estimated from the free vibration part of the records. The movies created focusing on the explosion and on the following seconds showed their response, which was dominated by the induced push in the direction of the explosion and opening and closing, i.e., breathing at the dome level. The phenomenon can be seen in Fig. 5.21 for the Sultanahmet mosque.

5.4.6 Long-Term Deformation Monitoring

Laser technology created remarkable opportunities for architectural and structural investigations. Traditional methods and technologies mostly involve point specific measurements. Laser technology, on the other hand, is capable of registering three-dimensional point clouds of objects to a high precision. This implies that it can be a tool for structural monitoring. Structural deformations/damage caused by earthquakes, or due to other static or dynamic mechanisms can be monitored with the help of laser technology and by comparative analyses of point cloud data. We have implemented this technology since 2015 in our structural monitoring activities. Historical monuments that we monitor are periodically scanned to assess the extent and progress of structural deformations (if there are any) that could potentially be the result of geotechnical, structural and/or material problems or be generated as a consequence of dynamic sources such as earthquakes.

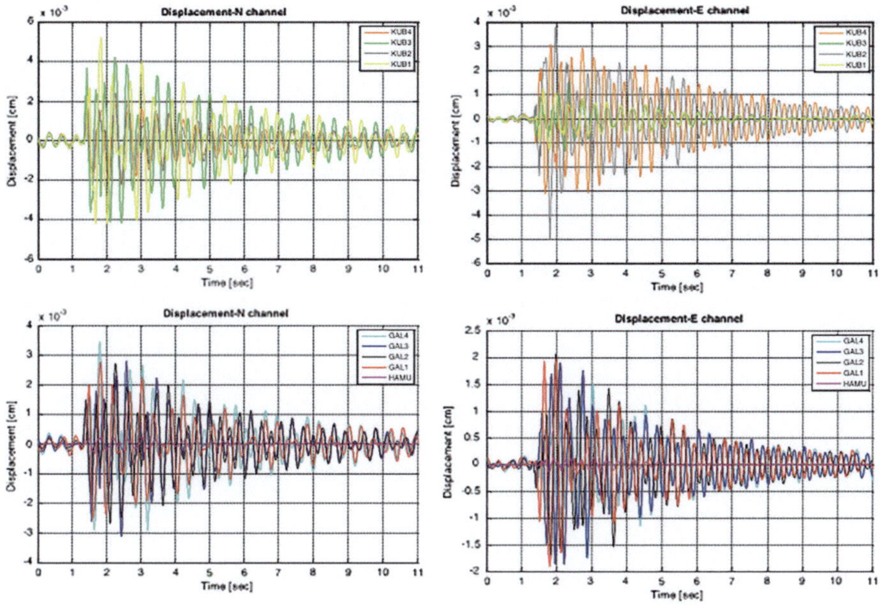

Fig. 5.21 Displacements on the Sultanahmet Mosque during the nearby explosion on 12 January 2016. Top row: dome base, bottom row: tops of the main piers

5.5 Conclusions

The number of structures with SHM systems is rapidly increasing. Istanbul is one of the major cities that has a large number of such systems.

The records from structures installed with SHM networks have significantly contributed to our understanding of structural dynamic response. Continuous monitoring enabled a fresh look at structural dynamics due to excitation sources other than earthquakes. Traditional location- or point-specific measurements are slowly being replaced by surface distributed systems and by systems using 3D laser technology. Among the new knowledge gained, we can include the surface-wave-induced, long-duration response of tall buildings, associated very low damping ratios, and the sensitivity of dynamic characteristics of structures to changes in environmental conditions.

Acknowledgements We would like to gratefully acknowledge the efforts and contributions of the following individuals from DEE-KOERI: our graduate students Mr. Emrullah Dar, Ms. Nesrin Yenihayat; our engineers and technicians Mr. Ahmet Korkmaz, Mr. Nafiz Kafadar, Ms. Gökçe Turan, Ms. Senem Korkmaz, and Mr. Emre Özdemir. The research described in this paper was supported by the Ministry of Development of the Republic of Turkey.

References

Çaktı E, Dar E (2015) Long-term dynamic response of Hagia Sophia in Istanbul to earthquakes and atmospheric conditions. In: Structural Health Monitoring, Proceedings, dpi-proceedings.com. https://doi.org/10.12783/shm2015/165

Safak E, Kaya Y, Skolnik D, Ciudad-Real M, Al Mulla H, Megahed A (2014) Recorded response of a tall building in Abu Dhabi from a distant large earthquake. In: Proceedings of the 10th U. S. national conference on earthquake engineering, Anchorage, Alaska, 21–25 July 2014

Satake N, Suda K-I, Arakawa T, Tamura Y (2003) Damping evaluation using full-scale data of buildings in Japan. J Struct Eng 129(4)

Tunc S (2016) Personal communication

Chapter 6
Earthquake Performance of Hagia Sophia

Mustafa Erdik

Abstract Hagia Sophia, during its history, was affected by many earthquakes, resulting in several reconstructions of parts of its main dome and main arches. Being one of the most important buildings of the mankind's common heritage and given Istanbul's high seismic risk, the worthiness of the present-day structure in a future major earthquake and what, if any, strengthening interventions will be needed to carry the edifice intact to the future were reviewed. The review encompasses studies conducted on structural and sub-structural assessments, material characterization, linear and nonlinear finite element dynamic analysis, nondestructive structural testing, ambient vibration testing, and the analysis of the strong motion data. The expected earthquake response, possible failure modes, and the retrofit proposals for the improvement of the earthquake performance of Hagia Sophia covered in different studies were comparatively and critically analyzed.

Keywords Hagia sophia · Earthquake · Analysis · Performance
Retrofit

6.1 Introduction

With its vast scale, immense cost, extraordinary speed of erection and stunning interior space, the creation of Hagia Sophia is unparalleled in architecture. Dedicated during the reign of Justinian in 537, after 5 years of construction, it has been an inspiring architectural and engineering marvel throughout the ages. The Anatolian architects/engineers, Anthemius of Tralles (todays Aydın, Turkey) and Isidorus of Miletus (todays Balat, Aydın, Turkey) created a building that was a big challenge given its size, the day's technology, and the construction materials brick and stone masonry (Fig. 6.1). It remained as the largest vaulted structure in the

M. Erdik (✉)
Boğaziçi University, İstanbul, Turkey
e-mail: erdik@boun.edu.tr

A. A. Kasimzade et al. (eds.), *Seismic Isolation, Structural Health Monitoring, and Performance Based Seismic Design in Earthquake Engineering*,
https://doi.org/10.1007/978-3-319-93157-9_6

Fig. 6.1 Hagia Sophia

world for some 800 years and it was the largest cathedral in the world for nearly a thousand years, until the completion of the Seville Cathedral in 1520.

Being one of the most important buildings of the mankind's common heritage and given Istanbul's high seismic risk, the worthiness of the present-day structure in a future major earthquake and what, if any, strengthening interventions will be needed to carry the edifice intact to the future millennia are important questions. This paper will review the studies conducted in response to these questions and will provide a summary of answers obtained.

6.2 Past Earthquake Damages and Structural Interventions

Hagia Sophia, during its history, was affected by many earthquakes, resulting in several reconstructions of parts of its main dome and main arches and repair of some structural elements. In 558, only 20 years after completion of construction, the central dome, including its supporting elements, partially collapsed due to an earthquake in 557. Nephew of Isidorus was assigned for the repair works and a new dome with an altered form was completed in 563. The important earthquake damages and the interventions are provided in Table 6.1.

Starting with the reconstruction of the main dome in 562 Hagia Sophia has experienced a multitude of structural interventions. After the collapse of the west main arch, sector of the main dome and the semidome in 989 extensive reconstructions were carried out till 994. Figure 6.2 provides renderings of the year 989 and year 1346 damages (Duppel 2009). Similar reconstructions on the eastern part were conducted on the eastern side between 1344 and 1354 after the collapse of east

Table 6.1 Structural history and past earthquake damages to Hagia Sophia

23 Feb 532: Construction started (Anthemius of Tralles and Isidorus of Miletus and about 10,000 workers)
27 Dec 537: Consecrated (Justinian I)
7 May 558: First collapse east arch, east semidome and east quadrant of main dome
24 Dec 562: East arch and east semidome rebuilt to original dimensions. New dome built 6m higher by Isidorus the younger
565–577: Southwest corner alterations due to building of patriarchal palace
768–769: Unspecified repairs
869: Damage to western arch caused by Earthquake and repairs
26 Oct 989: Second collapse of west arch, west semidome and west quadrant of main dome
993: West arch rebuilt to present larger size. Semidome, part of main dome and rebuilt to original dimensions by Tradat. Tympana under north and south main arches were built.
1317: Exerior buttresses added
19 May 1346: Third collapse east arch, east semidome east quadrant of main dome
1347: East Arch, East semidome rebuilt to original dimensions by Phakeolatos
1354: East quadrant of dome rebuilt to original dimensions by Astras and Peralta
1453: Sultan mehmet II builds the two western minarets
Sept. 10, 1509 Partial damage due to earthquake
May 10, 1556 Partial damage due to earthquake
1572–4: Architect Sinan (Under Sultan Selim II) builds two buttresses (One at the southwest corner, the other in the middle of the south) and repairs other buttresses
1574–6: Sultan Murat III Orders Architect Sinan to complete the repairs started by Selim II
1609: Sultan Ahmet I orders various repairs
Sept. 2, 1754 Partial damage due to earthquake
May 22, 1766 Partial damage due to earthquake
1847–9: Sultan Abdülmecid orders the fossati restoration
July 10, 1894 Light damage due to earthquake

main arch, sector of the main dome and the semidome. Buttresses to the south and southeast were built in early thirteenth century, later amended during the Ottomans. Four flying buttresses were added in fourteenth century to support the dome at the location of repair joints. During the reign of Selim II (1566–1577), the edifice was extensively strengthened with the addition of new and strengthening of existing buttresses to its exterior by the great architect Mimar Sinan. One of the most comprehensive restoration and retrofit of the Hagia Sophia was completed during the period 1847–1849 by the architects Fossati brothers. In addition to consolidating the dome and vaults and straightening columns, iron bars were placed around the square dome base and the four flying buttresses were removed. The last major earthquake that has caused light damage (probably reactivation of the old cracks) was 1894 earthquake (Fig. 6.3).

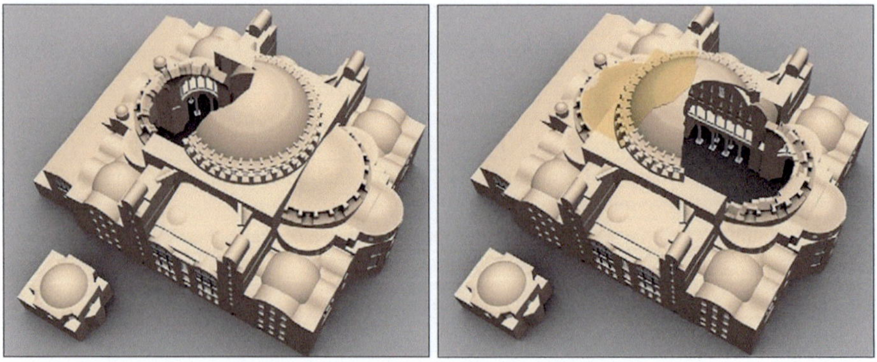

Fig. 6.2 Collapse in tenth and thirteenth centuries (after Duppel 2009)

Fig. 6.3 Sketch of damages at Hagia Sophia (looking east) in 1894 Earthquake

6.3 Current Structure

The period of systematic analytical study, assessment, and restoration of Hagia Sophia has been initiated in 1940 by the Byzantine Institute of the United States and the Dumbarton Oaks Field Committee. In this connection, R. L. Van Nice, R. J. Mainstone, and C. Mango have provided valuable information of the structure of the edifice (Mainstone 1988, 1992, 1993a, b, c; Van Nice 1964, 1965, 1986). R. L. Van Nice has carried out the first modern survey of Hagia Sophia and in 1965 and 1966 has released his survey drawings and documents. R. J. Mainstone in 1988 published his extensive studies as a book. In addition to several structural studies conducted by Japanese, Italian, and German researchers, since 1989, a consortium formed by Princeton and Bogazici Universities have undertaken extensive studies to understand the structural characteristics of Hagia Sophia. Isometric cutaway figures of Hagia Sophia depicting the current status and status prior to tenth century are presented in Figs. 6.4 and 6.5 provides plan and sections of the edifice.

The foundations of Hagia Sophia are well placed in terms of the surrounding geology. It rests upon natural rock, which is Folded Devonian schist at the crest of the Devonian anticline. Because this rock is a compressed and old formation, it was good for a structure on the scale of Hagia Sophia (Emerson and Van Nice 1943).

The main dome is spherically shaped and rests on a square dome base. Major elements include the four main piers supporting the corners of the dome base and the four main arches that spring from these piers and support the edges of the dome base. Entrance to nave of the structure is provided by through a narthex form west. The nave is shaped by a procession of piers and colonnades that separates it from aisles at the ground level and from the galleries above. The nave ends with an apse on the east side, covered by a barrel vault and semidome. The structure's nave has an uninterrupted free internal space of 30 m wide, 80 m long. At the corners of the nave, the colonnades become exedra, covered with semidomes above the gallery level. These two half domes are used to expand the nave space from west to east

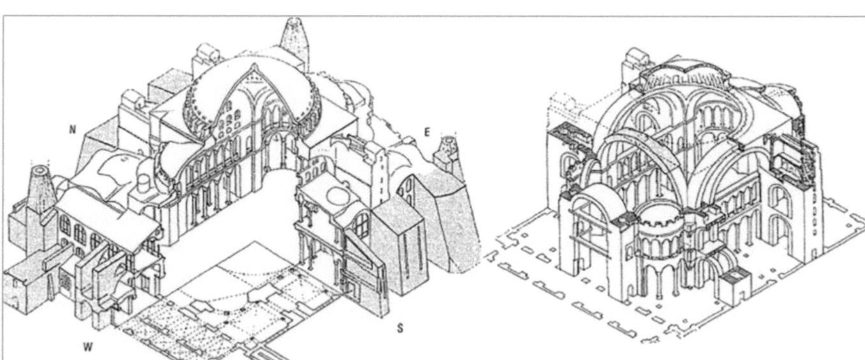

Fig. 6.4 Cutaway view of the current structure (left) and the original structure (right) (after R. J. Mainstone)

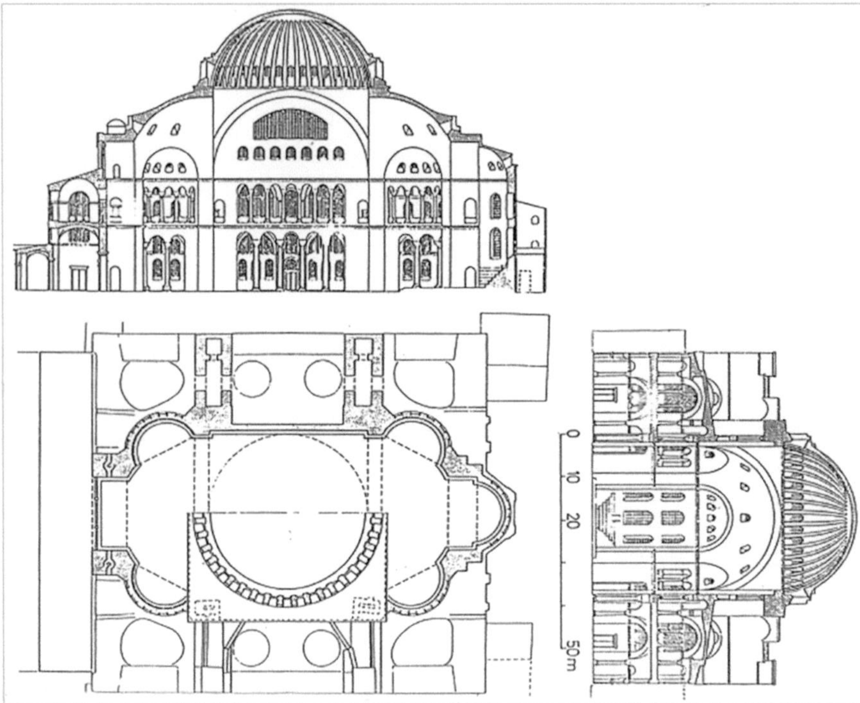

Fig. 6.5 Plan and N-S and E-W section views (after Van Nice 1965)

and also to counter support the outward thrust of the main dome. Four main piers, forming a square at the central of the structure, support arches above them. Four buttress piers are connected to the main piers by barrel vaults and arches passing transversely the aisles at ground and gallery level. Four great arches span 30 m between the main piers. Space above the main arches is filled by surcharge for the purpose of generating a square platform. This platform converted into a circle shaped domebase by four pendentives which have a form of equilateral triangles.

The north and south ones are actually two arches which are supported by tympana walls and the gallery and aisle colonnades under them. The two arches on the north (N) and south (S) sides are stiffened by additional arches with infill provided between the primary and secondary arches. The N and S arches have approximately the same dimensions and are symmetrically located.

The east and west arches are abutted by main semidomes, which are in turn are supported by both secondary piers and exedra semidomes. The E and W arches span an open area with buttress piers providing lateral support to the main piers in the N-S direction. A secondary colonnade system also provides some resistance to lateral motion of the main piers in the E-W direction. The E and W arches are not symmetric in dimension and are slightly asymmetric in location. This asymmetry is due to a reconstruction and enlargement of the W arch after a partial collapse in 989.

Main dome, supported by 40 ribs resting on this circle, has a maximum diameter of 31.24 m and reaches 55.6 m height above floor. Due to settlements, deformations, and consecutive repairs, the dome has lost its perfect circular base and has become somewhat elliptical with a diameter varying between 31.24 m and 30.86 m.

At the present time, the dome shows four lines of discontinuity following to the meridians which constitute the limiting points between the dissimilar parts reconstructed at different times. The North and South parts show two segments of what was once the sixth-century dome. On the West and East, interrupting the continuity of the first reconstruction more or less symmetrically, we can see the parts reconstructed in the tenth and fourteenth centuries respectively, whose area (similar in shape and size) is almost double that of the two parts remaining from the sixth century. We can also see irregular points and lines of discontinuity in the curvature of the dome, the arches, and the pendentives. This discontinuity reflects the junction lines between the parts which survived collapse (now permanently deformed) and the reconstructed parts, where it was evidently impossible to maintain continuity of the tangents which would have meant repairing the existing deformation, or altering the newly reconstructed geometry (Fig. 6.6).

Speed of construction (5 years) and the use of puzzolanic mortar (that needs years to set) have caused plastic deformations in the structure. These deformations can be correlated with the sequence of construction, depicted in Fig. 6.7

The dome, the galleries, and many of the structural elements, such as arches, columns, cornices and pillars, show irregularities in their geometry, reminders of the buildings many vicissitudes. Such deformities provide useful clues on which to base a reconstruction of the current situation. The partial collapses and rebuilding represent the main concerns regarding the behavior under the effects of seismic actions.

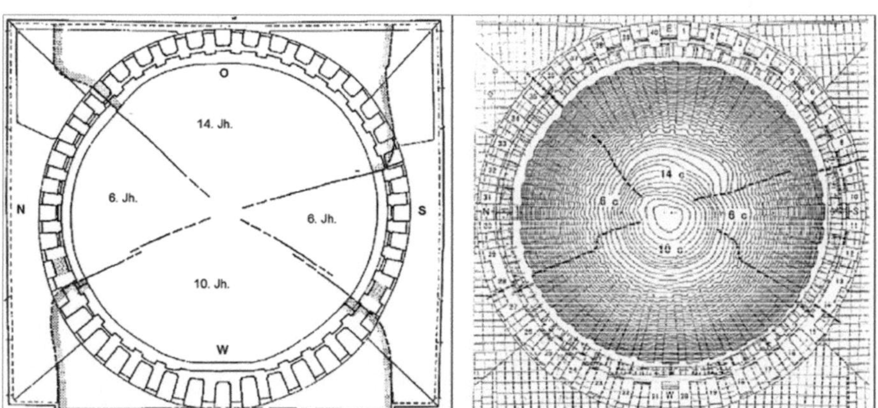

Fig. 6.6 Dome sectors from sixth, tenth, and fourteenth centuries (Mainstone 1988; Sato and Hidaka 2001)

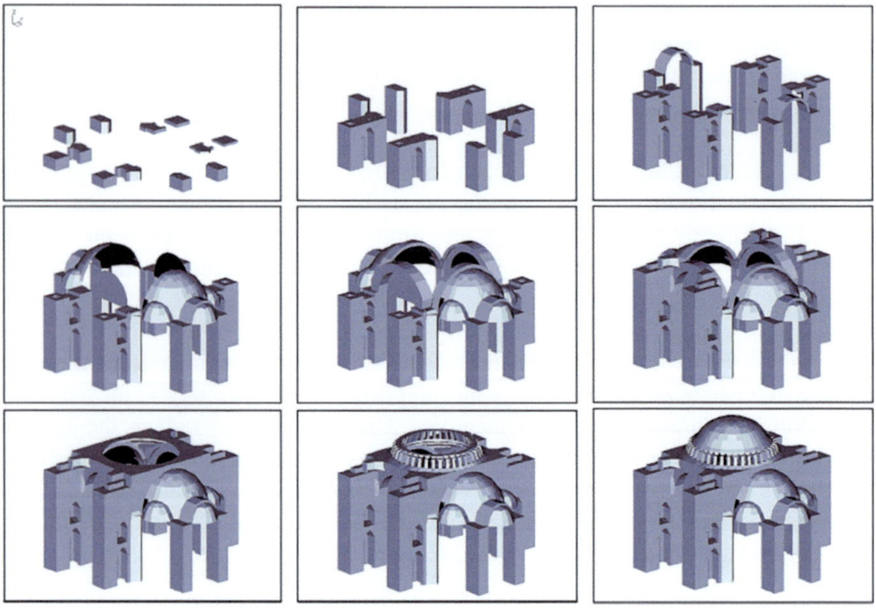

Fig. 6.7 Presumed Sequence of Construction (after A. S. Cakmak)

Along the north–south direction main piers have an approximately 85 cm outward inclination from ground level to top of the piers. Also for the east–west direction the piers exhibit 16 cm outward inclination. The same outward tendency in east–west direction can be observed especially for the secondary piers easily. On the main arches, vertical drops are obvious at the crowns. The crowns of north, south, east, and west arches have vertical displacements of 9, 18, 26, 13 cm in the direction of gravity, respectively. The east and west arches tilt inward 14 and 15 cm due to semidomes attached to them. The north and south arches' crowns also lead outward 60 and 57 cm, respectively. But the lower cornices on the north and south sides bow inward in contrast to the outward bowing of arches. This deformation can be explained by the tilt of piers. The columns at southeast and southwest corners incline outward diagonally. An isometric sketch of these deformations is presented in Fig. 6.8

6.4 Construction Materials

Main building materials used in the Hagia Sophia are stone, brick, and mortar. Most part of the structure is built with brick masonry, while the main piers and the lower parts of the external buttresses are made of stone block masonry. The main arches and dome are comprised of brick masonry.

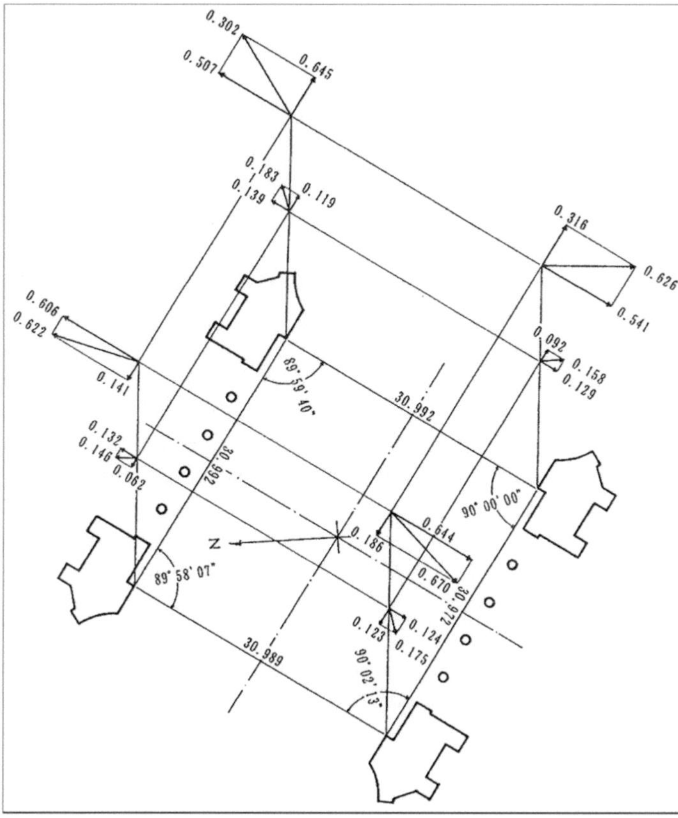

Fig. 6.8 Isometric view of the existing deformations (after Aoki et al. 1997)

Stone is either a limestone or a local type of granite. The blocks are about 0.45 m × 0.45 m ×1 m size. Brick elements are square shaped with a dimension of 35–40 cm and a thickness of 4–5 cm. However, the brick sizes in the main arches exceed 0.6 m.

The pozzolanic mortar used in the brick masonry may need approximately a curing period of one year to gain their full strength. The mortar thickness is on the order of that of the bricks (about 5–7 cm). In such a masonry construction, the bricks may have the role of providing stiffness rather than strength. The mortar samples taken from Hagia Sophia tested at National Technical University of Athens exhibit tensile strength varying between 0.5 and 1.2 MPa (Çakmak 1994). The ultrasonic test results reported provide the values of 4.2, 0.9, and 1.5 GPa, respectively for the elastic moduli of brick, mortar and the composite masonry. For stone masonry, the elastic modulus was found to be 15.2 GPa (Mark and Cakmak 1994)

6.5 Structural Analysis

Since 1989, the structural assessment and analysis investigations initiated jointly by Princeton and Boğaziçi Universities, with support from US National Science Foundation, have led to numerous investigations that throw light on the static and dynamic response of Hagia Sophia and on its earthquake performance (Erdik and Cakti 1990, 1992, 1993; Hill 1991; Durukal 1992; Mark and Cakmak 1992, 1994; Livingstone 1993; Mark et al. 1993; Erdik et al. 1993; Swan and Cakmak 1993; Livingstone et al. 1992, 1993; Cakmak et al. 1993, 1995a, b, 2009; Davidson 1993; Natsis 1994; Çakmak 1995; Durukal and Erdik 1994, 2003; Durukal et al. 1998a, 1998b, 2001, 2003; Erdik and Durukal 1996, 2003, 2007; Gürbüz et al. 1993; Keypour 2001; Moropoulou et al. 2002; Kırlangıç 2008).

A similar project on Hagia Sophia was conducted from 1990 to 1997 as a joint research between Japanese and Turkish scholars (several Japanese universities and Istanbul Technical University in Turkey). The results of the project were published as the following four parts (Aoki et al. 1997): (1) on the investigation of crack patterns (Aoki et al. 1992, 1993); (2) on the measurements of natural frequencies by ambient vibrations (Aoki et al. 1993); (3) on the estimation of the material properties; and (4) on finite element analysis of response (Kato et al. 1992; Aoki et al. 1997).

A research project entitled "Engineering Studies of the Main Dome and Main Pillars of the Hagia Sophia in Istanbul" was launched in 2004 by German Research Foundation to evaluate its structural response and stability in case of an earthquake, based on realistic data. The nondestructive geophysical studies conducted provided data on the geometric, structural, and material characteristics of the edifice (Duppel 2010a, b, 2011; Wenzel and Duppel 2010). High-resolution CAD and finite element models of Hagia Sophia were developed (Blankenhorn and Schweizerhof 2009) and comprehensive static and dynamic analyses were conducted.

The investigations conducted by Princeton and Boğaziçi Universities have served to illuminate the following structural aspects of Hagia Sophia:

- Structural system properties and dimensions
- Material properties (in situ nondestructive- and in situ shear testing, chemical and physical properties of mortar)
- Soil and foundation properties and the substructure (seismic refraction, seismic topography, geo-radar and micro-gravimetric methods)
- Existence and location of embedded iron ties and rings (magneto-metric procedures)
- Forces on the tie bars of secondary arches (dynamic methods)
- Modal frequencies and shapes (ambient vibration tests)
- Strong motion instrumentation and analysis of data
- Simulation of expected earthquake ground motions at the site
- 3-D finite element modeling calibrated to yield the measured dynamic properties

- Nonlinear earthquake response analysis and determination of failure mechanisms
- Proposals for strengthening
- Analysis of earthquake data from the 1999 earthquakes.

A finite element model of Hagia Sophia is created with the intention of understanding the interaction between the structural elements of the building, the reason for the deformation which can be observed, to understand the mode of failures during or after earthquakes in the past and to provide insight into the future behavior of the edifice in a big earthquake. The dimensions of the finite element model are based on Van Nice (1986) and Mainstone (1988). In modeling, an idealized geometry is adopted despite the fact the real structure is hardly regular. The creation of the model is described in detail by Hill (1991), Durukal (1992) Mark et al. (1993), Davidson (1993), Cakmak et al. (1995). For the finite element model of Hagia Sophia, frame, shell, and solid elements of are employed. Arches, surcharge, pendentives, and piers are modeled as solid elements; while domes and tympanum walls are represented by shell elements (Fig. 6.9). Ribs of the main dome, columns, and beams are modeled as frame elements. The total number of the elements used in the model is 7139 solids, 1012 shells, and 216 frames. The material types used in this finite element model is illustrated in Fig. 6.10.

This linear numerical model is calibrated for effective dynamic material properties using measured response during the ambient vibration testing and during earthquakes to provide reasonable estimates of overall dynamic characteristics including frequency and primary modes of response. The model is then studied in terms of the time-domain response of selected locations and in terms of stress analysis. Cakmak et al. (1995) have incorporated soil-structure interaction to this finite element model in an approximate way to improve the low-intensity predictions. The most recent study in these series of linear investigations conducted by

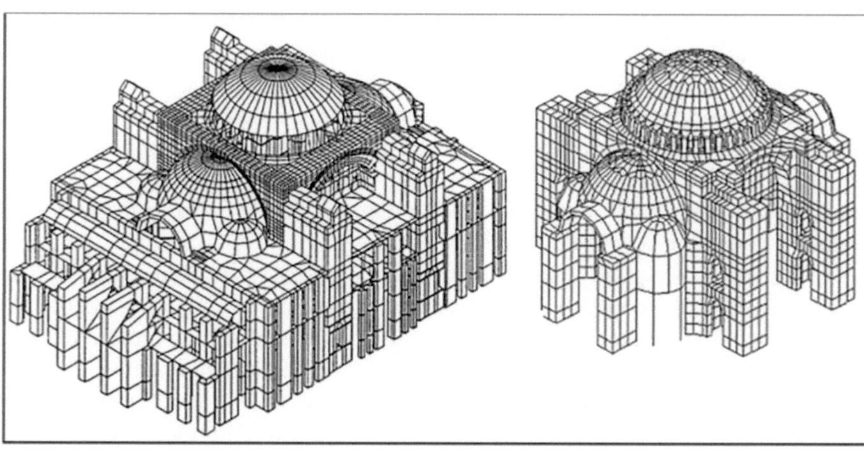

Fig. 6.9 Full and reduced finite element models

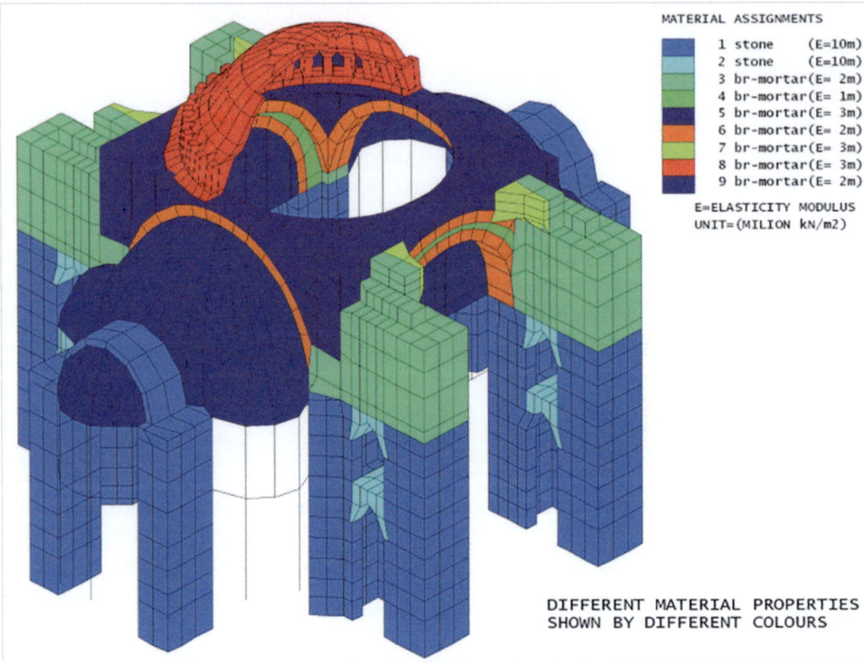

Fig. 6.10 Material types used in the model

Kırlangıç (2008), the finite element model was transferred to SAP2000 Version 10 and the previously proposed retrofit options were analytically tested.

Several scenario earthquakes were synthesized and applied as input to this finite element model to understand the stress levels throughout the structure. The stress analysis indicates that the top of the semidomes and the main arches on the east and west sides are the most vulnerable zones during earthquakes. In fact, these zones are those where past damages and collapses have concentrated. It appears that future suggestions regarding the strengthening of the structure should concentrate on these zones. The earthquake response analyses have shown that during a Mw = 7 earthquake that would emanate from the Main Marmara Fault about 20 km away, some stress concentrations that could lead to damage should be expected in the top portions of the main arches and the semi-domes.

The nonlinear structural modeling of Hagia Sophia has been conducted to provide insight to the collapse mechanism of the structure during a big earthquake and to simulate the earthquake records obtained on the structure in a more realistic way. In this connection, Swan and Cakmak (1993) applied an effective medium approach to model the elastoplastic masonry. The analysis is treated in two distinct stages: quasi-static to account for the deformations incurred by the structure under gravitational loading during its construction phase, and dynamic to account for the response to numerous seismic excitations. Keypour (2001) developed a simplified

3-D nonlinear finite element model of the structure with appropriate constitutive models and conducted transient dynamic analysis under selected ground motions. In the model nonlinear material types are used for the main arches, dome, and semidomes. For constitutive relationships a pressure dependent Von Mises model is used. Through this analysis, the regions of extensive nonlinear deformation and possible yield mechanisms have been identified. The yield propagation with respect to distinct time steps is presented in Fig. 6.11 for an excitation in the N-S direction.

Extensive investigations on the structural response of Hagia Sophia have also been carried out under the guidance of Prof. Giorgio Croci (Croci and Viscovic 1999; Croci 2006, 2007). A finite element model of Hagia Sophia including all the elements that are significant to represent in the real structure was developed. The model is supported with partial models with a much detailed mesh to investigate the dome. Static and nonlinear pushover type analyses have been carried out. The static analysis, taking into account only the weight of each section of the overall structure, has shown that the stresses remain at low levels. Maximum principal stresses under gravity loading are illustrated in Fig. 6.12. Static analysis shows that no critical zones exist in the main dome. At the base of the ribs, compression stresses range from 100 to 250 kPa and the lowering of the vertex is limited to just 3.5 mm.

The simulation of the earthquake is made by utilizing a response spectrum where the acceleration is independent from the period, with a value of 0.3 g (0.15 g at ground level with an amplification coefficient of 2 for the masonry structure). In the

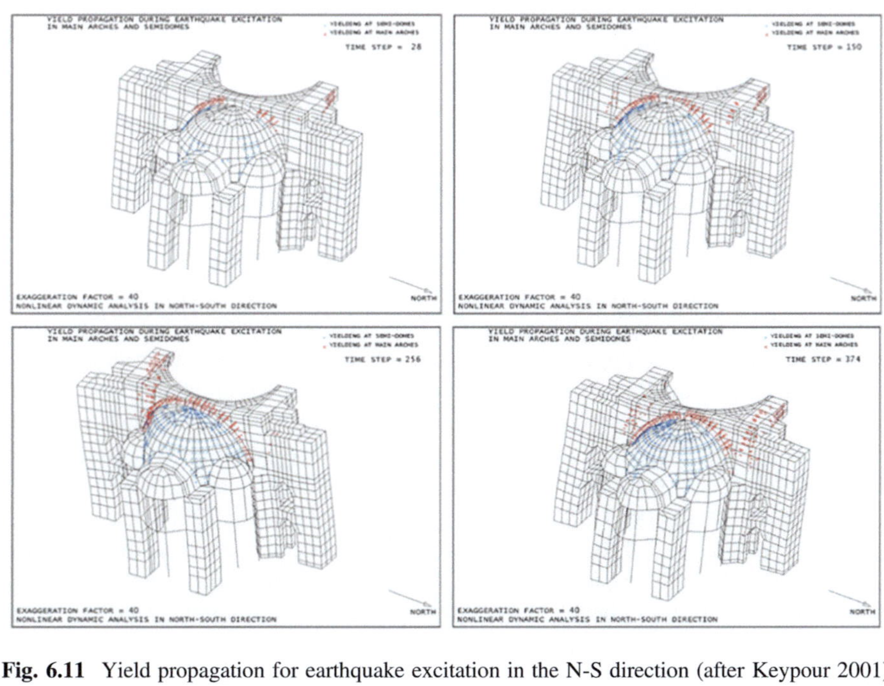

Fig. 6.11 Yield propagation for earthquake excitation in the N-S direction (after Keypour 2001)

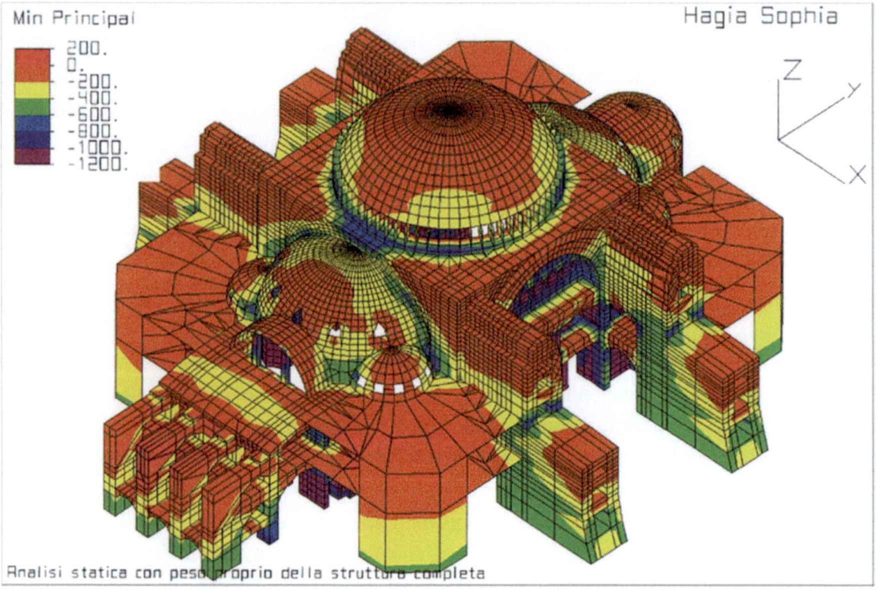

Fig. 6.12 Maximum principal stresses under gravity loading (after Croci 2006)

step-by-step nonlinear seismic (pushover) analysis, the effect of the earthquake is simulated by equivalent static forces increased of 0.1 g in each step, with a limit of 0.5 g. The seismic analysis indicates the crucial role of the secondary elements, such as the tympana, whose stiffness creates a pronounced "abutment" effect. This effect is so important when the earthquake acts in the longitudinal direction that the collapse of the tympana produces the collapse of the whole structure. Maximum principal tresses under pushover analysis in E-W direction are illustrated in Fig. 6.13.

The analysis of the structure in the N-S direction indicates tension concentrations produced in the pendentives and in the east and west arches (not stiffened by the presence of a tympanum) with substantial deformations. These deformations may invert the curvature at the key of the arch and therefore result in collapse. It is interesting to note that the damage produced in the model, simulating the structural behavior of the dome during an earthquake in direction of the semidome, well represents the situation that actually happened when the structure collapsed in the tenth and fourteenth centuries.

Aoki et al. (1992, 1993, 1997, 1998) and Sato and Hidaka (2001) have carried out structural analyses of the main dome. Structural characteristics of the dome are investigated through material and geometrical nonlinear analyses. The three-dimensional elastoplastic analysis is applied to the comprehensive structural system including the dome and supporting structures. The analytical models correspond to one quarter and one half of the actual structure, respectively, depending on the loading cases of dead and dead plus earthquake load. Through results of

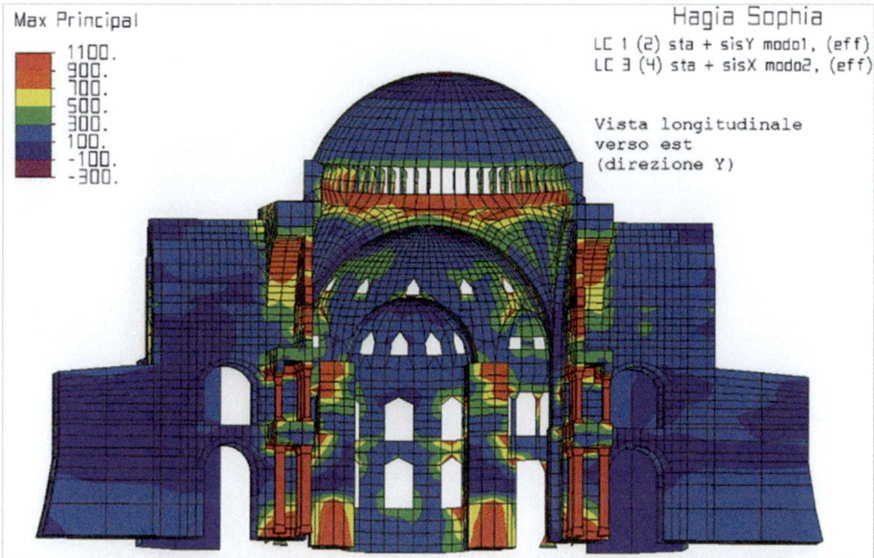

Fig. 6.13 Maximum principal stresses under pushover analysis in E-W axis (after Croci 2006)

finite element analyses, it has been shown that the oval displacement pattern of the base of the dome is in accordance with the actual state of the structure (Aoki et al. 1997). Results indicate that the double arches on the north and the south sides deflect outward due to the thrust of the dome. On the eastern and the western sides, the upper parts of the semidomes deflect inward, due to the rigidity discontinuity at the north–south great arches, while the main dome producing the outward thrust. These investigations have concluded that the main structural weakness of the whole structures exists in the east and west arches combined with the two adjacent half domes.

Significant structural analysis studies of Hagia Sophia include those conducted by Sahin and Mungan (2005) and Ozkul and Kuribayashi (2007a, b). Sahin and Mungan (2005) have investigated two alternative structural models of Hagia Sophia using dynamic finite element analysis that incorporates the crack propagation. The original model represents the structure as built, whereas a fictitious second one is obtained by strengthening the east and west arches to the level of the north and south arches. Ozkul and Kuribayashi (2007a, b) have modeled the structure with curved isoparametric trapezoidal finite elements with eight nodes and calculated the linear dynamic response of the structure under earthquake excitation.

Wenzel and Duppel (2008) and Duppel (2009) have conducted comprehensive engineering studies on the main dome and the main pillars of the Hagia Sophia. Duppel (2009) have developed a comprehensive finite element model and carried out an analysis of the collapse in the tenth century. The accurate high-resolution 3D finite element models developed (Blankenhorn and Schweizerhof 2009) were utilized in Almac et al. (2013) to assess the static and dynamic response of Hagia

Fig. 6.14 Finite Element Model developed by Blankenhorn and Schweizerhof (2009)

Sophia. An illustration of this finite element model is provided in Fig. 6.14. Under static gravity loading, the displacement scheme of the structure indicates that the crowns of west and east main arches deform inward, while the crowns of north and south arches lean outward. The thrust of the east and west main arches forces the main piers and buttress piers to incline outward and serves to create tension at their connections. As expected, the compressive stresses are located in the main piers at the ground level. These findings essentially duplicate and validate findings from previous investigations. To investigate the dynamic response of the edifice, Almac et al. (2013) impose a prescribed displacement time history to the structure through the nodes on the connection between ground and super-structure. Under N-S dynamic excitation, the slender tympanum walls under the north and south main arches indicate extensive out-of-plane oscillations. The stress distribution of the structure under lateral excitation demonstrates that the east and west main arches, together with adjacent portions of semi-domes and central dome, are the most vulnerable parts of the structure as illustrated in Fig. 6.15.

6.6 Ambient Vibration Testing and Strong Motion Instrumentation

Ambient vibration testing is an important element of structural studies of complex systems. The results provide clues about the linear dynamic characteristics of the structure, which in turn help to calibrate the numerical structural models and to design a reliable and optimal strong motion instrumentation system. The details of

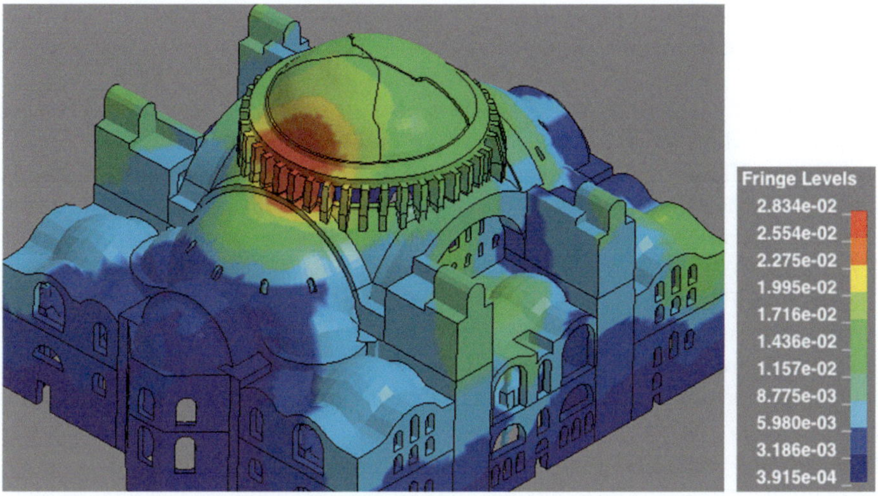

Fig. 6.15 Displacements on E-W direction under longitudinal excitation (Almac et al. 2013)

Table 6.2 Measured Modal Vibration Parameters

Modal frequencies and general mode shapes of Hagia Sophia determined by analysis of ambient vibration data		
Mode #	Modal frequency	Dominant motion
Mode 1.	1.85 Hz	Total lateral vibrations in E-W direction
Mode 2.	2.10 Hz	Total lateral vibrations in N-S direction
Mode 3.	2.35 Hz	Combined E-W lateral and torsional vibrations, Center of rotation located on the south side
Mode 4.	2.50 Hz	Combined N-S lateral and torsional vibrations Center of vibration is located on the west side
Mode 5.	2.70 Hz	Breathing mode
Mode 6.	2.80 Hz	Diagonally warping vibrations

the ambient vibration survey and the analysis are given by Erdik (1990) and Durukal (1992). Table 6.2 provides the results of this analysis in terms of modal frequencies of vibration and the identified mode shapes.

To assess whether there are permanent damages as a result of 1999 Kocaeli earthquake in Hagia Sophia, a preliminary post-earthquake ambient vibration testing was performed in April, 2000 (Durukal et al., 2003). A frequency drop of about 5.4% in the E-W direction and 4.3% in N-S direction was detected from the comparison of pre- and post-earthquake modal frequencies.

Aoki et al. (1993) have also conducted ambient vibration measurement at Hagia Sophia. The fundamental frequencies of vibration of Hagia Sophia are measured as 2.12 and 1.87 Hz in directions of N-S and E-W, respectively. As for the southeast brick and the southwest stone minaret, the fundamental frequencies in both directions are measured as 1 and 1.25 Hz, respectively. Through the comparison of ambient vibration measurements conducted on Hagia Sophia before and after the 1999 Kocaeli earthquake, Sato et al. (2008) have shown that the fundamental frequencies of vibration have decreased after the earthquake about 8.9% in E-W and 6.8% in N-S direction.

Stengel (2009) used frequency domain decomposition models to assess the modal frequencies and shapes of Hagia Sophia from ambient vibration data. The first and second modal frequencies of vibration were determined respectively as 1.743 and 2.036 Hz. The associated modal shapes on main piers and arches were illustrated in Fig. 6.16. The relatively much larger modal vibration amplitudes (in three directions) at the top of the SW pier (No. 6 in Fig. 6.16) is to be noted.

The 17 August 1999, Kocaeli, 12 November 1999, Düzce earthquakes and following aftershocks created a number of recordings in Hagia Sophia. Prior to the events of 1999, the Hagia Sophia arrays had provided data pertaining to events with magnitudes in between 3.9 and 4.8 Mb and epicentral distances in between 22 and 140 km. In earthquakes predating the 1999 events, maximum acceleration recorded in Hagia Sophia was 13 cm/s^2. During the Kocaeli earthquake (Mw 7.4 at 100 km) recorded peak accelerations reached 77 cm/s^2 in Hagia Sophia.

Based on the analysis of pre-1999 data, it has been determined that the first three modes of vibration of the structure involve coupled lateral-torsional motions. In the first and second modes, translations dominate respectively in the E-W and N-S directions. In the third mode, the dominant motion is torsional. The records have also indicated nonlinearites associated with the structure at surprisingly low levels of excitation. Mode shapes were observed in real time and found to be more complex than those of the finite element models.

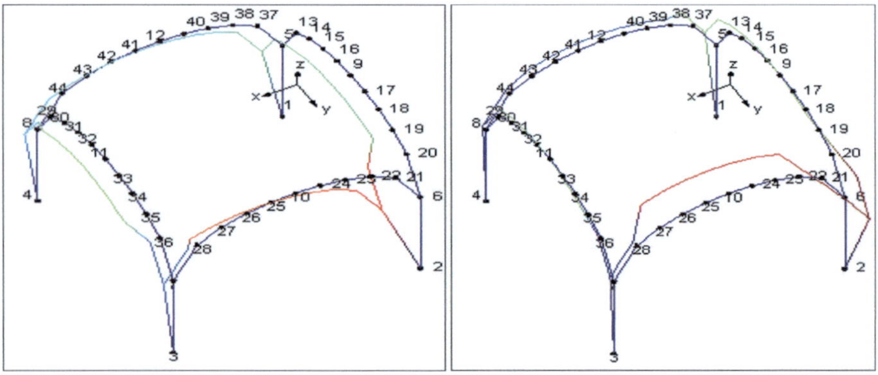

Fig. 6.16 The illustration of first and second mode shapes on main piers and arches (after Stengel 2009). X and Y directions indicate W and N directions. Colors on the modal shape show the degree of displacement (red maximum, blue minimum). A strong motion accelerometer network consisting of nine interconnected SSA-2 units was installed in Hagia Sophia in August 1991 (Erdik and Cakti 1993). The accelerometers, situated at three levels at structurally significant locations, are self-contained with triaxial recording capabilities (Fig. 6.17)

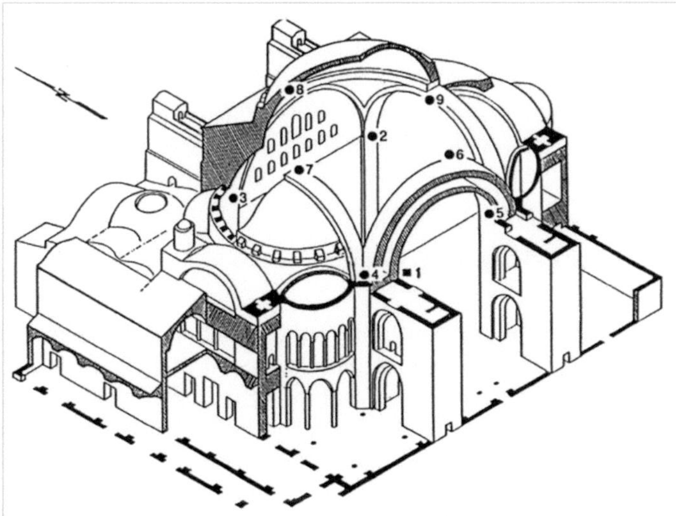

Fig. 6.17 Hagia Sophia strong motion array

Durukal et al. (2003) provide an analysis of strong motion data obtained at the Hagia Sophia strong motion array as a result of the 17 August 1999, Kocaeli (Mw 7.4) and 12 November 1999, Düzce, Turkey (Mw 7.2) earthquakes and their aftershock activity. The earthquake-induced variation of modal frequencies noted previously in Hagia Sophia (and also in Süleymaniye Mosque) was quantified by drops of 25 and 27%, respectively, in the first (E-W axis) and second (N-S axis) modal frequencies of vibration of Hagia Sophia, compared to the pre-earthquake ambient vibration survey results. These reduction levels in modal vibration frequencies indicate about 50% drop in the lateral stiffnesses during an earthquake ground motion with a PGA of 0.077 g measured at the ground floor. This loss of stiffness recovers gradually, reportedly within weeks.

The current findings from the Hagia Sophia strong motion array are encompassed in Dar (2015) and Çaktı and Dar (2015). Since the 1999 Kocaeli Earthquake, strong motion data from about 100 earthquakes were obtained. Their analysis indicates results similar to those obtained by Durukal et al. (2003). The drops in the first modal frequency of vibration (E-W axis) seem to increase linearly from about 3% at PGA levels of 0.001–0.002 g (measured at ground floor) to 9% at about 0.01 g. For the second modal frequency of vibration (N-S axis), the drop was about 4% at PGA levels less than 0.002 g and reaches to about 14% at 0.006 and 0.016 g. The amplification of the horizontal PGA between the ground floor of Hagia Sophia and the springing points of arches at the top of main piers (about 3–6 fold) and the top of main arches (about 3–10) remain almost the same for a range of ground floor PGAs that range from 0.0001 to 0.01 g, essentially indicating structural response in linear ranges of deformation. What is significant is the variation of this amplification factor. The highest amplification factors are observed at the top of the SW main pier (about

3–6) and at the top of the south and west main arches (about 7–11). Whereas the lowest amplification factors are observed at the springing points (about 2) and at the top of the north main arch (about 2–3). Similar trends on the vertical response were also observed. The amplification factor of vertical PGA is highest on SW pier (about 5–7) and at the top of the south and west main arches (about 8–12).

6.7 Expected Earthquake Performance

He expected earthquake performance of Hagia Sophia has been extensively dealt within the studies conducted by the Princeton and Boğaziçi University team through linear and nonlinear response analysis of the finite element model of the structure under exposure to a series of ground motion and simulated earthquake excitations due to different magnitude earthquakes originating on the Main Marmara Fault, passing about 20 km south of the edifice. The actual response of the structure measured by the existing strong motion array has also thrown light on these performance assessments.

Mark et al. (1993) have noted the predisposition in the numerical models of the east and west great arches that provide support to the central dome to warp out-of-plane under gravity loading. With additional out-of-plane motion caused by an earthquake (in this regard note also that the lowest vibration mode is east–west), the basis for the collapse of adjacent portions of the central dome (in the east and west) at different times throughout the building's history begins to get clearer.

Cakmak et al. (1993) have shown that significant E-W response motion occurs at the tops of all four main piers and at the tops of E and W arches, and that the vertical response is significant only at the tops of the E and W arches. The transfer functions show the tendency of the SW main pier and the S and W main arches that spring from it to respond at higher amplitudes than the other piers and arches. This behavior is probably associated with the foundation of the southwest main pier. Table 6.3 summarizes the maximum stresses in tshe critical crown region of the east and west arches corresponding to the static and earthquake loadings (Cakmak et al. 1995). These results give benchmarks for severe response characteristics and highlight the importance of dynamic response to past and potential failures of the primary support structure.

Although these results are based on linear analysis, qualitatively we can say that under the expected excitation of a magnitude Mw7.5 earthquake on the Main Marmara Fault, there will be extensive damage associated with the east and west main arches. Similar stress results have also been reported in Çamlıder (2009).

It has been repeatedly pointed out that (e.g., Erdik and Çaktı 1992; Erdik and Durukal 2007) the locations of stress concentrations from linear analysis, and yield propagation and numerical collapse scenarios from the nonlinear transient dynamic analysis, as well as the existing information regarding historical partial collapses, indicate the eastern and western main arches and adjacent portions of the main and corresponding semi-domes as the most vulnerable structural elements of Hagia

Table 6.3 Maximum Stresses Computed for Simulated Earthquakes (All stresses are in MPa, tensions and compressions in parentheses)

Load case	East arch		West arch	
	Mw = 6.5	Mw = 7.5	Mw = 6.	Mw = 7.5
Dead load	0.80 (3.30)	1.21 (3.61)	0.80 (3.30)	1.21 (3.61)
Earthquake	0.60 (0.60)	1.72 (1.72)	0.51 (0.51)	1.48 (1.48)
Total	1.40 (3.90)	2.52 (5.02)	1.72 (4.12)	2.73 (5.09)

Sophia. The yield propagation sequences provided in Fig. 6.11, taken from the results of the nonlinear analysis of Hagia Sophia, highlights this conclusion.

Studies conducted under the leadership of Professor Giorgio Croci (2006) have highlighted the prominent role played by the deformability of the main pillars and arches and have shown that the behavior of the structure differs according to the direction of the earthquake forces (i.e., depending on the direction, different degrees of acceleration produce different mechanisms of collapse, and different crack patterns). In particular, the damage caused by an earthquake in the longitudinal direction (E-W, parallel to the tympana), is represented by a theoretical crack pattern which does not correspond to the crack lines that mark the collapses in the tenth and fourteenth centuries. The effects of an earthquake in the transversal direction (N-S), instead, produce theoretical damage and collapse mechanisms very close to the historical records and what is possible to observe today. These analyses have also shown that the existing buttresses, which are much deformed and honeycombed with steps and passageways leading to the galleries, do not provide sufficient structural stiffness compared to that provided in the direction parallel to the main axis and given by the semidomes and the tympana. The transversal behavior (in the direction of the abutments) is worsened by the possible weakening of the transversal arches as a result of the possible detachment between the arches themselves and the semidome; that may produce “hammering” effects by the longitudinal component of an earthquake. The “hammering” effect is probably an additional factor, even if not essential because of the semidome’s small thickness compared to the transversal arch section. For this reason, the hammering more likely may cause serious damages to the semidome itself, and only local damages to the arch in its keystone intrados zone are expected. In any case, the disconnection of the transversal arches from their respective semidomes makes the arches more prone to deformation along their plane surface (the semidomes have a double curvature shape giving them more stiffness) and therefore the structure is weaker under transversal earthquake effects.

On the basis of the comparison of the displacements of two transverse (east and west) arches and double (north and south) arches, Aoki (1997, 1998) reports that the east and west arches are weaker than the north and south arches. These studies further indicate that the dome of Hagia Sophia seems to be vulnerable to severe earthquakes.

On the basis of the post-1999 earthquake ambient vibration measurements, Sato et al. (2008) claim that the average drift angle experienced by Hagia Sophia in the 1999 earthquake can be presumed to be approximately 1/250. Since, generally it is thought that a structure begins to receive damage at 1/100–1/200 drift angle levels, it is considered that the Hagia Sophia stood against the 1999 earthquake motion without great margin of safety, which (they claim) is also evidenced by the significant decrease of natural frequency of vibration.

Under the action of gravity and earthquake, Ozkul and Kuribayashi (2007a, b) have found that the top section of the main arches and the main arches neighboring the east and west semi-domes are the most critical regions in the structure. In a strong earthquake, the east and west semi-domes can receive damage, get detached from the east and west arches and can no longer provide support to take the trust of the main dome. This condition may and can lead to the collapse of the eastern and western sectors of the main dome as it was experienced in the past earthquakes.

6.8 Earthquake Retrofit and Strengthening

The earthquake retrofit proposals and applications for Hagia Sophia (for that matter for all historical edifices) should certainly possess: retractability (reversibility), unobtrusiveness (minimum needed), and compatibility with existing materials. The use of ring beams and exterior gravity walls ('counterfort'), and external buttresses is the main choice for retrofitting of old masonry structures. Another traditional technique to increase the strength of masonry walls is to strengthen the connection between different structural elements with reinforced injections or with stiffening elements (i.e., longitudinal and/or transverse reinforcing elements, tie bars and posttensioning).

Existing information regarding historical partial collapses, as well as locations of stress concentrations from linear analysis, and yield propagation and numerical collapse scenarios from the nonlinear transient dynamic analysis provide relevant evidence that the detachment of the eastern and western semidomes from the eastern and western main arches constitutes the most important collapse mechanism in Hagia Sophia in the event of a large earthquake (as it was the case in tenth- and fourteenth-century earthquakes). One possible preventive action against this mechanism is to install tension links by means of suitable anchorages to avoid detachment (and hammering) to ensure a uniform dynamic behavior of semidomes and the eastern and western main arches. Similarly, the crowns of the main arches may need to be strengthened by posttensioned internal tendons.

As it has been pointed out, the eastern and western main arches and adjacent portions of the main and corresponding semi-domes as the most vulnerable structural elements of Hagia Sophia. Extensive structural analysis conducted also indicates that the most plausible mode of damage to be expected under exposure to a large magnitude earthquake would be the weakening of the east and west arches lost of their contact with the main and semidomes that would eventually lead to a partial

Fig. 6.18 Earthquake strengthening proposal (Erdik and Çaktı 1992)

collapse. One possible action against this scenario is presented in Fig. 6.18, where establishment links by means of suitable anchorages are proposed to provide and integrity of the structural elements at these junctions and thus to ensure uniform dynamic behavior of semidomes and the eastern and western main arches (Erdik and Çaktı 1992).

Similar findıngs regarding the possible earthquake failure modalities of Hagia Sophia were later reported by Aoki et al. (1997), Croci (2006) and Almac et al. (2013) among others. On the basis of the results obtained from 3D elastoplastic model, Aoki et al. (1997) states that: "the structural weakness of the whole structures exist in the east and west arches combined with the two adjacent half domes" and to mitigate damage in the event of an earthquake "the west and east great arches, the two half domes and the adjacent supporting structures have to be structurally repaired". The results of dynamic response analysis conducted by Almac et al. (2013) using high-resolution finite element models indicate that "the east and west main arches, together with adjacent portions of semi-domes and central dome are the most vulnerable areas of the structure". Almac et al. (2013) also recommend the modification of the finite element model to allow for the repeated separation and contact (i.e., pounding) between the interface of these structural parts with contact elements to capture the reaction forces during dynamic simulations.

Kırlangıc (2008) has compared the responses of existing and retrofitted structure to simulated earthquake scenarios. The main objective of the retrofit was to keep the tensile stresses in the structure under 1 MPa under the earthquake excitation. Two retrofitting methods have been investigated: inclusion of posttensioned steel bars at the main dome base along the main arches and fiber reinforced polymers (FRPs) to improve the tensile strength of the semidomes. Results of the first retrofit method indicate the effective lowering achieved in tensile stresses. However, extensive "bowing" type displacements observed in the east and west arches due to post-tensioning remains to be subject in need of more elaboration.

6.9 Conclusions

There have been several studies of the dynamic response of the Hagia Sophia. The studies have successfully enlightened the global response and earthquake performance characteristics. The uncertainties and the difficulties of measurement of macroscale material properties provide limitations on such quantitative assessments of the earthquake performance of the edifice. The complexity of the modal shapes indicates the complexity and irregularity of the structure.

The time-domain response characteristics of the analytical model models were successful in simulating the horizontal response levels obtained from earthquake recordings. However, the finite element models cannot adequately represent the complex reality of the building, nor take into account the stresses accumulated due to collapses and deformations of the past.

The excessive vertical vibrations at the crowns of the arches, observed in actual earthquake response, could not be simulated successfully. The analytical results are rather low compared to the actual recordings. Similarly due to the fact that an ideal symmetrical model is created, the excessive vibrations of the SW piers could not be simulated as well.

The existing buttresses, which are much deformed and emptied with steps and passageways leading to the galleries, do not provide sufficient structural stiffness in the N-S direction compared to that provided in the E-W direction given by the semidomes and the tympana. The effects of an earthquake in the N-S direction can theoretically produce damage and collapse mechanisms very close to the historical records.

Previous collapses of the east and west arches occurred through excessive outward movements of their springing and excessive local compressions at the hinge regions at crown and haunches. The analytical and empirical studies have also shown the east and west arches as the weakest structural elements. In this connection, the western arch is in a better shape than the eastern arch due to its increased size (strengthening) after the tenth-century earthquake.

A significant decrease in the natural frequencies of vibration and increase of effective damping ratios has been observed in transient basis even during low-level excitations (hence strains). Decrease in the frequency of vibration is indicative of decrease in stiffness of the masonry (i.e., mortar). These co-earthquake excitation decreases were in a temporary manner with subsequent recovery. However, the measured decreases of pre- and post-Kocaeli earthquake frequencies of vibration have taken much longer to recover (years) and with some amount yet to be recovered.

The continuous monitoring of the movements at critical points and the opening across important cracks of the structure is important for the decisions on the health of the structure. Similarly, the measurement of the structural vibrations, on-going since 1992, should continue with increased station numbers.

On the basis of dynamic response analysis, the east and west main arches, together with adjacent portions of semi-domes and central dome are the most vulnerable areas of the structure.

A first retrofit option appears to be the tying the eastern semidome to eastern arch. In addition, other strengthening measures such as the re-institution of ties and consolidation of infill walls may need to be undertaken.

Besides all these structural interventions, the most important requirement for the protection of Hagia Sophia is its proper maintenance, including the periodic replacement of lead shields, repair of deteriorated masonry and re-pointing of joints.

Acknowledgements My sincere thanks are to Prof. Ahmet Cakmak of Princeton University who has dedicated his time and energy to the protection of cultural heritage in Istanbul and initiated my involvement on the earthquake protection of Hagia Sophia.

I acknowledge the help provided by the Directorate of the Hagia Sophia Museum, the Directorate of Restoration and Monuments, the Central Laboratory for Restoration, and support provided by the Ministry of Culture.

Special thanks and gratitude are due to my colleagues whose finding constitute the backbone of this review study.

References

Almac U, Schweizerhof K, Blankenhorn G, Duppel C, Wenzel F (2013) Structural behaviour of Hagia Sophia Under dynamic loads. In: Vienna congress on recent advances in earthquake engineering and structural dynamics, Vienna, Austria, p 475

Aoki T (1998) Collapse mechanism of Hagia Sophia—a case of dead load (in Japanese). J Struct Constr Eng AIJ 508:79–85

Aoki T, Kato S, Hidaka K (1992) Crack pattern in hagia sophia dome and its comparison with those from analysis, In: J, Proceedings of the IASS-CSCE international symposium, Toronto, vol 1, pp 730–741

Aoki T, Kato S, Ishikawa K (1993) Structural characteristics of the dome of hagia sophia from measurement of micro tremor, Proceedings 5th conference on soil dynamics and earthquake engineering

Aoki T, Kato S, Ishikawa K, Hidaka K, Yorulmaz M, Cili F (1997) Principle of structural restoration for hagia sophia dome. In: STREMAH V—international conference on structural studies, repairs and maintenance of historical buildings, San Sebastian, vol 3, pp 467–476

Aoki T, Shiro K, Koichiro I, Kenichiro H (1997b) Deformation and seismic strength of the dome of Hagia Sophia (in Japanese). J Struct Eng 43B:641–646

Aoki T, Kato S, Ishikawa K, Hidaka K, Yorulmaz M, Çili F (1997) Principle of Structural Restoration for Hagia Sophia Dome. In: Proceedings of STREMAH international symposium, pp 467–476

Blankenhorn G, Schweizerhof K (2009) Die Erdbebengefährdung der Hagia Sophia in Istanbul, Verifizierung und Validierung numerischer Rechenmodelle für dynamische Beanspruchungen Institutsbericht 2009 Institut für Mechanik, Berichtszeitraum, Oktober 2008 bis September 2009

Çakmak AŞ (1995) Interdisciplinary study of dynamic behavior and earthquake response of Hagia Sophia. Soil Dyn Earthquake Eng 14:125–133

Cakmak AS, Davidson R, Mullen CL, Erdik M (1993) Dynamic analysis and earthquake response of Hagia Sophia. In: STREMA'93, Proceedings of the 3rd International Conference on structural studies, repairs and maintenance of historical buildings, Bath, UK, CML Publications, Southampton, pp 67–84

Cakmak AS, Moropoulou A, Mullen CL (1995a) Interdisciplinary study of dynamic behavior and earthquake response of Hagia Sophia. Soil Dyn and Earthquake Eng 14:125–133

Cakmak AS, Natsis MN, Mullen CL (1995b) Foundation effect on the dynamics of Hagia Sophia. Soil Dyn Earthquake Eng 14:125–133

Çakmak AS, Taylor RM, Durukal E (2009) The structural configuration of the first dome of Justinian's Hagia Sophia (A.D. 537–558): an investigation based on structural and literary analysis. Soil Dyn Earthquake Eng 29(4):693–698

Çaktı E, Dar E (2015) Ayasofya'nin Depremler ve Atmosfer Koşulları Altında Uzun Dönemli Dinamik Davranışı, Sekizinci Ulusal Deprem Mühendisliği Konferansı, 11 Mayıs-14 Mayıs, 2015, İstanbul

Croci G (2006) Seismic behavior of masonry domes and vaults: Hagia Sophia in Istanbul and St. Francis in Assisi, Keynote Address K8, First European conference on earthquake engineering and seismology, Geneva, Switzerland, 3–8 September 2006

Croci G (2007) Santa Sophia, Istanbul, Il Comportamento Strutturale. Recupero e Conservazione 78:36–41

Croci G, Viskovic A (1999) Hagia Sophia in Istanbul: study, research and mathematical models for the analysis of the structural behaviour program of investigation, Report, Istanbul, Turkey

Dar E (2015) Long-term dynamic Response of Hagia Sophia in Istanbul to Earthquakes and Atmospheric Conditions, MSc thesis, Graduate Program in Earthquake Engineering Boğaziçi University

Davidson RA (1993) The mother of all churches: a static and dynamic structural analysis of Hagia Sophia, Unpublished senior thesis, civil engineering and operations research department, Princeton University
Duppel C (2009) Ingenieurwissenschaftliche Untersuchungen an der Hauptkuppel und den Hauptpfeilern der Hagia Sophia in Istanbul. Univ., Diss, Karlsruhe, p 2009
Duppel C (2010a) Ingenieurwissenschaftliche untersuchungen an der Hagia Sophia in Istanbul—teil 1: das konstruktionsgefüge. Bautechnik, 87, H. 11:708–716
Duppel C (2010b) Ingenieurwissenschaftliche untersuchungen an der Hagia Sophia in Istanbul—teil 2: zum tragverhalten. Bautechnik, 87, H. 12: 790–799
Duppel C (2011) On the structural characteristics and present stability of the Hagia Sophia in Istanbul. In: WCCE-ECCE-TCCE joint conference 2, seismic protection of cultural heritage
Durukal E (1992) Structural identification and seismic vulnerability assessment of Hagia Sophia, MSc thesis, Bogazici University
Durukal E, Erdik M (1994) Comparison of system identification techniques: a case study for Hagia Sophia. In: Savidis SA (ed) Proceeding the second international conference on earthquake resistant construction and design, vol 2, pp 993–1000
Durukal E, Erdik M (2003) Proteggere Hagia Sophia dai terremoti: un sfida per l'ingegneria e la conservaziona. In: Earthquake Protection of Hagia Sophia: a challenge for engineering and conservation. Disegnare Idee Immagini, Anno XIV, n. 26, Luglio 2003, pp 8–19
Durukal E, Çakmak A, Yüzügüllü O, Erdik M (1998a) Assessment of the earthquake performance of Hagia Sophia. In: PACT 56 1998 Proceedings-compatible materials for the protection of European cultural heritage, pp 49–57
Durukal E, Yüzügüllü, Ö, Beyen K (1998b) Non-destructive testing techniques of structural materials in historical structures. In: PACT 55 1998 Proceedings-compatible materials for the protection of european cultural Heritage, Technical Chamber of Greece, Athens, pp 221–229
Durukal E, Erdik M, Cimilli S (2001) Strong motion networks: a tool for the assessment of earthquake response of historical monuments. In: Erdik M et al (eds) Strong motion instrumentation for civil engineering structures, Kluwer
Durukal E, Cimilli S, Erdik M (2003) Dynamic response of two historical monuments in Istanbul deduced from the recordings of Kocaeli and Duzce earthquakes. Bull Seismological Soc Am 93(2):694–712
Emerson W, Van Nice RL (1943) Haghia Sophia, Istanbul: preliminary report of a recent examination of the structure. Am J Archaeology 47:403–436
Erdik M, Cakti E (1990) Istanbul Ayasofya Muzesi Yapisal Sisteminin ve Deprem Guvenliginin Saglanmasina Yonelik Tedbirlerin Tespiti. Kandilli Observatory and Earthquake Research Institute, Bogazici University, Istanbul, Ikinci Asama Raporu
Erdik M, E Cakti (1992) Determination of earthquake risk and restoration work strengthening project of Hagia Sophia (Research Report, in Turkish). Kandilli Observatory and Earthquake Research Institute, Bosphorus University
Erdik M, E Cakti (1993) Instrumentation of aya sofya and the analysis of the response of the structure to an earthquake of 4.8 Magnitude. In: Proceedings of STREMA '93. In: Brebbia CA, Frewer RJB (eds) 3rd international conference, Bath, UK, June 16–18, pp 99–114
Erdik M, Durukal E (1996) Use of Strong motion data for the assessment of the earthquake response of historical monuments, In: Proceedings of the 11th world conference on earthquake engineering, Acapulco, Mexico, June 23–28
Erdik M, Durukal E (2007) Earthquake performance of Hagia Sophia. In: Proceedings 8th conference of the hellenic society for theoretical and applied mechanics (HSTAM), Patras
Erdik M, Durukal E, Yuzugullu B, Beyen K, Kadakal U (1993) Strong-motion instrumentation of Aya Sofya and the analysis of response to an earthquake of 4.8 magnitude. In: Brebbia CA (ed) Structural repair and maintence of historical buildings III
Gürbüz C, Deal M, Bouwer J, Bekler T, Çakmak AS, Erdik M (1993) Technical report on the seismic refraction studies in Hagia Sophia (in turkish). Bogaziçi University, Kandilli Observatory and Earthquake Research Institute

Hill KE (1991) Hagia Sophia: static analysis of a finite element model, Unpublished Senior thesis, Deportment of civil engineering and operations research, Princeton University

Kato S, Aoki T, Hidaka K, Nakamura H (1992) Finite element modeling of the first and second dome of Hagia Sophia. The Hagia Sophia: from the age of justinian to the present. Cambridge University Press, New York, pp 103–119

Keypour H (2001) Comprehensive nonlinear dynamic analysis of Hagia Sophia, PhD thesis. Boğazici University. Istanbul

Kırlangıç S (2008) Evaluation of earthquake performance of Hagia Sophia, MSc thesis submitted to department earthquake engineering, Boğaziçi University

Livingston RA, Stutzman PE, Mark Erdik M (1993) Preliminary analysis of the masonry of the Hagia Sophia Basilica, Proceedings of STREMA '93, 3rd international conference, Bath, UK, June 16–18

Livingstone RA (1993) Materials analysis of the masonry of the Hagia Sophia Basilica, Istanbul. In: Structural repair and maintenance of historical buildings III. In: Brebbia CA, Frewer RJB (eds) Proceedings of the third international conference held at bath in June 1993, Southampton 1993, pp 15–31

Livingstone RA, PE Stutzman, R Mark and M Erdik (1992) Preliminary analysis of the masonry of the Hagia Sophia Basilica, Istanbul, In: Vandiver PB, Druzik JR, Wheeler GS, Freestone IC (eds) Materials issues in art and archaeology III, proceedings of the 1992 MRS symposium, materials research society, Pittsburgh, pp 721–738

Mainstone RJ (1988) Hagia sophia architecture, structure and liturgy of justinian's Great Church, Thames and Hudson, ISBN 0-500-34098-6

Mainstone RJ (1992) Questioning Hagia Sophia. In: Mark R, Cakmak AS (eds) Hagia Sophia from the age of justinian to the present. Cambridge, pp 158–178

Mainstone RJ (1993) Materials analysis of the masonry of the Hagia Sophia Basilica, Istanbul. In: Brebbia CA (ed) Structural repair and maintence of historical buildings III

Mainstone RJ (1993) The structural conservation of Hagia Sophia. In: Brebbia CA, Frewer RJB (eds) Structural repair and maintenance of historical buildings III, Proceedings of the third international conference held at bath in June 1993, Southampton 1993, pp 3–14

Mainstone RJ (1993c) Present state of the Hagia Sophia monument with its recommendations for preservation and restoration. UNESCO, Paris, p 1996

Mark R, Cakmak AS (eds) (1992) Hagia Sophia from the age of justinian to the present. Cambridge University Press, New York

Mark R, Cakmak AS (1994) Mechanical tests of material from the Hagia Sophia Dome. Dumbarton Oaks Papers 48(1994):277–279

Mark R, Cakmak AS, Hill KE, Davidson R (1993) Structural analysis of Hagia Sophia: a historical perspective, proceedings of STREMA '93, 3rd international conference, Bath, UK, June 16–18

Mark R, Cakmak AS, Hill K, Davidson R (1993) Structural analysis of Hagia Sophia: a historical perspective, Proceedings 6th International Confernce on soil dynamics and earthquake engineering, pp 867–880

Moropoulou A, Cakmak AS, Biscontin G, Bakolas A, Zendri E (2002) Advanced byzantine cement based composites resisting eartquake stresses: the crushed brick/lime mortars of Justinian's Hagia Sophia. Constr Build Mater 16(8):543–552

Natsis MN (1994) The enigma of Hagia Sophia: a dynamic structural analysis of justinian's great church, senior thesis, Princeton University

Ozkul TA, Kuribayashi E (2007a) Structural characteristics of Hagia Sophia: I-A finite element formulation of dynamic analysis. Build Environ 42:1212–1218

Ozkul TA, Kuribayashi E (2007b) Structural characteristics of Hagia Sophia: II-A finite element formulation of dynamic analysis. Build Environ 42:2100–2106

Şahin M, "Dynamic Response of Hagia Sophia Considering Cracks", PhD thesis, Mimar Sinan University, 2002

Sahin M, Mungan I (2005) Dynamic performance of the roof of Hagia Sophia considering cracking. Int J Space Struct 20(3):135–141

Sato T, Hidaka K (2001) Deformation of the upper structure of Hagia Sophia. In: Hagia Sophia surveying project conference, proceedings of the conference. Tokyo 2001, pp 37–64

Sato T, Nakamura Y, Saita J (2008) The change of the dynamic characteristics using microtremor. Proc., The 14th World Conference on Earthquake Engineering, 12–17 Oct 2008, Beijing, China

Stengel D (2009) System identification for four types of structures in istanbul by frequency domain decomposition and stochastic subspace identification methods, diploma study submitted to submitted to the kandilli observatory and earthquake research Institute, Boğaziçi University and to the Department of Reinforced Concrete Structures, University of Karlsruhe

Swan CC, Cakmak AS (1993) Nonlinear quasi-static and seismic analysis of the Hagia Sophia using an effective medium approach. Soil Dyn Eq Engrg 12(5):259–271

Van Nice RL (1964) St. Sophia's structure: a new assessment of the half domes. Architectural Forum: The Magazine Of Building 121(2):45–49

Van Nice RL (1965, 1986) St Sophia in Istanbul: an architectural survey. Dumbarton Oaks Dumbarton Oaks Center for Byzantine Studies, Washington

Wenzel F, Duppel C (2008) Engineering studies on the main dome and the main pillars of the Hagia Sophia in Istanbul, IABSE Reports, vol 90

Wenzel F, Duppel C (2010) Investigations into the construction and repair history of the Hagia Sophia. Annual of Ayasofya Museum, 13

Chapter 7
Recent Studies on Earthquake Performance Assessment of Hagia Sophia in Istanbul

Eser Çaktı, Emrullah Dar and Gülen Uncu

Abstract Over the last 25 years there have been a series of academic efforts to understand the particulars associated with earthquake performance and vulnerability of Hagia Sophia. Linear and nonlinear structural analyses; literary investigations; non-destructive tests and investigations involving its construction materials, foundations and main structural elements; ambient vibration tests; monitoring its earthquake response with the help of an accelerometric network were among them. More recently the foci of efforts were the analysis of long-term dynamic response of Hagia Sophia to earthquakes and variations in atmospheric conditions, shake table testing of a scaled model of the structure, analysis of recently added tiltmeter recordings, evaluation of strengthening alternatives and monitoring of its static deformations via laser technology. This contribution presents an overall picture of the research efforts carried out so far on earthquake response analysis of Hagia Sophia, emphasizing the studies over the last 5 years.

Keywords Hagia Sophia · Structural health monitoring · Earthquake response
Environmental effects · Shake table tests

7.1 Introduction

Hagia Sophia in Istanbul is one of the most prominent architectural and structural creations in the history of mankind (Fig. 7.1). Built in the sixth century, it has a remarkable size and structural configuration. Over its 1500 yearlong history, Hagia Sophia experienced many earthquakes, as a result of which several of its parts were damaged and its main dome collapsed partially along with the adjacent parts of the main arches and semi-domes three times. Its existence in a part of the world marked

E. Çaktı (✉) · E. Dar · G. Uncu
Department of Earthquake Engineering, Kandilli Observatory and Earthquake Research Institute, Boğaziçi University, Çengelköy, Istanbul, Turkey
e-mail: eser.cakti@boun.edu.tr

A. A. Kasimzade et al. (eds.), *Seismic Isolation, Structural Health Monitoring, and Performance Based Seismic Design in Earthquake Engineering*,
https://doi.org/10.1007/978-3-319-93157-9_7

Fig. 7.1 Hagia Sophia in Istanbul, Turkey

with very high earthquake hazard levels will ever be tested by seismic activity. This contribution aims at presenting an overall picture of the research efforts carried out so far on earthquake response analysis of Hagia Sophia, emphasizing the studies over the last 5 years.

7.2 Building History

Hagia Sophia was included in the world heritage list of UNESCO in 1985. The planners of Hagia Sophia, Isidore of Miletus and Anthemius of Tralles designed it as a combination of a centralized structure and a longitudinal basilica. The construction period (532–537) lasted slightly over 5 years, which is a relatively short period of time given its size, structural complexity and the use of masonry as the construction material. The longitudinal length of the structure is 100 m, while its width is 69.5 m. The height between the central dome and the ground level is 55.60 m. The diameter of the dome is 31.87 m in the north–south direction and 30.86 m in the east–west direction. Hagia Sophia served as the central church of the Byzantine Empire for more than 900 years. After the conquest of Istanbul by the Ottomans in 1453, it was converted into a mosque. In the republican era Hagia Sophia was converted to a museum in 1935. It serves to this day as a museum attracting visitors from all over the world as a cultural heritage building. Hagia Sophia experienced many earthquakes in its history resulting in damages and partial collapses. The first destructive damage occurred on the eastern side of the main dome in 553 during an earthquake. As a consequence, the main dome was rebuilt six meters higher than the original dome (Müller-Wiener 2007). The second severe earthquake occurred in 989 and caused damage in the west main arch and collapse of the adjacent parts of the main dome. Then a wider and thicker arch was built (Van Nice 1963). In 1349, the east main arch and the adjacent parts of the dome collapsed for the last time and rebuilt (Mainstone 1998). The structure never collapsed as a whole.

7.3 Structural Analyses

Understanding the static and dynamic behavior properties of Hagia Sophia is the primary objective of all analytical and experimental studies conducted in the past and in the present. It is only with collective information generated as a result of such investigations, that we will be able to answer the question on the performance of the structure in a future large earthquake near Istanbul with some confidence. To this aim a series of analytical studies have been conducted on the dynamic behavior of Hagia Sophia up to now. Static and dynamic linear structural modeling and analyses started in early 90s. Under the supervision of Ahmet Cakmak from Princeton University, Hill (1991) created the earliest finite element model of Hagia Sophia for static analysis. Durukal (1992), Davidson (1993) and Natsis (1994) extended and improved the SAP90 model created by Hill (1991). Durukal (1992) carried out dynamic analysis and validated the model with real earthquake recordings under the supervision of Mustafa Erdik of Boğaziçi University. Davidson (1993) looked at the effect of construction stages and phases. Natsis (1994) studied the soil–structure interaction, cracking in the arches and the effect of material anisotropy in the main piers. Kokkinos (2013) and Almaç et al. (2013) modeled.

Hagia Sophia using Abaqus and LS-Dyna respectively. The finite element model of Hagia Sophia can be seen in Fig. 7.2. The structural behavior was also investigated using software with nonlinear analysis capabilities by Rhodes (1994), Worcester (1995) and Smith (1996).

7.4 Material Properties

Assessment of material properties of Hagia Sophia is crucial for explaining existing static deformations and past collapses, and for estimating its future structural behavior. Most of the interior surfaces in the nave and narthexes are covered with marble, mosaics, and plaster. Yet, some parts of the main dome remain exposed allowing a view of the brick + mortar fabric. It is parts of the structure like the staircase- and passage interiors, as well as the recently restored western exterior, where one could observe the original construction materials. To give an overview,

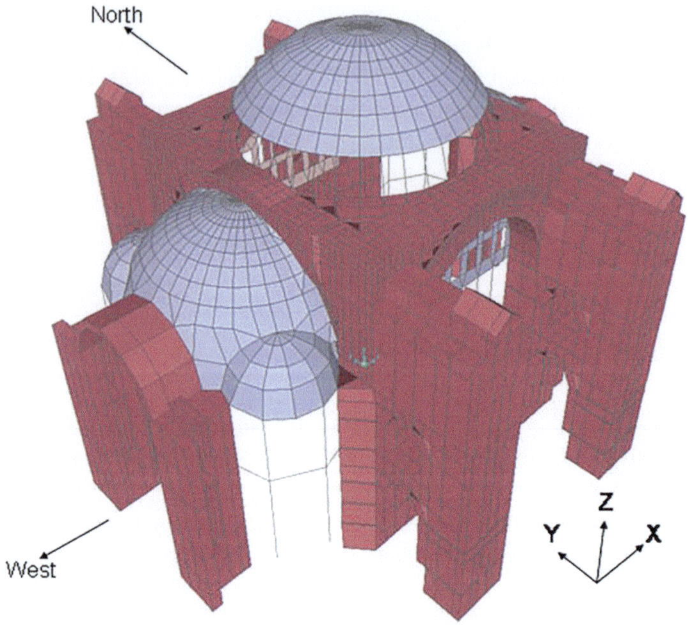

Fig. 7.2 Finite element model of the primary structural system of Hagia Sophia

the main piers and the secondary piers are constructed of stone masonry, while all structural elements springing from them or above them are of brick masonry. Physical experiments on mortar samples, in situ ultrasonic tests and inverse engineering approach using the structural models and analyses were used in the assessment of material properties (Livingston et al. 1992; National Geographic 2010; Moropoulou et al. 2002).

7.5 Structural Monitoring

7.5.1 Accelerometric Network

The first accelerometric network was established in August 1991 (Erdik et al. 1993). The system was completely renewed and updated in 2008 to meet the technological developments (Çaktı and Dar 2015). The network consists of nine accelerometers. Acceleration sensors have three components and currently operate at 100 Hz sampling frequency. One of the accelerometers is located on the ground level, four of them are installed at the top of four main load-bearing piers and the remaining four accelerometers are stationed at the crowns of the four main arches, which bear the main dome. Figure 7.3 displays the location of the sensors and the station codes on a schematic representation of the structural elements of the building.

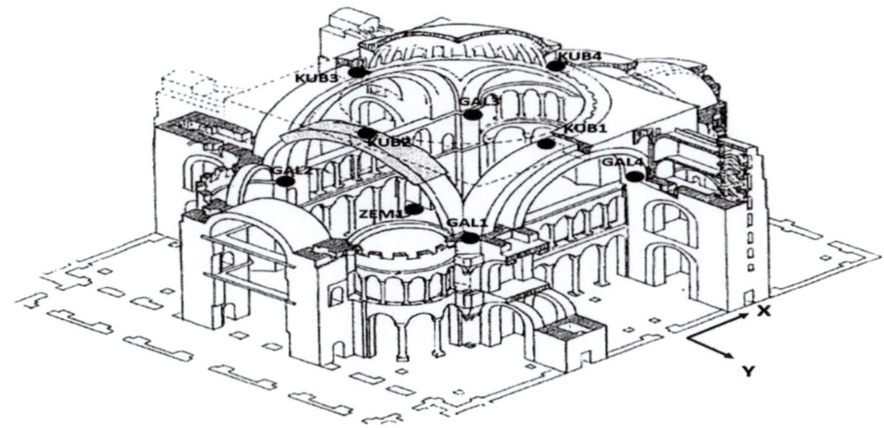

Fig. 7.3 Accelerometric monitoring system in Hagia Sophia; Stations GAL1, GAL2, GAL3 and GAL4 include two-way tiltmeters in addition to accelerometers. The isometric view is from Mainstone (2006)

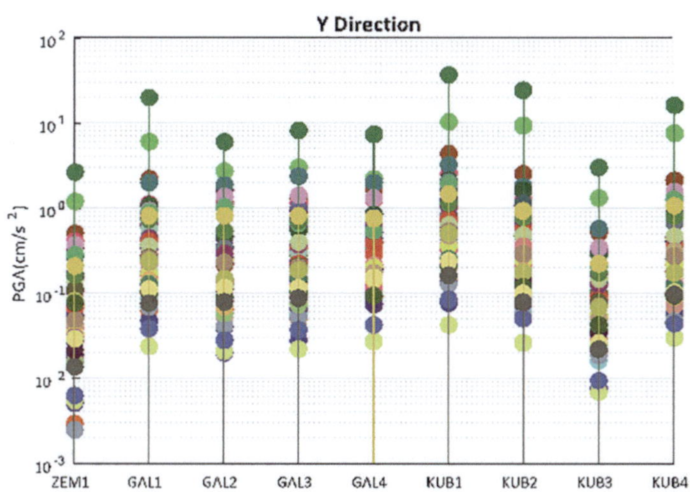

Fig. 7.4 Peak accelerations recorded at Hagia Sophia stations between years 2008 and 2015

Since November 2008, the network operates in real-time mode. About a hundred earthquakes have been recorded in Hagia Sophia between 2008 and 2015. The range of recorded peak ground accelerations can be seen in Fig. 7.4. Collected data are used in the identification of dynamic response parameters of the structure such as modal shapes, modal frequencies, and modal damping. They are also used in the study of time variation of these parameters, as well as in research on identification

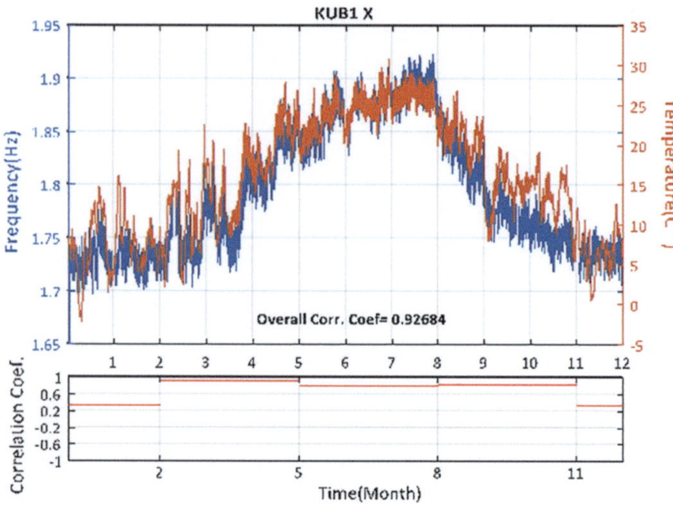

Fig. 7.5 Variation of the first modal frequency of Hagia Sophia and the temperature in 2013

and characterization of damage-related structural parameters (Durukal et al 2003; Çaktı and Dar 2015 among others).

Atmospheric conditions affect modal vibration frequencies and damping coefficients of a structure. Daily and seasonal variations in the temperature for example induce continuous cyclic variation in the dynamic response. Hagia Sophia is no exception to this phenomenon. Figure 7.5 displays the correlation between Hagia Sophia's first modal frequency and temperature over the year 2013.

7.5.2 *Tiltmeters*

In addition to acceleration sensors, four two-component tiltmeters are installed at stations GAL1, GAL2, GAL3, and GAL4. Tiltmeter is a measuring device that is used to monitor angular rotations of a structural element or of the ground. The recorded tilt of a structural element is contributed by the tilt due to loads, by the change in the tiltmeter output due to change in temperature of the gage, and by the change in tilt of the element due to temperature. Assuming that all tiltmeter channels are exposed to the same temperature (all sensors are located in the interior at the same elevation) and that we look at event specific data implying that all sensors will record similar levels of tilt change of the main piers due to temperature, we can attempt an inter-station comparative interpretation of the tiltmeter data. In earthquakes with magnitudes larger than 5.0, typical recorded maximum rotations were in the order of 0.0004–0.0006 degrees. Largest recorded tilt is close to 0.005 degrees and is registered during a magnitude 6.5 event. The rotations recorded by station GAL1 are consistently higher in both directions during all earthquakes.

7.5.3 Long-Term Deformation Monitoring

Laser technology created remarkable opportunities for architectural and structural investigations. We have started to work in Hagia Sophia in 2016 as part of a larger campaign for monitoring long-term deformations of a series of monumental structures in Istanbul with the help of laser technology. The measurements will be carried out periodically over an extended period of time. The aim is to assess the extent and progress of structural deformations (if there are any) of Hagia Sophia which could potentially be the result of geotechnical, structural and/or material problems or be generated by dynamic sources such as earthquakes or traffic.

7.6 Subsurface Conditions

Hagia Sophia is situated on the first hill of the historical peninsula of Istanbul, where the geology is dominated by the so-called Trakya formation. According to the study by the Istanbul Metropolitan Municipality the site conditions of Hagia Sophia can be classified as NEHRP C type (IMM 2009). Extended areas on the first hill of Istanbul are covered with artificial fill as a consequence of being exposed to continuous settlement for over 2500 years. Investigations on the subsurface conditions and foundations of Hagia Sophia are limited in number, extent and depth. The geophysical study by Gürbüz et al. (1994) estimates that the northern main piers extend down for about 1 m, whereas the southern main piers continue for about 4 m before reaching engineering bedrock. Geo-radar and reflection studies by Anatolian Geophysical (2009a, b) identify following shear wave velocity ranges within the building: 750–1000 m/s at 10 m depth and 900–1800 m/s at 20 m depth.

7.7 Strengthening Alternatives

Studies attempting to develop strengthening alternatives are rare. There are several reasons for that. The first reason is related to the fact that we are still far away from understanding Hagia Sophia's static and dynamic behavior completely. Overwhelming majority of existing studies rely on linear modeling, which on its own is not sufficient. In the 25 years since the beginning of structural analyses of Hagia Sophia, there have been significant developments in terms of computer power and software, which make analytical investigations on strengthening and behavior by nonlinear analysis more feasible now than ever. Another reason should be that strengthening a building of the caliber of Hagia Sophia is an immense task in terms of governance, decision-making and consensus of stakeholders. Existing studies are mostly on the conceptual level (Erdik, personal communication) or in the form of preliminary feasibility analysis of FRP applications and post tensioning (Kırlangıç 2008).

Fig. 7.6 Scaled model of Hagia Sophia on the shake table

7.8 Shake Table Model

For a documentary film on Hagia Sophia (Glassman et al 2015) we developed and built its scaled model (Fig. 7.6). The model was designed on the basis of architectural drawings and material properties of the building observing similitude laws. It was constructed in stages probably similar to the actual construction sequence followed by Anthemius of Tralles and Isidorus of Miletus. The scaled model was subjected to shake table testing in the shake table laboratory of Boğaziçi University's Department of Earthquake Engineering. The experimental program involved sequential testing under regularly increased acceleration records from the Kocaeli earthquake and continued until collapse.

7.9 Conclusions

More studies, investigations, analyses, experiments are needed concerning Hagia Sophia without any doubt. Further investigations on material properties using non-destructive techniques; three dimensional imaging, and models of subsurface conditions beneath and around the building; structural analyses and nonlinear models to replicate existing permanent deformations; nonlinear dynamic modeling to understand the building's existing and future earthquake behavior; improvement and enhancement of the current structural health monitoring system with sensors for strain measurements, climate conditions and static deformations; and assessment of collapse mechanisms would be invaluable contributions to Hagia Sophia.

References

Almaç U, Schweizerhof K., Blankenhorn G, Duppel C, Wenzel F (2013) Structural behaviour of Hagia Sophia under dynamic loads. Paper presented at Vienna congress on recent advances in earthquake engineering and structural dynamics, Vienna, 28–30 August 2013

Anatolian Geophysical (2009a) Ayasofya Müzesi Zemin Araştırması için Sismik Etüd, İstanbul Rölöve ve Anıtlar Müdürlüğü için Hazırlanmış Rapor, 20 Haziran, 2009

Anatolian Geophysical (2009b) Ayasofya Müzesi Yapı İncelemesi için Yer Radarı Etüdü, İstanbul Rölöve ve Anıtlar Müdürlüğü için Hazırlanmış Rapor, 20 Haziran, 2009

Ancient Megastructures: Istanbul's Hagia Sophia (2010) National Geographic

Çaktı E, Dar E (2015) Long-term dynamic response of Hagia Sophia in Istanbul to earthquakes and atmospheric conditions. Paper presented at 10th international workshop on structural health monitoring (IWSHM), Stanford, USA, 1–3 Sept 2015

Davidson RA (1993) The mother of all churches: a static and dynamic structural analysis of Hagia Sophia. Dissertation, Princeton University

Durukal E (1992) A Study on structural identification and seismic vulnerability assessment of Aya Sofya. Dissertation, Boğaziçi University

Durukal E, Cimilli S, Erdik M (2003) Dynamic response of two historical monuments in İstanbul deduced from the recordings of Kocaeli and Düzce Earthquakes. Bulletin of Seismological Society of America 93(2)

Erdik M. Personal communication, Boğaziçi University, Istanbul

Erdik M, Durukal E, Yüzügüllü G, Beyen K, Kadakal U (1993) Strong-motion instrumentation of Aya Sofya and the analysis of response to an earthquake of 4.8 Magnitude. Soil dynamics and earthquake engineering VI, Computational Mechanics Publications. New York, pp 899–914

Glassman G, Julien O, Klein L (2015) Hagia Sophia: istanbul's mystery, nova. http://www.imdb.com/title/tt4254156/fullcredits?ref_=tt_ov_dr#directors Accessed at August 2016

Gürbüz et al (1994) Kandilli observatory and earthquake research institute. Department of Geophysics, Internal Report

Hill KE (1991) Hagia Sophia: static analysis of a finite element model. Dissertation, Princeton University

IMM (2009) Istanbul'un Olası Deprem Kayıpları Tahminlerinin Güncellenmesi. Istanbul Metropolitan Municipality, Project Report in Turkish

Kırlangıç AS (2008) Re-evaluation of earthquake performance and strengthening alternatives of Hagia Sophia. Dissertation, Boğaziçi University

Kokkinos K (2013) Dynamic analysis of Hagia Sophia in Abaqus: a study on a historical masonry structure under earthquake loading. Dissertation, The University of New South Wales

Livingston A, Stuzman E, Mark R, Erdık M (1992) Preliminary analysis of the masonry of Hagia Sophia Basilica. Materials Issues in Art and Archaeology III, Mat. Res. Soc. Pittsburgh, pp 721–736

Mainstone RJ (1998) Hagia Sophia: architecture. Structure and Liturgy of Justinian's Great Church. Thames and Hudson, London

Moropoulou A, Cakmak A, Bıscontın G, Bakolas A, Zendrı E (2002) Advanced Byzantine cement based composites resisting earthquake stresses: the crushed brick/lime mortars of Justinian's Hagia Sophia. Constr Build Mater 16:543–552

Müller-Wiener W (2007) İstanbul'un Tarihsel Topografyası: 17. Yüzyıl Başlarına kadar Byzantion—Konstantinopolis. Yapı Kredi Yayınları, İstanbul

Natsis MN (1994) The enigma of Hagia Sophia: a dynamic structural analysis of Justinian's great church. Dissertation, Princeton University

Rhodes H (1994) Hagia Sophia: a structural and symbolic monument. Dissertation, Princeton University

Smith J (1996) Hagia Sophia: advancement of nonlinear modeling. Dissertation, Princeton University

Van Nice RL (1963) The structure of St. Sophia. Architectural Forum 118:45–138

Worcester J (1995) Hagia Sophia: a nonlinear static analysis with cracking and phasing. Dissertation, Princeton University

Chapter 8
Evaluation of Historical Merzifon Dönertaş Mosque with a Single Dome in Terms of Its Structure

Burçin Şenol Şeker, Sertac Tuhta and Varol Koç

Abstract The mosques are the most important pieces of Islamic culture in terms of their architectural and structural properties. The mosques and complexes are built in the center of living areas and constituted important structures in the surroundings in Islamic culture. Therefore, it is very common to observe the mosques in almost every city center. In this study, Merzifon Dönertaş Mosque, which is located in the city, center of Merzifon and which takes an important part among Amasya's historical monuments is investigated by means of static and dynamic analyses by the use of finite element method. The material properties and formulas, given in the literature, have been considered in order to obtain the structural behavior of the structure. The results of the analyses show that the most critical parts of the mosque.

Keywords Structural performance · Mechanical tests · Finite element method Structural behavior

8.1 Introduction

The City Merzifon, which covers a large area is a place of settlement with a historical structure, which is situated at the crossroads of highways in Middle Black sea region. Now it is a town of Amasya, and it is bordered by Gümüşhacıköy to the

B. Ş. Şeker (✉)
Department of Civil Engineeing, Amasya University,
Merzifon Vocational School, Amasya, Turkey
e-mail: senolseker@yahoo.com

S. Tuhta · V. Koç
Department of Civil Engineering, Ondokuz Mayis University,
Faculty of Engineering, Atakum, Samsun, Turkey
e-mail: stuhta@omu.edu.tr

V. Koç
e-mail: kvarol@omu.edu.tr

A. A. Kasimzade et al. (eds.), *Seismic Isolation, Structural Health Monitoring, and Performance Based Seismic Design in Earthquake Engineering*,
https://doi.org/10.1007/978-3-319-93157-9_8

east, Vezirköprü to the West and the City Çorum to the southwest. Throughout history, Kimmers, Phyrgians, Lydians, and Persians reigned in the town, whose history goes back to the age of bronze. Before Turks conquered this region, Eastern Roman Empire used to reign there. Then, the reign of Seljuks started in the region, and as a result of Mongol occupation, it was now under the rule of Ilkhands for some time. With the conquest of the city by Sultan Yıldırım Bayezit in 1393, Ottoman rule began (Çerkez 2005). The fast urbanization of the places conquered, together with the rule of the Ottoman Empire, resulted in the construction of new buildings across the city. The Dönertaş Mosque is one of the buildings constructed in this period. In this study, this historical Mosque in Merzifon was examined in terms of its structural movements in detail. For the present historical structure of this town, which is situated on a first-degree seismic belt on the North Anatolia fault zone, to be passed down to the next generation, it is important to determine the movements of such buildings during an earthquake and find out the parts that are influenced most by the earthquake as the main block. Due to the fact that analytical examinations are time-consuming and require many suppositions, this examination was conducted using the software system ANSYS (ANSYS 2014), one of the programmes using finite element analysis technique, which is convenient for buildings with large blocks. Our aim is to detect the parts of the building, whose main supporting blocks were built of stone and whose dome was built of bricks, where tension and deformation were intense, through static and dynamic analysis.

8.2 The Architectural and Historical Features of the Mosque

The building has a square plan with a northax on the northern side. The minaret is on the eastern side, connected to the main block. On the main block, which consists of a main masonry wall planned in the form of a square, stands an octagonal frame with beveled corners. The frame, whose supporting arches on the four corners are squinched at the back, and which is built of ashlars with smooth surfaces covering it from behind rising in stages like stairs, has a structure that looks like a triangle. Both this part and the domes are covered with leads. The entrance door faces north with two rectangular windows on both sides. The upper edges of the main sanctuary door and the two windows are level at the top. The side windows on the west side at the bottom of the main outer wall have iron bars on them in a way that cover both windows joining them together. These windows also have jambs on either side. The windows on the upper part have the shape of a pointed arch, and they are positioned in a different way compared to the ones below them. The building materials used for construction are smooth ashlars for the main block and bricks for the dome and minaret's. There are also marble pillars in the northax. The mosque was built in 1262, and it has been restored in the following years. There is not much information about its history (Fig. 8.1).

Fig. 8.1 General appearance of the Mosque

8.3 Structural Analysis

8.3.1 Three-Dimensional Numerical Model

Before the analysis stage regarding the construction, it was modeled three-dimensionally on the computer in order to see general movements of the construction clearly Fig. 8.2.

Thanks to this modeling, the mosque can be simulated through the computer programme using finite elements method, and necessary features of materials can be assigned according to the requirements of each section of the mosque. On the one hand, complexity of its geometry makes it difficult to model such masonry historical buildings, and on the other hand, the fact that the materials used during the construction is not homogeneous in every part of it necessarily requires some

Fig. 8.2 Finite element model of the Mosque

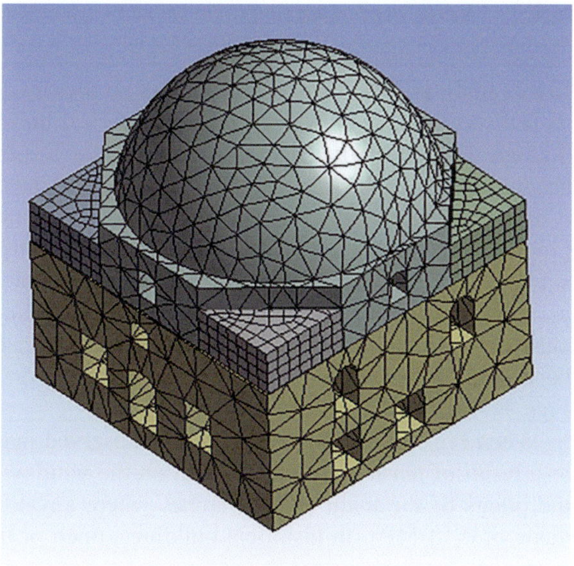

Table 8.1 Physical and mechanical features of the materials used for the Mosque

Department of Mosque	Modulus of elasticity (MPa)	Poisson rate	Unit volume weight (kg/m^3)	Average compressive strength (MPa)	Average the tensile strength (MPa)
The main mass (Wall and last part of community)	4803	0.17	2350	48.03	7.50
Domes	1605	0.15	2032	16.05	2.65

suppositions. Because of this reason, before making analysis, the model mosque was accepted to have been built of a mixture of different materials, which was a good way for macro modeling technique. In addition, only the main block was modeled in order to observe the movement, tension, and deformation in the main block and to avoid making it a big model. Solid 186 element, one of the solid element types, which is available in ANSYS model library, and which only has translation degree of freedom in all three dimensions, has been used in the analysis. The features of the materials used for the construction was determined making use of the results of the experiments done on the sample materials taken from a mosque in the town which had been restored before. The results of the experiment are given in Table 8.1 Moreover, for the mechanical features and limit values of the material to be assigned to the interior walls of the mosque, we made use of Regulation about the Constructions to be Built in Earthquake Zone, Chap. 5. The finite element model of the mosque is shown in Fig. 8.2.

8.3.2 *Self-load Analysis*

In this analysis, deformation and stresses on the construction due to the influence of its self-weight have been determined. Historical buildings are designed from top to bottom. The effect of the main dome on the bearing walls has been taken into account, and then the sizes of other parts under it have been determined accordingly (Çamlıbel 1998). The bottom surface of the main outer walls have been accepted as ground fixed, bricks have been assigned to the parts that constitute the dome and squinches of the corners, and mechanical features of stone materials have been assigned to other parts. To be able to see the distribution of deformation and stress clearly in such buildings with complex geometry, colored distribution scale has been used. Distribution of stress and deformation on the construction are shown in Figs. 8.3, 8.4 and 8.5.

When Fig. 8.2 is examined, it can be observed that maximum stresses that occur as a result of tension are mostly both on the windows of the bottom section and at the points of the main bearing arches where keystones are used, and it takes the value of 0.13 MPa. In historical buildings, open or hidden small arches were built

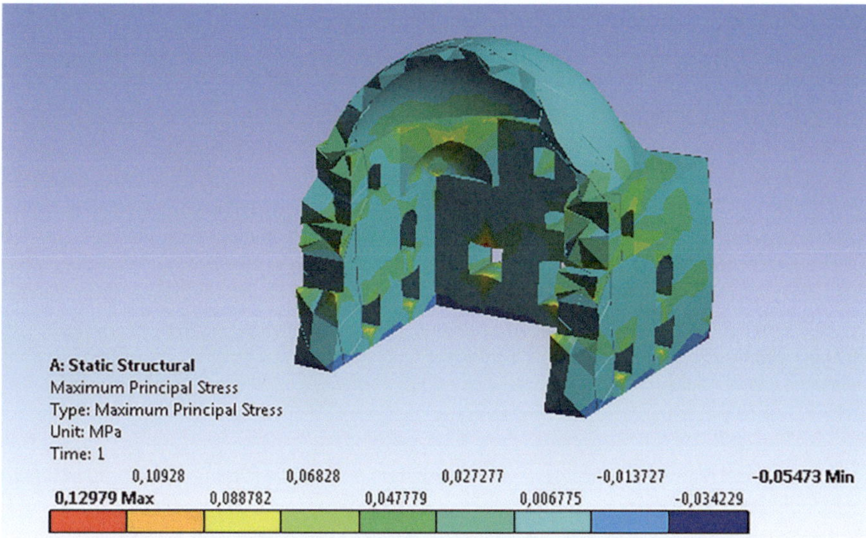

Fig. 8.3 Mosque I. Noble stress distribution

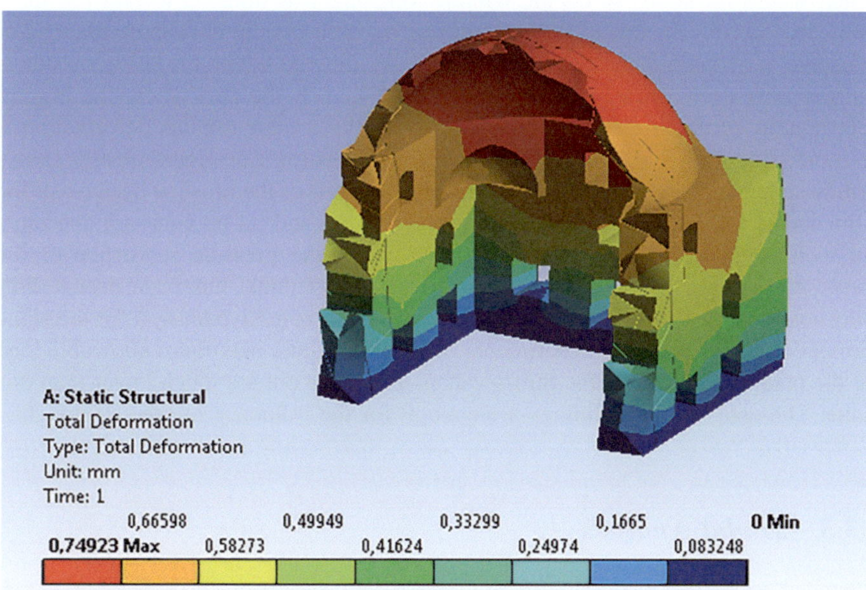

Fig. 8.4 Total replacement distribution of the Mosque

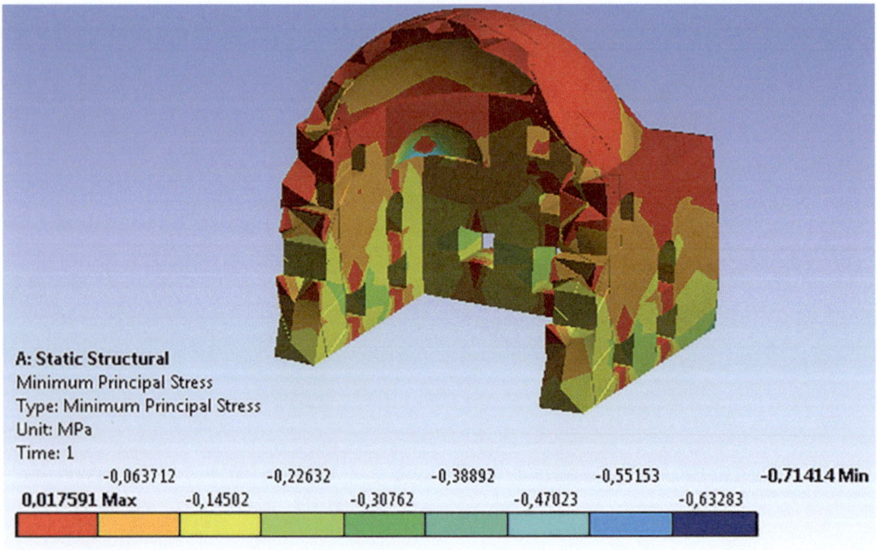

Fig. 8.5 Mosque 3. Noble stress distribution

over the windows, or tension elements such as timbers were used so that the load which would normally be on the window opening was directed toward the main walls. In order for the arches in the main bearing arch system to resist to the load of the dome, it becomes clear from the analysis that the part where the keystone stands and the arch itself must be thick enough. In Fig. 8.3, the part where the biggest deformation occurred was the dome, and the highest value for this deformation is 0.75 mm. Figure 8.4 shows maximum pressure tension. The figure clearly shows that these tensions increase gradually toward the floor of the mosque because of the influence of the load caused by the upper structure, and its maximum value turns out to be 0.71 MPa. In Code TE 2007, 25% of the pressure resistance of the construction material, which makes up the masonry units, has been accepted as allowable compressive stress. Resistance value of compressive tension for the stone element obtained from the experiments is 48 MPa. Thus, maximum allowable load for the main walls is 12 MPa. Stress values that come out are much lower than this value. The sizes of the buildings are enough for the influence of the self-weight.

8.3.3 Modal Analysis

In this analysis, free vibration shapes of the building have been determined considering mechanical and dynamic features of the construction. The magnitude corresponding each period has been dealt with as mode shapes, which show vibration shapes of the construction under an external influence. Even though

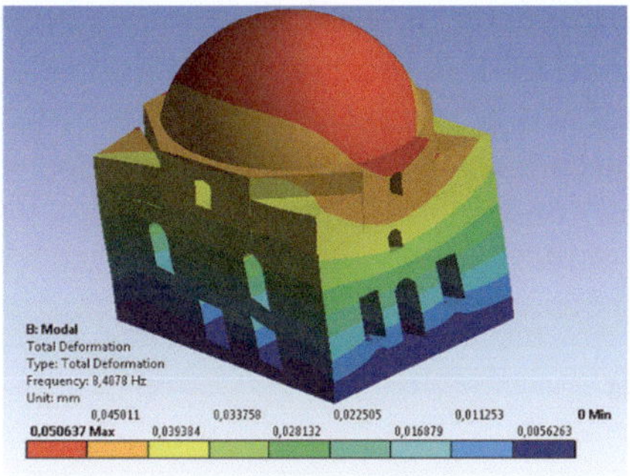

1.Mode Shape (translation in the direction of Kible)

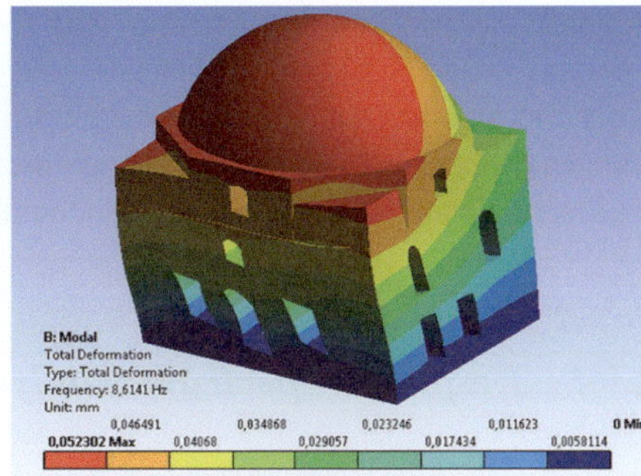

2. Mode Shape (Translation in east-west direction)

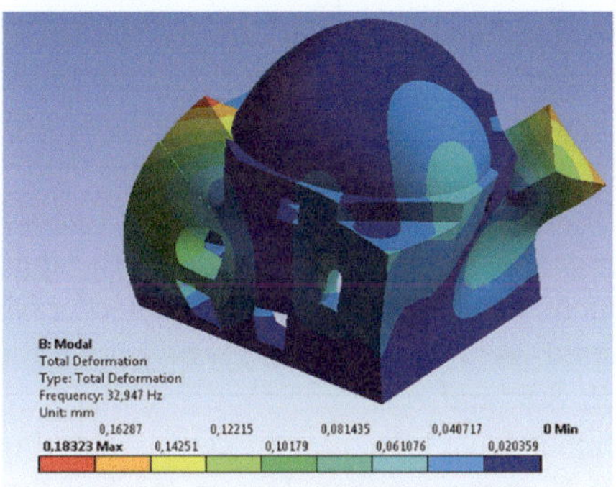

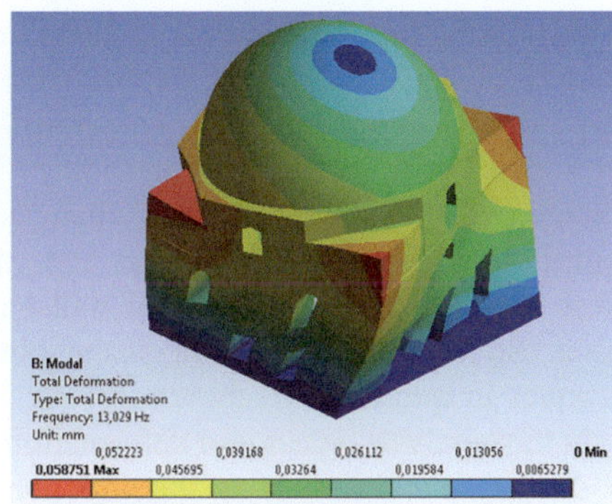

3. Mode Shape (Torsion) 17. Mode Shape

Fig. 8.6 Mode shapes

external load changes, mode shape does not change, for it is connected with the structural geometry and rigidness. Thanks to the mode shapes obtained it is possible to find out the part of the construction that could hardly resist in case of an earthquake. The mode shapes obtained through analysis for the mosque in question is shown in Fig. 8.6, and the mass ratio added to these shapes is given in Table 8.2.

When the result of the analysis are examined, it can be seen that effective mode come out as translation in both directions. It is seen that in these kinds of constructions built symmetrically in general, torsional effects are reduced to minimum,

Table 8.2 Mode frequencies and mass ratios that influence the mosque

Mode	Frequency (hz)	Mass participation rate (X direction)	Mass participation rate (Y direction)
1	8.4078	0.747	0.22E−2
2	8.6140	0.25E−2	0.7098
3	13.0293	0.19E−6	0.18E−1
17	32.9468	0.25E−1	0.14E−1

preventing unwanted constraining. The fact that the total mass fraction involved in torsional mode is not more than 3% is a proof for this situation. This information about the construction of historical buildings can be obtained through experience. The result obtained shows that both architectural knowledge and knowledge of engineering have been taken into account in constructions similar to mosques.

8.3.4 Response Spectrum Analysis

In this kind of analysis, based on the principle of statistical combination of the modes obtained as a result of model analysis, changeable suppositions are required (valid). Using the spectrum curve given in the earthquake regulations, the stress and changes in shape caused by the influence of the mode related to taking each mode shape period into account are calculated. Then, the values obtained are added up statistically using SRSS method (Square Root of the Sum of Squares). The values obtained are the total stress and changes in shape that will happen thanks to the contribution of each mode. Stress and changes. The values obtained are the total stress and changes in shape that will happen thanks to the contribution of each mode. Stress and changes in shape obtained as a result of analysis about the mosques in question are shown in Figs. 8.7, 8.8, and 8.9.

In this analysis, stress and deformation values have been determined considering mode shapes and periods in the direction of Y of the mosque. Due to the fact that

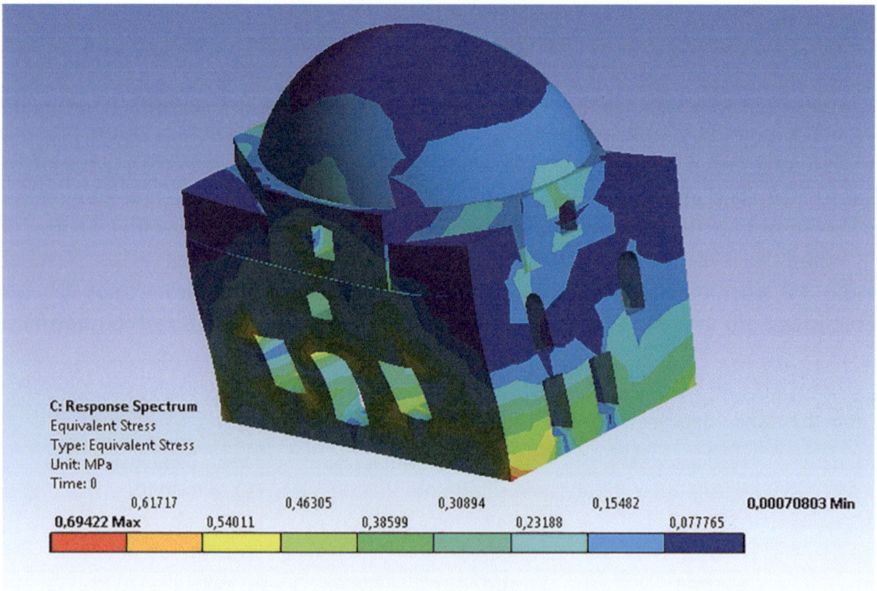

Fig. 8.7 Mosque equal (possible) stress distribution

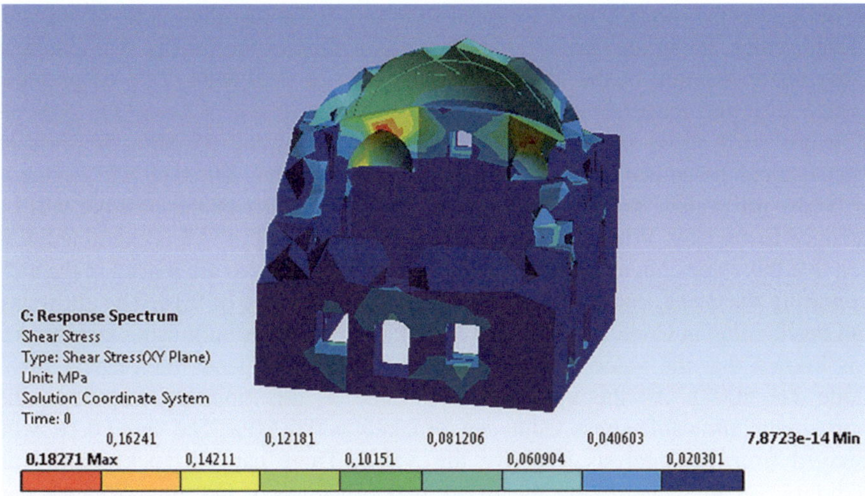

Fig. 8.8 Mosque shear stress distribution

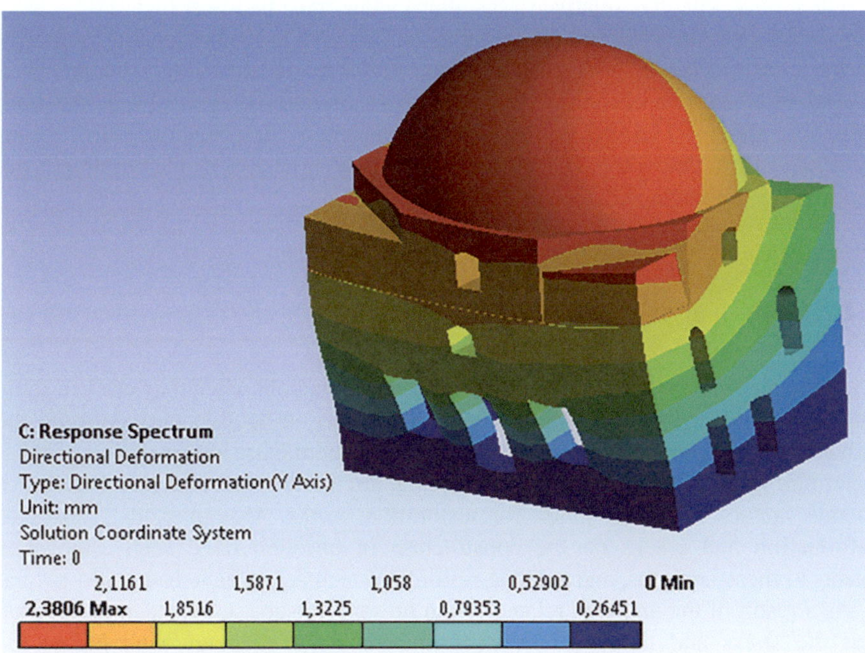

Fig. 8.9 Deformation of the mosque in the direction of Y

the mosque is the model with a complex structure, stress distribution is shown with colored scales. It can also be observed that equal stresses are mostly at the bottom of the main structure of the mosque with a value of 0.69 MPa. 25% of pressure resistance of the materials used to build walls takes place in Code TE 2007 as allowable wall stress (Code TE 2007) Thus, 25% of the 48 MPa of pressure resistance can be considered as MPa allowable stress value. The 0.69 MPa value is far below this value. So it is not expected that these parts of the mosque will be damaged in anyway. When the shear stress distribution in Fig. 8.8 is examined, it is seen that the values intensify at the point where the key stones are placed in the arch supporting the dome, and it has reached the maximum value of 0.18. The allowable wall crack value in Code TE 2007 is 0.15 MPa for the walls built using clay bricks. It is known that the value of allowable shear tension depends on vertical stress (Code TE 2007). As the vertical stress value at this point is very low, the approximate allowable stress value can be taken as 0.5 MPa. The value 0.18 MPa obtained from the analysis is above this value. There can be cracks during an earthquake at these points on the upper part of the arch. Opening of crack can change based on the magnitude and the period of the earthquake. If this value is high, it can cause damage to the upper structure (the dome). Deformation values in Fig. 8.9 are mostly on the main dome. Deformations with a maximum value of 2.38 mm can be compared with the formulation given in Code TE 2007.

The highest effective relative (translation) value must be lower than $0.02 \times h/R$. Here, h refers to the height of the building and R refers to load reduction coefficient of the construction in an earthquake, for masonry construction its value is 2. Considering that top elevation of the dome is approximately 11.44 m, the limit value is $0.02 \times 11{,}440/2 = 114.4$ mm. Maximum deformations that occur (2.38 mm) is far below this value. Based on possible translations the building is not expected to get damaged.

8.4 Conclusions

It is of vital importance to analyze historical buildings both when they are in a static state and under the effect of an earthquake in order to be able to understand the movements of these kinds of buildings, and help them exist longer period of time. Determining the parts of the constructions that are likely to get damaged because of outside factors will be a guide for restoration works. Moreover, the causes of deformation and cracks on the construction in question have been established thanks to these analyses, and suitable restoration techniques have been used to treat it. As a result of the analyses related to the present mosque, it has been established that the stress and deformations occurring in static state are well below the allowable values. The parts of the construction that lack any supporting columns, if possible, should be built in a geometrical shape like an arch so that it can reduce possible stress to minimum. In general, the sizes of mosques are sufficient for static state.

Dynamic analysis is a kind of effect that brings about quite different results compared to static analysis in terms of its effects and results. With this analysis about the construction in question, mode shapes and periods of free vibration have been determined for the first time. In the reflection, mode shapes in both orthogonal directions turned out to be 70% with more mass rate as translation. This suggests that there will be little torsional effect on the construction in case of an earthquake and that there will not be anything that will cause complete destruction of the masonry construction. It can be said that the tension elements to be used at certain levels on the main walls will absorb the pulling stress that will occur because of the influence of this translation. Most probably, the same logic lies behind using bonding timbers in historical buildings. The shear stress values, the most important effect for masonry buildings, occur on the upper points of the arches supporting the main dome, and it is important that the hanger points of these supporting arches are thick and big enough in size. Any damage to this part may cause the main dome to collapse. Yet, based on the duration and effect of the earthquake, the main dome runs the risk of collapse when the tension and deformation values exceed the elasticity limit. It is considered that the results of this study will be useful in passing these kinds of buildings down to the next generations.

References

ANSYS Finite Element Analysis Program USA 2014

Çamlıbel N (1998) Sinan Mimarlığında Yapı Strüktürünün Analitik İncelenmesi. Doçentlik Tezi, İTÜ, Fen Bilimleri Enstitüsü, İstanbul

Çerkez M (2005) Merzifon'da Türk Devri Mimari Eserleri. Doktora Tezi, Ankara Üniversitesi Sosyal Bilimler Enstitüsü, Ankara

Code TE (2007) Specifications for the buildings to be constructed in disaster areas. Ministry of Public Works and Settlement, Ankara, Turkey

Chapter 9
Analytical and Experimental Modal Analysis of a Model Cold Formed Steel (CFS) Structures Using Microtremor Excitation

Azer A. Kasimzade, Sertac Tuhta, Gencay Atmaca and Sevda Ozdemir

Abstract In this study was investigated the possibility of using the recorded micro-tremor data on ground level as ambient vibration input excitation data for investigation and application Operational Modal Analysis (OMA) on the bench-scale earthquake simulator (The Quanser Shake Table) for model cold formed steel (cfs) structures. As known OMA methods (such as EFDD, SSI and so on) are supposed to deal with the ambient responses. For this purpose, analytical and experimental modal analysis of a model (cfs) structure for dynamic characteristics was evaluated. 3D Finite element model of the building was evaluated for the model (cfs) structure based on the design drawing. Ambient excitation was provided by shake table from the recorded micro tremor ambient vibration data on ground level. Enhanced Frequency Domain Decomposition is used for the output-only modal identification. From this study, best correlation is found between mode shapes. Natural frequencies and analytical frequencies in average (only) 2.99% are differences.

Keywords Experimental modal analysis · Modal parameter · EFDD
Shake table

A. A. Kasimzade (✉) · S. Tuhta · G. Atmaca · S. Ozdemir
Department of Civil Engineering, Faculty of Engineering,
Ondokuz Mayis University, Atakum, Samsun, Turkey
e-mail: azer@omu.edu.tr

S. Tuhta
e-mail: stuhta@omu.edu.tr

G. Atmaca
e-mail: gencayatmaca@hotmail.com

S. Ozdemir
e-mail: sevdaozdemir@hotmail.com

A. A. Kasimzade et al. (eds.), *Seismic Isolation, Structural Health Monitoring, and Performance Based Seismic Design in Earthquake Engineering*,
https://doi.org/10.1007/978-3-319-93157-9_9

9.1 Introduction

There are many varieties of the structural and architectural structures in the world. Common features of these structures, it can be managed to survive under static and dynamic loads. Structures under dynamic loads and vibrations occurred impact consists of vibrations that do not require or require intervention on the structure brings many damage occurs. In this case, the vibration should be known and may occur in nature and will be focused on the effects generated by these vibrations. In recent years, several earthquakes have occurred in the world and are given as a result of heavy losses. If the resulting perceived to pose several problems for the countries of heavy losses, the structure of the receipt of the knowledge of the current situation and how important it is understood that the necessary measures. In this case, experimental determination of the behavior they showed against vibrations from the structures and obtained the theoretical and the creation of finite element model to represent the actual structure by comparing the experimental value are emerging requirements. As known forced (shaker, impact, pull back or quick release tests) and ambient vibration techniques are available for vibration testing of large structures. Force vibration methods more complex and are generally more expensive than ambient vibration tests. Ambient vibration testing (also called Operational Modal Analysis) is the most economical non-destructive testing method to acquire vibration data from large civil engineering structures for Output-Only Model Identification. General characteristics of structural response (appropriate frequency, displacement, velocity, acceleration rungs), suggested measuring quantity (such as velocity or acceleration) depends on the type of vibrations (Traffic, Acoustic, Machinery inside, Earthquakes, Wind) are given in Vibration of Buildings (1990).

This structures response characteristics gives a general idea of the preferred quantity and its rungs to be measured. A few studies the analysis of ambient vibration measurements of buildings from 1982 until 1996 are discussed in Ventura and Schuster (1996). Last 10 years Output-Only Model Identification studies of buildings are given in appropriate references structural vibration solutions. For the modal updating of the structure it is necessary to estimate sensitivity of reaction of examined system to change of parameters of a building, Kasimzade (2006), Kasimzade and Tuhta (2012). System identification is the process of developing or improving a mathematical representation of a physical system using experimental data investigated in HO and Kalman (1966), Kalman (1960), Ibrahim and Miculcik (1977), Ibrahim (1977), Bendat (1998), Ljung (1999), Juang (1994), Van Overschee and De Moor (1996), and system identification applications in civil engineering structures are presented in works Trifunac (1972), Brincker et al. (2000), Roeck (2003), Peeters (2000), Cunha et al. (2005), Wenzel and Pichler (2005),Ventura and Brincker (2015), Kasimzade and Tuhta (2005, 2006, 2007a, b, c, 2009, 2012). Extracting system physical parameters from identified state-space representation was investigated in references. Alvin and Park (1994), Balmes (1997), Juang et al. (1988), Juang and Pappa (1985), Lus et al. (2003), Phan et al.

(2003), Sestieri and Ibrahim (1994), Tseng et al. (1994a, b). In engineering structures there are three types of identification: modal parameter identification; structural-modal parameter identification; control-modal identification methods are used. In the frequency domain the identification is based on the singular value decomposition of the spectral density matrix and it is denoted Frequency Domain Decomposition (FDD) and its further development Enhanced Frequency Domain Decomposition (EFDD). In the time domain there are three different implementations of the Stochastic Subspace Identification (SSI) technique: Unweighted Principal Component (UPC); Principal component (PC); Canonical Variety Analysis (CVA) is used for the modal updating of the structure Friswell and Mottershead (1995), Marwala (2010). It is necessary to estimate sensitivity of reaction of examined system to change of random or fuzzy parameters of a structure. Investigated measurement noise perturbation influences to the identified system modal and physical parameters. Estimated measurement noise border, for which identified system parameters are acceptable for validation of finite element model of examine system. System identification is realized by observer Kalman filter (Juang et al. 1993) and Subspace Overschee and De Moor (1996), algorithms. In special case observer gain may be coincide with the Kalman gain. Stochastic state-space model of the structure are simulated by Monte-Carlo method.

The Quanser Shake Table is a bench-scale earthquake simulator ideal for teaching structural dynamics, control topics related to earthquake, aerospace and mechanical engineering and widely used in applications. In this study was investigated of possibility using the recorded micro tremor data on ground level as ambient vibration input excitation data for investigation and application Operational Modal Analysis (OMA) on the bench-scale earthquake simulator (The Quanser Shake Table) for model (cfs) structures.

For this purpose, analytical and experimental modal analysis of a model (cfs) structure for dynamic characteristics was evaluated. 3D Finite element model of the building was evaluated for the model (cfs) structure based on the design drawing. Ambient excitation was provided by shake table from the recorded micro tremor ambient vibration data on ground level. Enhanced Frequency Domain Decomposition is used for the output-only modal identification.

9.2 Modal Parameter Extractions

The (FDD) ambient modal identification is an extension of the Basic Frequency Domain (BFD) technique or called the Peak-Picking technique. This method uses the fact that modes can be estimated from the spectral densities calculated, in the case of a white noise input, and a lightly damped structure. It is a non parametric technique that determines the modal parameters directly from signal processing. The FDD technique estimates the modes using a Singular Value Decomposition

(SVD) of each of the measurement data sets. This decomposition corresponds to a Single Degree of Freedom (SDOF) identification of the measured system for each singular value (Brincker et al. 2000).

The Enhanced Frequency Domain Decomposition technique is an extension to Frequency Domain Decomposition (FDD) technique. This t technique is a simple technique that is extremely basic to use. In this technique, modes are easily picked locating the peaks in Singular Value Decomposition (SVD) plots calculated from the spectral density spectra of the responses. FDD technique is based on using a single frequency line from the Fast Fourier Transform analysis (FFT), the accuracy of the estimated natural frequency based on the FFT resolution and no modal damping is calculated. On the other hand, EFDD technique gives an advanced estimation of both the natural frequencies, the mode shapes and includes the damping ratios (Jacobsen et al. 2006). In EFDD technique, the single degree of freedom (SDOF) Power Spectral Density (PSD) function, identified about a peak of resonance, is taken back to the time domain using the Inverse Discrete Fourier Transform (IDFT). The natural frequency is acquired by defining the number of zero crossing as a function of time, and the damping by the logarithmic decrement of the correspondent single degree of freedom (SDOF) normalized auto correlation function Peeters (2000).

In this study modal parameter identification was implemented by the Enhanced Frequency Domain Decomposition. The relationship between the input and responses in the EFDD technique can be written as, In this method, unknown input is represented with $x(t)$ and measured output is represented with $y(t)$

$$\left[G_{yy}(j\omega)\right] = [H(j\omega)]^{*}[G_{xx}(j\omega)][H(j\omega)]^{T} \tag{9.1}$$

where $G_{xx}(j\omega)$ is the $r \times r$ Power Spectral Density (PSD) matrix of the input. $G_{yy}(j\omega)$ is the $m \times m$ Power Spectral Density (PSD) matrix of the output, $H(j\omega)$ is the m x r Frequency Response Function (FRF) matrix, and * and superscript T denote complex conjugate and transpose, respectively. The FRF can be reduced to a pole/residue form as follows:

$$[H(\omega)] = \frac{[Y(\omega)]}{[X(\omega)]} = \sum_{k=1}^{m} \frac{[R_k]}{j\omega - \lambda_k} + \frac{[R_k]^{*}}{j\omega - \lambda_k^{*}}, \tag{9.2}$$

where n is the number of modes λ_k is the pole and, R_k. is the residue. Then Eq. (9.1) becomes as

$$G_{yy}(j\omega) = \sum_{k=1}^{n}\sum_{s=1}^{n}\left[\frac{[R_k]}{j\omega - \lambda_k} + \frac{[R_k]^{*}}{j\omega - \lambda_k^{*}}\right] G_{xx}(j\omega)\left[\frac{[R_s]}{j\omega - \lambda_s} + \frac{[R_s]^{*}}{j\omega - \lambda_s^{*}}\right]^{\overline{H}} \tag{9.3}$$

where s the singular values, superscript is H denotes complex conjugate and transpose. Multiplying the two partial fraction factors and making use of the

Heaviside partial fraction theorem, after some mathematical manipulations, the output PSD can be reduced to a pole/residue form as follows;

$$\left[G_{yy}(j\omega)\right] = \sum_{k=1}^{n} \frac{[A_k]}{j\omega - \lambda_k} + \frac{[A_k]^*}{j\omega - \lambda_k^*} + \frac{[B_k]}{-j\omega - \lambda_k} + \frac{[B_k]^*}{-j\omega - \lambda_k^*}, \tag{9.4}$$

where A_k is the k th residue matrix of the output PSD. In the EFDD identification, the first step is to estimate the PSD matrix. The estimation of the output PSD known at discrete frequencies is then decomposed by taking the SVD (singular value decomposition) of the matrix;

$$G_{yy}(j\omega_i) = U_i S_i U_i^{\overline{H}}, \tag{9.5}$$

where the matrix $U_i = [u_{i1}, u_{i2}, \ldots, u_{im}]$ is a unitary matrix holding the singular vectors u_{ij} and s_{ij} is a diagonal matrix holding the scalar singular values. The first singular vector u_{ij} is an estimation of the mode shape. PSD function is identified around the peak by comparing the mode shape estimation u_{ij} with the singular vectors for the frequency lines around the peak. From the piece of the SDOF density function obtained around the peak of the PSD, the natural frequency and the damping can be obtained.

9.3 Description of Model (CFS) Structure

The Quanser shake Table II is a uniaxial bench-scale shake table. This unit can be controlled by appropriate software was illustrated in Fig. 9.1a–c. It is effective for a wide variety of experiments for civil engineering structures and models. Shake table specifications are as follows Quanser (2008) (Table 9.1):

Model (cfs) structure is 1.03 m height. Thickness of elements is 0.001588 m. The structure dimensions are shown in Fig. 9.2.

9.4 Analytical Modal Analysis of Model (CFS) Structure

A finite element model was generated in SAP2000 (1997). Beams and columns were modeled as 3D beam-column elements (in Fig. 9.3 shown by the black color). Structure modeled as an absolutely rigidity floor (rigid diaphragm). The selected structure is modeled as a space frame structure with 3D elements. Beams and columns were modeled as 3D beam-column elements which have degrees of freedom. At the base of the structure in the model, the ends of every element were fixed against translation and rotation for the 6-DOF then creating finite element model of the structure in SAP2000. The following assumptions were taken into account. Model

(a) (b) (c)

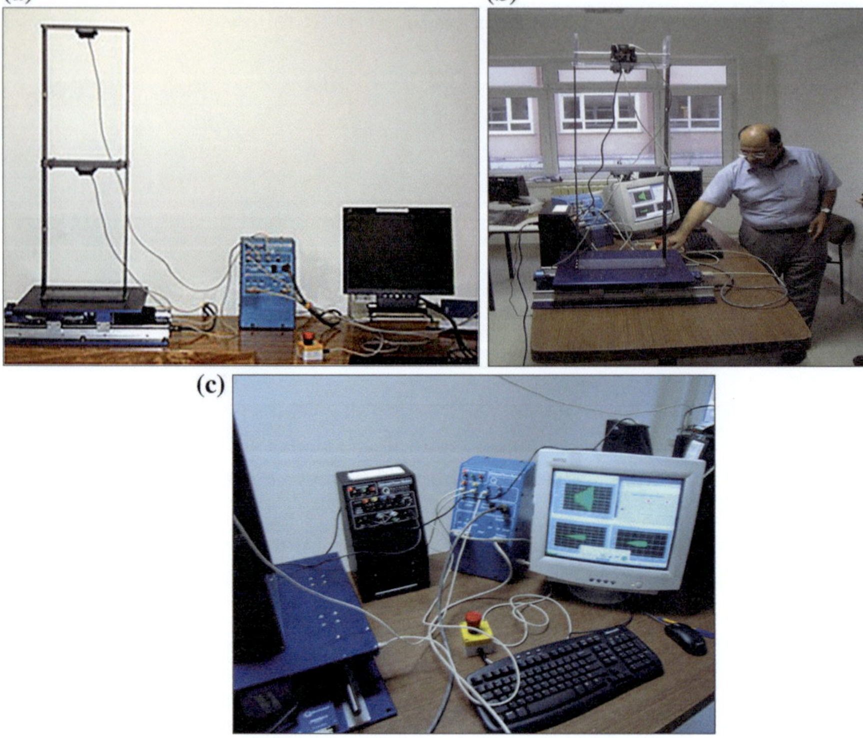

Fig. 9.1 **a–c** Illustration of model (cfs) structure and shake table

(cfs) structure is modeled using an equivalent thickness and shell elements with isotropic property. All supports are modeled as fully fixed. The members of steel frame are modeled as rigidly connected together at the intersection points. In modeling of beams and columns the modulus of elasticity E = 2.000E11 N/m^2, Poisson ratio $\mu = 0.3$, mass per unit volume ρ = 78,500 N/m^3.

Natural frequencies and vibration modes are concerned a significant impact on the dynamic performance of buildings is an important dynamic properties. A total of five natural frequencies of the structure are attained which range between 2 and 9 Hz. The first five vibration mode of the structure is shown in Fig. 9.4. Analytical modal analysis results at the finite element model are shown in Table 9.2.

Table 9.1 Shake table specifications

Dimensions ($H \times L \times W$)	61 cm × 46 cm × 13 cm
Total mass	27.2 kg
Payload area ($L \times W$)	46 cm × 46 cm
Maximum payload at 2.5 g	7.5 kg
Maximum travel	±7.6 cm
Operational bandwidth	10 Hz
Maximum velocity	66.5 cm/s
Maximum acceleration	2.5 g
Lead screw pitch	1.27 cm/rev
Servomotor power	400 W
Amplifier maximum continuous current	12.5 A
Motor maximum torque	7.82 N.m
Lead screw encoder resolution	8192 counts/rev
Effective stage position resolution	1.55 μm/count
Accelerometer range	±49 m/s^2
Accelerometer sensitivity	1.0 g/V

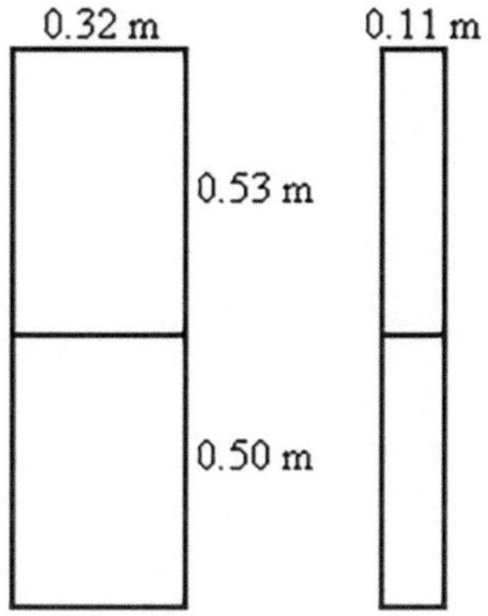

Fig. 9.2 View of model (cfs) structure and shake table

9.5 Experimental Modal Analysis of Model (CFS) Structure

Ambient excitation was provided by the recorded micro-tremor data on ground level. Three accelerometers (with both x and y directional measures) were used for the ambient vibration measurements, one of which were allocated as reference sensor always located in the first floor (they are shown by the red line in Fig. 9.5a, b). Two accelerometers were used as roving sensors (they are shown by the black line in Fig. 9.5a, b). The response was measured in two data sets (Fig. 9.5a, b). For two data sets were used 3 and 5 degree of freedom records respectively (Fig. 9.5a, b). Every

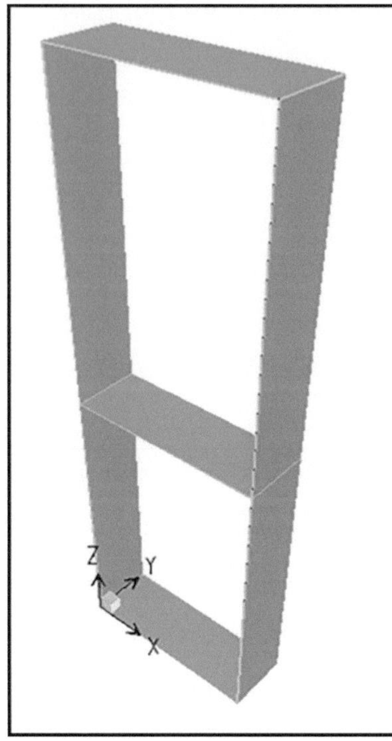

Fig. 9.3 Finite element model of model (cfs) structure

data set (Fig. 9.5a, b) was measured 100 min. The selected measurement points and directions are shown in Fig. 9.5a, b.

The data acquisition computer was dedicated to acquiring the ambient vibration records. In between measurements, the data files from the previous setup were transferred to the data analysis computer using a software package. This arrangement allowed data to be collected on the computer while the second, and faster, computer could be used to process the data in site. This approach maintained a good quality control that allowed preliminary analyses of the collected data. If the data showed unexpected signal drifts or unwanted noise or for some unknown reasons, was corrupted, the data set was discarded and the measurements were repeated (Fig. 9.6).

Before the measurements could begin, the cable used to connect the sensors to the data acquisition, equipment had to be laid out. Following each measurement, the roving sensors were systematically located from floor to floor until the test was completed. The equipment used for the measurement includes three quanser accelerometers (with both x and y directional measures) and geosiguni-axial accelerometer, matlab data acquisition toolbox (wincon). For modal parameter estimation from the ambient vibration data, the operational modal analysis (OMA) software ARTeMIS Extractor (1999) is used.

Fig. 9.4 Analytically identified mode shapes of model (cfs) structure

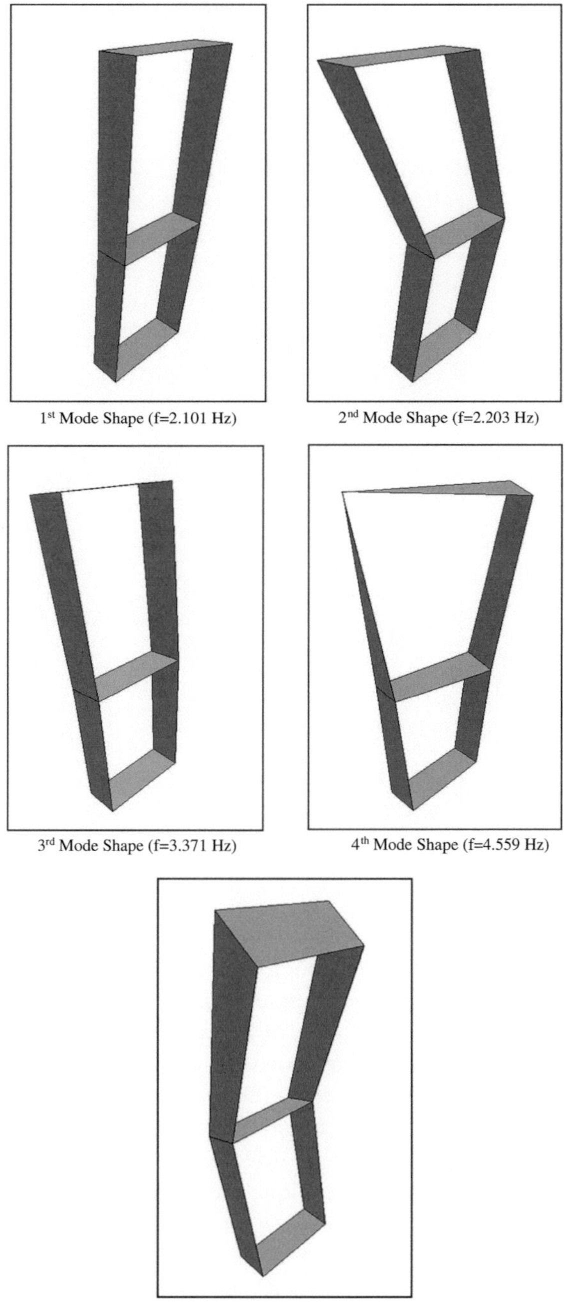

1st Mode Shape (f=2.101 Hz)

2nd Mode Shape (f=2.203 Hz)

3rd Mode Shape (f=3.371 Hz)

4th Mode Shape (f=4.559 Hz)

5th Mode Shape (f=5.971 Hz)

Table 9.2 Analytical modal analysis result at the FE model

Mode number	1	2	3	4	5
Frequency (Hz)	2.101	2.203	3.371	4.559	5.971

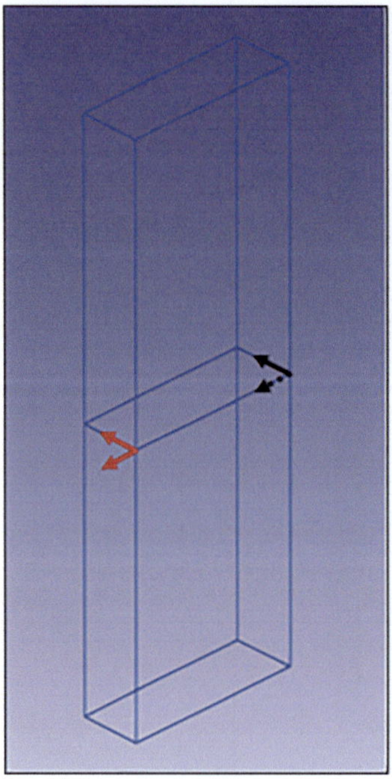

(a) First setup

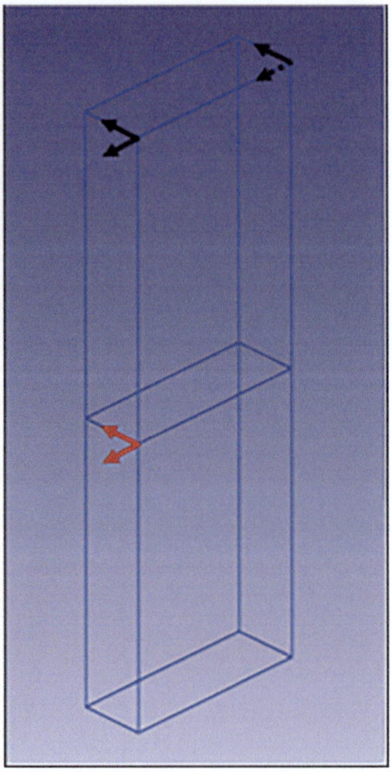

(b) Second setup

Fig. 9.5 Accelerometers location of experimental model in the 3D view

The simple peak-picking method (PPM) finds the eigenfrequencies as the peaks of nonparametric spectrum estimates. This frequency selection procedure becomes a subjective task in case of noisy test data, weakly excited modes and relatively close eigenfrequencies. Also for damping ratio estimation the related half-power bandwidth method is not reliable at all. Frequency domain algorithms have been the most popular, mainly due to their convenience and operating speed.

Singular values of spectral density matrices, attained from vibration data using PP (Peak Picking) technique are shown in Fig. 9.7. Natural frequencies acquired from the all measurement setup are given in Table 9.3. The first five mode shapes extracted from experimental modal analyses are given in Fig. 9.8. When all measurements are examined, it can be seen that there are best accordance is found between experimental mode shapes. When the analytically and experimentally identified modal parameters are checked with each other, it can be seen that there is a best agreement between the mode shapes in experimental and analytical modal analyses (Table 9.4).

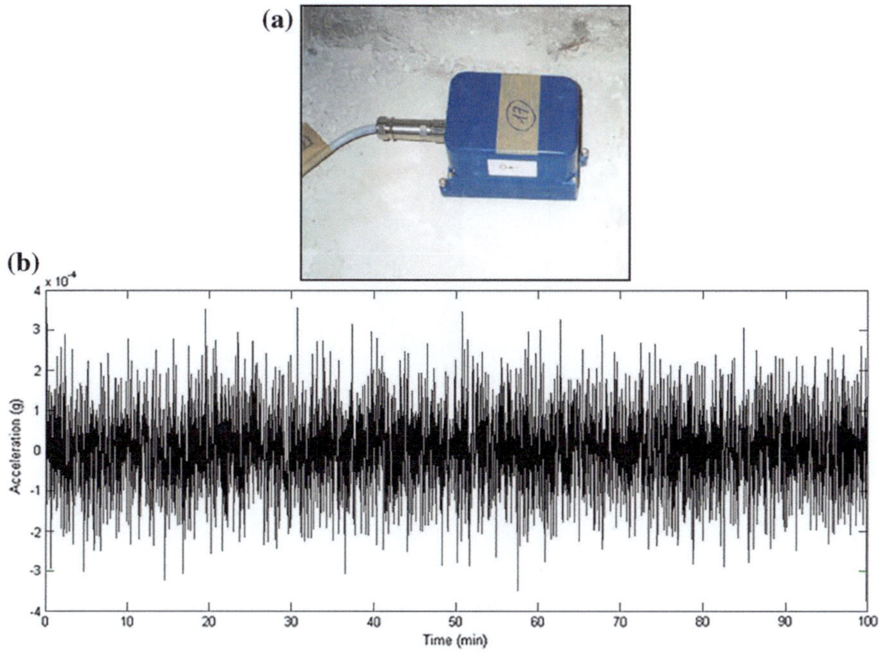

Fig. 9.6 **a** Ambient vibrations recorded by the accelerometer, **b** Ambient excitation data from the recorded micro tremor data on ground level used in the shake table

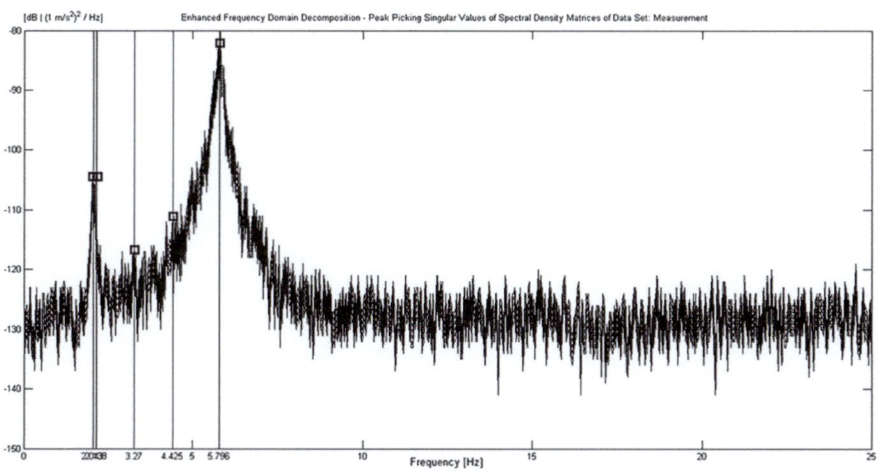

Fig. 9.7 Singular values of spectral density matrices

Table 9.3 Experimental modal analysis result at the model (cfs) structure

Mode number	1	2	3	4	5
Frequency (Hz)	2.040	2.138	3.270	4.425	5.796
Modal damping ratio (ξ)	0.925	0.606	0.467	1.448	1.275

Fig. 9.8 Experimentally identified mode shapes of model (cfs) structure

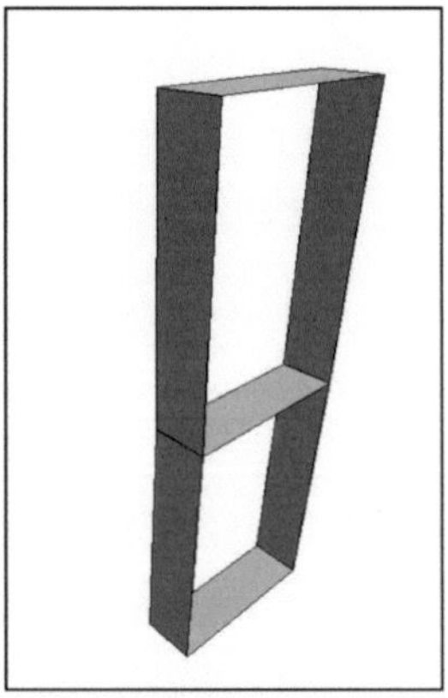

1st Mode Shape (f=2.040 Hz, ξ=0.925)

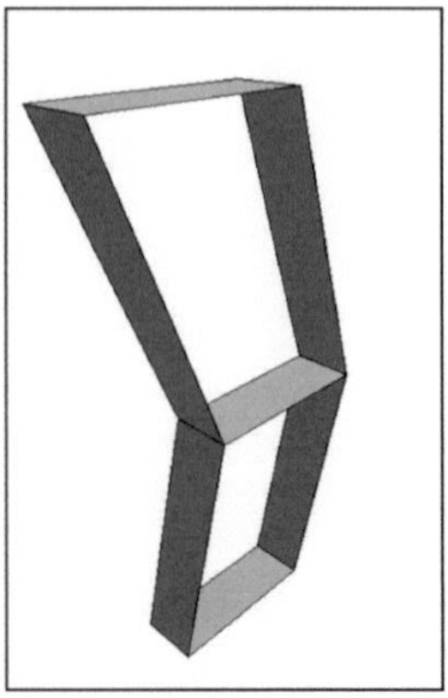

2nd Mode Shape (f=2.138 Hz, ξ=0.60

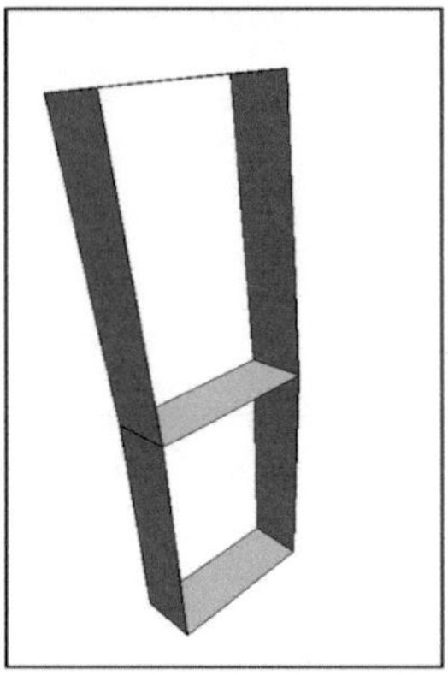

3rd Mode Shape (f=3.270 Hz, ξ=0.467)

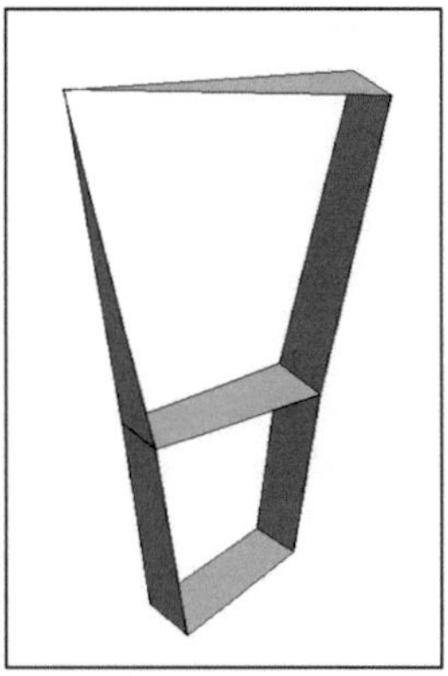

4th Mode Shape (f=4.425 Hz, ξ=1.448

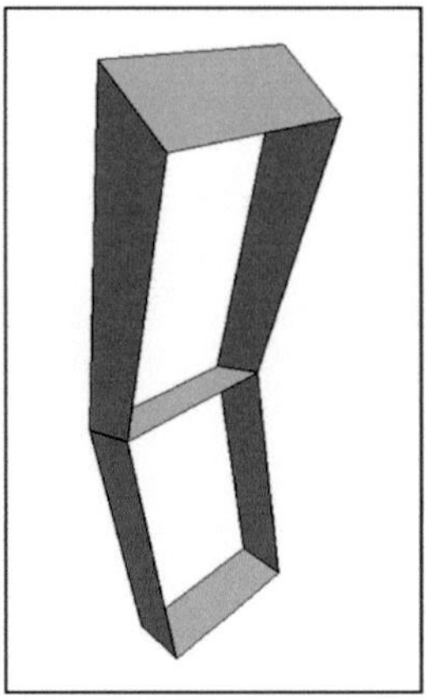

5th Mode Shape (f=5.971 Hz)

Table 9.4 Comparison of analytical and experimental modal analysis results

Mode number	1	2	3	4	5
Analytical frequency (Hz)	2.101	2.203	3.371	4.559	5.971
Experimental frequency (Hz)	2.040	2.138	3.270	4.425	5.796
Difference (%)	2.903	2.950	2.996	2.939	2.930

9.6 Conclusions

In this paper, analytical and experimental modal analysis of model (cfs) structure was presented. Comparing the result of study, the following observation can be made:

From the finite element model of model (cfs) structure a total of five natural frequencies were attained analytically, which range between 2 and 6 Hz. 3D finite element model of model (cfs) structure is constructed with SAP2000 software and dynamic characteristics are determined analytically. The ambient vibration tests are conducted under provided by shake table from ambient vibration data on ground level. Modal parameter identification was implemented by the Enhanced Frequency Domain Decomposition (EFDD) technique. Comparing the result of analytically and experimentally modal analysis, the following observations can be made:

From the finite element model of the model (cfs) structure, the first five mode shapes are attained analytically that range between 2 and 6 Hz.

From the ambient vibration test, the first five natural frequencies are attained experimentally, which range between 2 and 6 Hz.

When comparing the analytical and experimental results, it is clearly seen that there is best agreement between mode shapes.

Analytical and experimental modal frequencies differences between 2.903 and 2.996%.

Presented investigation results are shown and confirm of possibility using the recorded micro tremor data on ground level as ambient vibration input excitation data for investigation and application Operational Modal Analysis (OMA) on the bench-scale earthquake simulator (The Quanser Shake Table) for model (cfs) structures and shed light on the development of related research.

References

Alvin KF, Park KC (1994) Second-order structural identification procedure via state-space-based system identification. AIAA J 32(2):397–406

ANSI S2.47-1990 Vibration of buildings-Guidelines for the measurement of vibrations and evaluation of their effects on buildings

ARTeMIS Extractor, Structural Vibration Solutions, Aalborg, Denmark (1999)

Balmes E (1997) New results on the identification of normal modes from experimental complex modes. Mech Syst Signal Process 11(2):229–243

Bendat JS (1998) Nonlinear systems techniques and applications. Wiley, Hoboken

Brincker R, Ventura C (2015) Introduction to operational modal analysis. Wiley, Hoboken

Brincker R, Zhang L, Andersen P (2000) Modal identification from ambient responses using frequency domain decomposition. In: Proceedings of the 18th International modal analysis conference (IMAC), San Antonio, Texas, USA

Cunha A, Caetano E, Magalhaes F, Moutinho C (2005) From input-output to output-only modal identification of civil engineering structures. In: 1st International operational modal analysis conference (IOMAC), Copenhagen, Denmark

Friswell M, Mottershead JE (1995) Finite element model updating in structural dynamics. Springer Science-Business Media

HO B, Kalman RE (1966) Effective construction of linear state-variable models from input/output functions. at-Automatisierungstechnik, 14(1–12):545–548

Ibrahim SR (1977) Random decrement technique for modal identification of structures. J Spacecraft Rockets 14(11):696–700

Ibrahim SR, Miculcik EC (1977) A method for the direct identification of vibration parameters from the free response. Shock Vib Bull 47(4):183–194

Jacobsen NJ, Andersen P, Brincker R (2006) Using enhanced frequency domain decomposition as a robust technique to harmonic excitation in operational modal analysis. In: International conference on noise and vibration engineering (ISMA), Leuven, Belgium

Juang JN (1994) Applied system identification. Prentice Hall

Juang JN, Pappa RS (1985) An eigensystem realization algorithm for modal parameter identification and model reduction. J Guid Control Dyn 8(5):620–627

Juang JN, Cooper JE, Wright JR (1988) An eigensystem realization algorithm using data correlations (ERA/DC) for modal parameter identification. Control-Theory Adv Technol 4 (1):5–14

Juang JN, Phan M, Horta LGand, Longman RW (1993) Identification of observer/kalman filter markov parameters-theory and experiments. J Guid Control Dyn 16(2):320–329

Kalman RE (1960) A new approach to linear filtering and prediction problems. J Basic Eng 82 (1):35–45

Kasimzade AA (2006) Coupling of the control system and the system identification toolboxes with application in structural dynamics. In: International control conference (ICC2006), Glasgow, Scotland, UK

Kasimzade AA, Tuhta S (2005) Ambient vibration test on detection of dynamic properties of reinforced concrete building. In: 1st International operational modal analysis conference (IOMAC), Copenhagen, Denmark

Kasimzade AA, Tuhta S (2007a) Ambient vibration analysis of steel structure. In: 2nd international operational modal analysis conference (IOMAC), Copenhagen, Denmark

Kasimzade AA, Tuhta S (2007b) Ambient vibration analysis of steel structure. Experimental vibration analysis of civil engineering structures (EVACES'07), Porto, Portugal, October

Kasimzade AA, Tuhta S (2007c) Particularities of monitoring, identification, model updating hierarchy in experimental vibration analysis of structures. Experimental vibration analysis of civil engineering structures (EVACES'07), Porto, Portugal

Kasimzade AA, Tuhta S (2009) Optimal estimation the building system characteristics for modal identification. In: 3rd International operational modal analysis conference (IOMAC), Porto Novo, Ancona, Italy

Kasimzade AA, Tuhta S (2012) Stochastic parametrical system identification approach for validation of finite elements models. TWMS J Pure Appl Math 3(1):11–32

Ljung L (1999) System Identification: theory for the user. Prentice Hall, Upper Saddle River

Lus H, De Angelis M, Betti R, Longman RW (2003) Constructing second-order models of mechanical systems from identified state space realizations. Part I: theoretical discussions. J Eng Mech 129(5):477–488

Marwala T (2010) Finite element model updating using computational intelligence techniques: applications to structural dynamics. Springer Science-Business Media

Peeters B (2000) System identification and damage detection in civil engineering. Ph.D. dissertation, Katholieke Universiteit Leuven, Leuven, Belgium

Phan MQ, Longman RW, Lee SC, Lee JW (2003) System identification from multiple-trial data corrupted by non-repeating periodic disturbances. Int J Appl Math Comput Sci 13(2):185–192

Quanser (2008) Position control and earthquake analysis. Quanser Shake Table II User Manual, Nr 632, Rev 3.50, QuanserInc, Markham, Canada

Roeck GD (2003) The state-of-the-art of damage detection by vibration monitoring: the SIMCES experience. J Struct Control 10(2):127–134

SAP2000, Integrated finite element analysis and design of structures, computers and structures Inc, Berkeley, California, USA (1997)

Sestieri A, Ibrahim SR (1994) Analysis of errors and approximations in the use of modal coordinates. J Sound Vib 177(2):145–157

Trifunac MD (1972) Comparisons between ambient and forced vibration experiments. Earthquake Eng Struct Dynam 1(2):133–150

Tseng DH, Longman RW, Juang JN (1994a) Identification of the structure of the damping matrix in second order mechanical systems. Spaceflight Mechanics, pp 167–190

Tseng DH, Longman RW, Juang JN (1994b) Identification of gyroscopic and nongyroscopic second order mechanical systems including repeated root problems. Spaceflight Mechanics, pp 145–165

Van Overschee P, De Moor BL (1996) Subspace identification for linear systems: theory-implementation-applications. Springer Science-Business Media

Ventura CE, Schuster ND (1996) Structural dynamic properties of a reinforced concrete high-rise building during construction. Can J Civ Eng 23(4):950–972

Wenzel H, Pichler D (2005) Ambient vibration monitoring. Wiley, Hoboken

Chapter 10
Optimal Estimation Example the Dynamic Parameters from Ambient Vibration for Modal Identification

Azer A. Kasimzade, Muxlis Haciyev, Sertac Tuhta and Gencay Atmaca

Abstract A novel approach of system characteristic matrix's correction in modal identification from ambient vibration is presented. As a result of this approach, actual system characteristic matrices are determined more accurately with minimum error. It is reflected on to updating system parameters more reliable. In first approximation, actual system characteristic matrices determined by singular value decomposition of block Hankel matrix, which build from the response correlation matrix. In second approximation, to make the system characteristic matrices optimal definite, for black-box modeling the input–output relation of the system used Kalman theory. Covariance of the nonmeasurable process noise and measurement noise matrixes are contained in Riccati equation are determined by expressing Hankel matrix's multiplicities from eigensolution of the system state matrix obtained in previous iteration. Another word process and measurement noises covariance matrixes indirectly is constructed only from measured output data. These iterations are repeated until satisfying estimated error. As a result of these iterations, actual system characteristic matrices are determined more accurately with minimum error. Then, from determined system characteristic, matrices are extracted system modal parameters. These system modal parameters are used for the system modal updating for which direct and iterative methods are applied. Supporting to this algorithm realized code maybe interfaced with finite element codes.

Keywords Structural parameters · Ambient vibration · System identification Kalman filter

---

A. A. Kasimzade (✉) · M. Haciyev · S. Tuhta · G. Atmaca
Department of Civil Engineering, Faculty of Engineering,
Ondokuz Mayis University, Atakum/Samsun, Turkey
e-mail: azer@omu.edu.tr

S. Tuhta
e-mail: stuhta@omu.edu.tr

G. Atmaca
e-mail: gencayatmaca@hotmail.com

A. A. Kasimzade et al. (eds.), *Seismic Isolation, Structural Health Monitoring, and Performance Based Seismic Design in Earthquake Engineering*,
https://doi.org/10.1007/978-3-319-93157-9_10

10.1 Introduction

When forced excitation tests are very difficult or only response data are measurable, while the actual loading conditions are unknown, operation modal analysis (output-only modal identification techniques) remains the only technique for structural identification. The main advantage of this method is that no special, artificial type of excitation has to apply to the structure to determine its dynamic characteristics (Kasimzade and Tuhta 2012; Ge and Lui 2005; Wei 1990; Raol and Madhuranath 1996). Furthermore, if a structure has high period (more than one second) modes, it may be difficult to excite it with a shaker, whereas this is generally no problem for drop weight or ambient sources. However, if mass-normalized mode shapes are required, ambient excitation cannot be used. Output-only model identification techniques efficiently use with model updating tools to develop reliable finite element models of structures. Last year's Output-Only Model Identification studies of systems and results are given in appropriate references structural vibration solutions, partly in (Cunha et al. 2005). For the modal updating of the structure it is necessary to estimate sensitivity of reaction of examined system to change of parameters of a building (Kasimzade 2006; Kasimzade and Tuhta 2005, 2007a).

System identification is the process of developing or improving a mathematical representation of a physical system using experimental data (Bendat 1998; Ibrahim and Mikulcik 1997; Juang 1994; Peeters 2000; ARTeMIS 1999; Ljung 1999; Van Overschee and De Moor 1996). In engineering structures, there are three types of identification that are used: modal parameter identification; structural-modal parameter identification; control-model identification methods (Andersen et al. 2007; Brownjohn and Carden 2007; Gawronski 2004; Johansson 1993). In the frequency domain, the identification is based on the singular value decomposition of the spectral density matrix and it is denoted Frequency Domain Decomposition (FDD) and its further development Enhanced Frequency Domain Decomposition (EFDD). In the time domain, there are three different implementations of the Stochastic Subspace Identification (SSI) technique, which are used: Unweighted Principal Component (UPC); Principal component (PC); Canonical Variety Analysis (CVA).

Below, a novel approach of system characteristic matrix's correction in modal identification from ambient vibration is presented. In this algorithm, the first approximation actual system characteristic matrices' is determined by the data-driven stochastic subspace identification method. For an optimal estimation state vector, another word to make the system characteristic matrices optimal definite in second approximation are obtained by applying the steady-state Kalman filter to the stochastic state-space model equation (Kalman 1960; Labarre et al. 2004; Kasimzade et al. 2006; Kasimzade and Tuhta 2007a, b, 2009). All off calculations are repeated until satisfying estimation error condition.

10.2 Determination of Approximate Values of System Characteristic Matrices

As mentioned above, the purpose of operational modal analysis is to determine modal parameters of a real system, which is assumed to be linear time invariant, measuring only its response at specific locations, being the exciting load unknown.

In the vibration analysis identification of actual system (black-box model) supporting to experimental measurements determination n approximate value of system characteristic matrices $\left[\widehat{A}\right], \left[\widehat{B}\right], \left[\widehat{C}\right]$ instead of its appropriate unknown real matrices value are appeared, first of all. Analysis sequences for the evaluation of this aim are shown below.

The equation of motion of the continuous system are arranged

$$[m]\{\ddot{u}(t)\} + [c]\{\dot{u}(t)\} + [k]\{u(t)\} = \{F(t)\} = [d]\{f(t)\}\frac{n!}{r!(n-r)!} \tag{10.1}$$

These are transformed to the state-space former of first-order equations, i.e., a continuous-time state-space model of the system are evaluated as (Kowalczuk and Kozlowski 2000).

$$\{\dot{z}(t)\} = [A_c]\{z(t)\} + [B_c]\{f(t)\} \tag{10.2a}$$

$$[A_c] = \begin{bmatrix} [0] & [I] \\ -[m]^{-1}[k] & -[m]^{-1}[c] \end{bmatrix} \tag{10.2b}$$

$$[B_c] = \begin{bmatrix} [0] \\ [m]^{-1}[d] \end{bmatrix};\ \{z(t)\} = \begin{bmatrix} u(t) \\ \dot{u}(t) \end{bmatrix} \tag{10.2c}$$

If the response of the dynamic system is measured by the m_1, output quantities in the output vector $\{y(t)\}$ using sensors (such as accelerometers, velocity, displacements, etc.,), for system model represented by the Eqs. (10.2a–c), appropriate measurement-output equation become as

$$\begin{aligned}\{y(t)\} &= [C_a]\{\ddot{u}\} + [C_v]\{\dot{u}\} + [C_d]\{u\} \\ &= [C]\{z(t)\} + [D]\{f(t)\}\end{aligned} \tag{10.3a}$$

$$[C] = [[C_d] - [C_a][m]^{-1}[k] \quad [C_v] - [C_a][m]^{-1}[c]] \tag{10.3b}$$

$$[D] = [C_a][m]^{-1}[d] \tag{10.3c}$$

where $[m], [c], [k]$ mass, damping, stiffness matrices of the structure are building by finite element method (Kasimzade 2006); $\{u\}$ is the vector of displacement; $[A_c]$ is an $n_1(=2n_2;\ n_2$ is the number of in depended coordinates) by n_1 state matrix; $[d]$ is an n_2 by r_1 input influence matrix, characterizing the locations and type of known inputs $\{f(t)\}$; $[C_a], [C_v], [C_d]$ are output influence matrices for acceleration, velocity, displacement for using sensors (such as accelerometers, tachometers, strain gages, etc.,), respectively; $[C]$ is an $m_1 \times n_1$ output influence matrix for the state vector $\{z\}$ and displacement only; $[D]$ is an $m_1 \times r_1$ direct transmission matrix; r_1 is the number of inputs; m_1 is the number of outputs.

In the output-only modal analysis environment, the main assumption is that input force $\{F(t)\} = [d]\{f(t)\}$ comes from white noise or time impulse excitation. Under this hypothesis, discrete-time stochastic state-space model may be written as

$$\{z_{k+1}\} = [A]\{z_k\} + [B]\{f_k\} + \{w_k\} \tag{10.4}$$

$$\{y_k\} = [C]\{z_k\} + [D]\{f_k\} + \{v_k\} \tag{10.5}$$

where,

$\{z_k\} = \{z(k\Delta t)\}$ is the discrete-time state vector;

$\{w_k\}$ is the process noise due to disturbance and modeling imperfections;

$\{v_k\}$ is the measurement noise due to sensors' inaccuracies;
$\{w_k\}$, $\{v_k\}$ vectors are nonmeasurable, but assumed that they are white noise with zero mean. If this white noise assumption is violated, in other words, if the input contains also some dominant frequency components in addition to white noise, these frequency components cannot be separated from the eigenfrequencies of the system and they will appear as poles of the system matrix $[A]$.

As shown from measurement-output equation (10.3a–c), it indirectly is dependent on system model (2) and contain appropriate system mass, damping, rigidity matrices $[m], [c], [k]$, respectively. For this reason, to carry out measurement by the relation (10.5), it will be require known system model (2) with matrices $[m], [c], [k]$ previously for **zero** approximation. For **zero** approximation, these known matrices are denoted as $[m], [c], [k]$ and include them $[A_c], [B_c], [A], [B], [C], [D]$, respectively.

In the real structures, exited by ambient vibration, the input $\{f(t)\}, \{f_k\}$ remains unmeasured and, therefore, it disappears from Eqs. (10.2a–c)–(10.5), respectively. Then, to take into consideration this fact, the input is implicitly modeled by the noise terms $\{w_k\}$, $\{v_k\}$ and mentioned relation became as

$$\{z_{k+1}\} = [A]\,\{z_k\} + \{w_k\} \tag{10.6a}$$

$$\{y_k\} = [C]\,\{z_k\} + \{v_k\} \tag{10.7a}$$

The main goal of the operational modal analysis, identification the modal parameters of the system from the output vector $\{y_k\}$, measured by the sensors located on the structure, whose dimensions are $n_{\text{sensors}} \times n_{\text{samples}}$ where n_{sensors} and n_{samples} are the number of sensors and samples, respectively. In the practical engineering of stochastic methods, the signal $\{y_k\}$ (it may be displacement, velocity or acceleration) given by sensors can be sampled at discrete time intervals depending on the characteristic of the computer processing.

Measurements are realized by the formulas (10.6a, 10.7a), and then defined system matrices $\left[\widehat{A}\right]$, $\left[\widehat{C}\right]$ by the next sequences. Measuring the output vector $\{y_k\}$ and $\{y_s\}_{\text{ref}}$ (it is a subset of the output vector $\{y_{k+s}\}$) by the sensors located on the $s+k = 1 \div (s_* + k_*)$ characteristic and $s = 1 \div s_*$ reference points appropriately of the building structure, then denoted the expected value operator as $E(\ldots)$, the correlation function calculated as

$$[R_k] = E\left(\{y_{k+s}\}\{y_s\}_{\text{ref}}^T\right) = \frac{1}{s_* - k} \sum_{s=0}^{s_*-k-1} \{y_{k+s}\}\{y_s\}_{\text{ref}}^T \tag{10.8}$$

From which, the following block Hankel matrix build as

$$\left[H_{p,q}\right] = \begin{bmatrix} [R_1] & [R_2] & \ldots & [R_q] \\ [R_2] & [R_3] & \ldots & \left[R_{q+1}\right] \\ \ldots & \ldots & \ldots & \ldots \\ \left[R_p\right] & \left[R_{p+1}\right] & \ldots & \left[R_{p+q-1}\right] \end{bmatrix} \tag{10.9}$$

The factorization of the block Hankel data matrix realized by using singular value decomposition,

$$[H(0)] = [U]\left[\sum\right][V]^T \tag{10.10}$$

Removing eigen values nearly zero from matrix $[\sum]$, consequently defined its rank (n) and in accordance with matrices $[U_n]$, $\left[\sum_n\right]$, $[V_n]$ are evaluated, respectively.

The **first** approximation of the system matrices $\left[\widehat{A}\right]$, $\left[\widehat{B}\right]$, $\left[\widehat{C}\right]$ are calculated as

$$\left[\widehat{A}\right] = \left[\sum_n\right]^{-1/2} [U_n]^T [H(1)] [V_n] \left[\sum_n\right]^{-1/2} \tag{10.11}$$

$$\left[\widehat{B}\right] = \left[\sum_n\right]^{1/2} [V_n]^T [E_r] \tag{10.12}$$

$$\left[\widehat{C}\right] = [E_m]^T [U_n] \left[\sum_n\right]^{1/2} \tag{10.13}$$

$$[E_r]^T = \left([I_{r_1}] \quad [0_{r_1}] \quad \ldots \quad [0_{r_1}] \right), [E_{m_1}]^T = \left([I_{m_1}] \quad [0_{m_1}] \quad \ldots \quad [0_{m_1}] \right)$$

where $[H(1)]$ a shifted is block Hankel matrix; r_1 and m_1 are the number of inputs and outputs, respectively; $[I_i]$ is an identity matrix of order i; $[0_i]$ is a null matrix of order i.

10.3 System Characteristic Matrices Optimal Definite

Here, for reach theoretical target to find the best (or optimal) estimate $\{\widehat{z}_k\}$ in the sense that the estimation error $\{e_k\} = \{z_k\} - \{\widehat{z}_k\}$ is as small as possible, achievement of to make definite **first** approximate values of the building system characteristics $\left[\widehat{A}\right]$, $\left[\widehat{C}\right]$ to the stochastic black–box models' is applied Kalman theory. Theoretically, the Kalman filter is very attractive because it has a closed-form solution (i.e., Riccati equation) for its gain matrix. However, the Kalman filter requires information, including of covariance $E(\{w_k\}\{w_k\}^T) = [q\delta(k-j)] = [Q]$ of the non-measurable process noise $\{w(k)\}$ and covariance $E(\{v_k\}\{v_k\}^T) = [r\delta(k-j)] = [R]$ of the non-measurable measurement noise $\{v(k)\}$ (Nelson 2000). For this reason, it is necessary to estimate these matrices $([Q], [R])$ indirectly from measured output data $\{y_k\}$ and $\{y_s\}_{\text{ref}} = \{y_{k+s}\}$. One of the crucial points (Kasimzade 2006) in presented method, determination these matrices at least approximately $[Q] \cong [\overline{Q}]$, $[R] \cong [\overline{R}] = [U_n]$ by expressing Hankel matrix's multiplicities from eigensolution of the system state matrix $\left[\widehat{A}\right]$ as shown below (In references (Juang 1994) by Observer/Kalman Filter Identification method to avoid this problem to make an effort estimate more approximately Kalman filter gain directly from experimental data (Markov parameters) without estimating the covariance (P) of the process and measurement noises and solving Riccati equation).

Shortly, in this optimal make definition stage of matrices $\left[\widehat{A}\right]$, $\left[\widehat{C}\right]$ for the **first** approximation, the system model represented by the relation (10.6b, 10.7b) become as

$$\{z_{k+1}\} = \left[\widehat{A}\right]\{z_k\} + \{w_k\} \tag{10.6b}$$

$$\{y_k\} = \left[\widehat{C}\right]\{z_k\} + \{v_k\} \tag{10.7b}$$

Under above assumptions and represented system model, the solution of this problem may be reached if the error covariance $[P] = E(\{e_k\}\{e_k\}^T)$ will satisfy discrete algebraic Riccati equation. Sequences of solution of formulated problem are given below.

Solving eigen problem of the matrix $\left[\widehat{A}\right]$, consequently

$$([\Lambda],[\Psi]) = \mathrm{eig}\left(\left[\widehat{A}\right]\right) \tag{10.8}$$

And expressing it in the form $[\widehat{A}] = [\Psi][\Lambda][\Psi]^T$, then the Hankel matrix's multiplicities are express as

$$[H_0] = \left[[U_n]\left[\sum_n\right]^{1/2}[\Psi]\right]\left[[\Psi]^{-1}\left[\sum_n\right]^{1/2}[V_n]^T\right] \approx [\overline{P}][\overline{Q}] \tag{10.9}$$

where $[\overline{R}] = [U_n]$, $[\overline{P}] = \left[[U_n][\sum_n]^{1/2}[\Psi]\right]$, $[\overline{Q}] = \left[[\Psi]^{-1}[\sum_n]^{1/2}[V_n]^T\right]$

Supporting this result is evaluated Riccati equation

$$[P] = \left[\widehat{A}\right][P]\left[\widehat{A}\right]^T - \left[\widehat{A}\right][P]\left[\widehat{C}\right]^T\left[[\overline{R}] + \left[\widehat{C}\right][P]\left[\widehat{C}\right]^T\right]^{-1}\left[\widehat{C}\right][P]\left[\widehat{A}\right]^T + [\overline{Q}] \tag{10.10}$$

The existence of the Riccati equation solution is only possible, if the correlation function is positive definite. There are a few proposals in literature to guarantee a solution. But the existing experiences, it remains an open problem in large scale stochastic realization theory.

If a solution of the Riccati equation exists, after defining it $[P] = \left[\widehat{P}\right]$, one can obtain Kalman gain of the building structure model

$$\left[\widehat{K}\right] = \left[\widehat{A}\right]\left[\widehat{P}\right]\left[\widehat{C}\right]^T\left[[\overline{R}] + \left[\widehat{C}\right]\left[\widehat{P}\right]\left[\widehat{C}\right]^T\right]^{-1} \tag{10.11}$$

Kalman filter equation is evaluated

$$\{\widehat{z}_{k+1}\} = \left[\left[\widehat{A}\right] - \left[\widehat{K}\right]\left[\widehat{C}\right]\right]\{\widehat{z}_k\} + \left[\left[\widehat{B}\right] - \left[\widehat{K}\right][D]\right]\{f_k\} + \left[\widehat{K}\right]\{y_k\} \tag{10.12}$$

With the output measurement $\{y_k\}$ satisfying

$$\{y_k\} = \left[\widehat{C}\right]\{\widehat{z}_k\} + \{\varepsilon_k\} \tag{10.13}$$

The output residual $\{\varepsilon_k\}$ satisfies $\{\varepsilon_k\} = [\widehat{C}]\{e_k\} + \{v_k\}$.

Comparing building system modeling by the Kalman filter equations (10.12, 10.13) with the first step system modeling equations, so

$$\{z_{k+1}\} = \left[\widehat{A}\right]\{z_k\} + \{w_k\}; \quad \{y_k\} = \left[\widehat{C}\right]\{z_k\} + \{v_k\}$$

The error is evaluated as

- the state estimation error

$$\{e_k\} = \{z_k\} - \{\widehat{z}_k\}; \quad \{E(e_k)\} = \{0\} \tag{10.14}$$

- error dynamics (the state estimation error at steady-state reduces to)

$$\{e_{k+1}\} = \left[\left[\widehat{A}\right] - \left[\widehat{K}\right]\left[\widehat{C}\right]\right]\{e_k\} - \left[\widehat{K}\right]\{v_k\} + \{w_k\} \tag{10.15}$$

- output residual

$$\{\varepsilon_k\} = [\widehat{C}]\{e_k\} + \{v_k\}; \quad \{E(\varepsilon_k)\} = \{0\} \tag{10.16}$$

If condition (10.16) is not satisfied, all off calculations are repeated until satisfying this condition. For example, if not satisfied condition (10.16), in **second** approximation case relation (10.6a, 10.7a) became as

$$\{z_{k+1}\} = \left[\widehat{A}\right]\{z_k\} + \{w_k\}; \quad \{y_k\} = \left[\widehat{C}\right]\{z_k\} + \{v_k\} \tag{10.17}$$

Measurements in **second** approximation case are realized by these formulas, and then will defined system characteristic matrices $\left[\overset{\frown}{A}\right]$, $\left[\overset{\frown}{C}\right]$, by the operations (10.8–10.13).

Then, for the optimal make definition of the system matrices $\left[\overset{\frown}{A}\right]$, $\left[\overset{\frown}{C}\right]$ relation (10.6b, 10.7b) became as

$$\{z_{k+1}\} = \left[\overset{\frown}{A}\right]\{z_k\} + \{w_k\}; \quad \{y_k\} = \left[\overset{\frown}{C}\right]\{z_k\} + \{v_k\} \tag{10.18}$$

In all off relation (10.8–10.16) instead of matrices $\left[\widehat{A}\right]$, $\left[\widehat{C}\right]$, will take place matrices $\left[\overset{\frown}{A}\right]$, $\left[\overset{\frown}{C}\right]$, respectively for this **second** approximation case. As a result of this iterations [when satisfied condition (10.16)], are definite real building system parameters, another word, system identified completely and obtained system matrices will mark as $[A]$, $[C]$ for the illustrating of operations in next section (Kasimzade and Tuhta 2012; Phan et al. 2003). Obtained modal parameters-damping, period (which contain the eigenvalue matrix), mode shape using as a reference modal "experimental" data (in mean that they are defined supporting experimental result) the analytical (finite element model) (Chen 2001) stiffness, damping, and mass matrices are corrected by the known direct (Caesar 1986)

Fig. 10.1 Mobile structural monitoring system and steel frame benchmark structure

and iterative updating methods (Link 1993) using convergence criteria (Dascotte and Vanhonacker 1989). For speed up updating procedure, marked updating parameters previously are fuzzyfied in the user definable intervals, these parameters using full factorial and orthogonal array testing on the base finite element method (Kasimzade 2006; Kasimzade and Tuhta 2005, 2007a, b, 2009, 2012; Kasimzade et al. 2006b). Results of those research partly was verified on the steel frame benchmark model (Fig. 10.1).

10.4 Analytical Modal Analysis of Model Aluminum Structure

A finite element model was generated in SAP2000 (1997). Beams and columns were modeled as 3D beam-column elements (in Fig. 10.2 shown by the black color). Structure modeled as an absolutely rigidity floor (rigid diaphragm). The selected structure is modeled as a space frame structure with 3D elements. Beams and columns were modeled as 3D beam-column elements which have degrees of freedom. At the base of the structure in the model, the ends of every element were fixed against translation and rotation for the 6-DOF then creating finite element model of the structure in SAP2000. The following assumptions were taken into account. Model aluminum structure is modeled using an equivalent thickness and shell elements with isotropic property. All supports are modeled as fully fixed. The members of aluminum frame are modeled as rigidly connected together at the intersection points. In modeling of beams and columns, the modulus of elasticity $E = 6.963\text{E}10\ \text{N/m}^2$, Poisson ratio $\mu = 0.33$, mass per unit volume $\rho = 27{,}145\ \text{N/m}^3$.

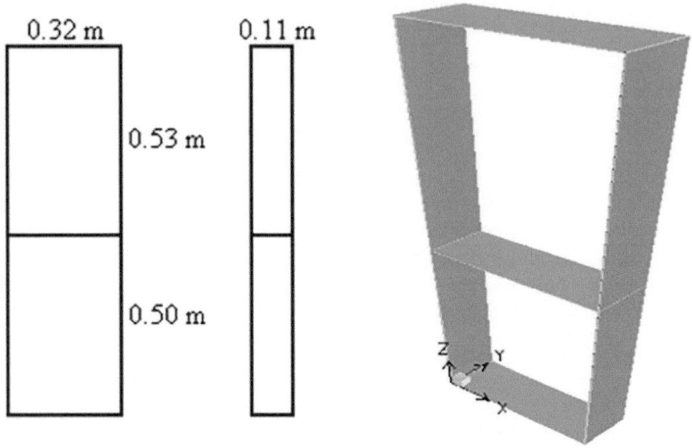

Fig. 10.2 Finite element model of model aluminum structure

Table 10.1 Analytical modal analysis result at the FE model

Mode number	1	2	3	4	5
Frequency (Hz)	1.651	4.816	7.619	8.083	9.004

Natural frequencies and vibration modes are concerned, a significant impact on the dynamic performance of buildings is an important dynamic property. A total of five natural frequencies of the structure are attained which range between 1 and 10 Hz. Analytical modal analysis results at the finite element model are shown in Table 10.1.

10.5 Experimental Modal Analysis of Model Aluminum Structure

Three accelerometers (with both *x* and *y* directional measures) were used for the ambient vibration measurements, one of which were allocated as reference sensor always located in the first floor (they are shown by the red line in Fig. 10.3a, b). Two accelerometers were used as roving sensors (they are shown by the black line in Fig. 10.3a, b). The response was measured in two data sets (Fig. 10.3a, b). For two data sets were used 3 and 5 degrees of freedom records, respectively (Fig. 10.3a, b). The selected measurement points and directions are shown in Fig. 10.3a, b.

When all measurements are examined, it can be seen that the best accordance is found between experimental frequencies. When the analytically and experimentally

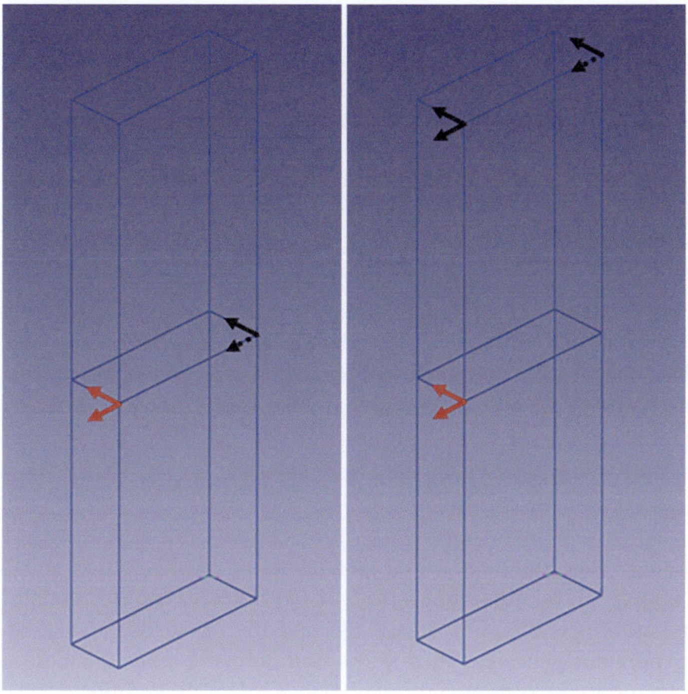

(a) First setup (b) Second setup

Fig. 10.3 Accelerometers location of experimental model in the 3D view

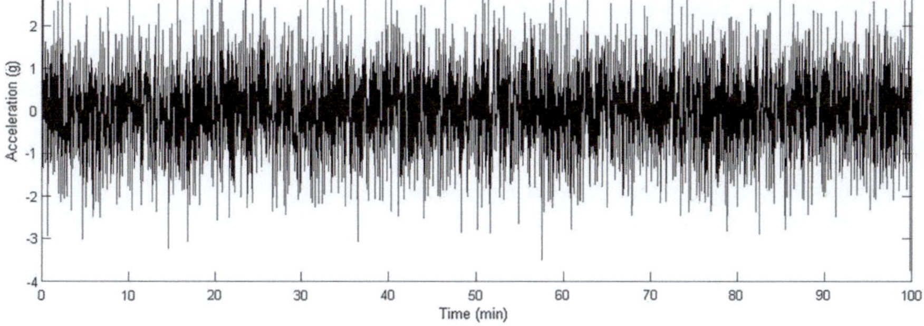

Fig. 10.4 Input excitation-time

identified dynamic parameters are checked with each other, it can be seen that there is a best agreement between the natural frequencies in experimental and analytical modal analyses (Tables 10.1 and 10.2).

Table 10.2 Experimental modal analysis result at the model aluminum structure

Mode number	1	2	3	4	5
Frequency (Hz)	1.601	4.674	7.388	7.853	8.745
Modal damping ratio (ξ)	1.947	1.880	1.395	2.178	1.472

10.6 Conclusion

System characteristic matrix's correction in modal identification from ambient vibration is presented. In this algorithm, the first approximation actual system characteristic matrices' determined by the data-driven stochastic subspace identification method. For an optimal estimation state vector, another word to make the system characteristic matrices optimal definite in second approximation are obtained by applying the steady-state Kalman filter to the stochastic state-space model equation. All off calculations are repeated until satisfying estimation error condition.

As a result of this approach, actual system characteristic matrices are determined more accurately with minimum error. It is reflected on to updating system parameters more reliable. For speed up updating procedure, marked updating parameters previously are fuzzyfied in the user definable intervals these parameters using full factorial and orthogonal array testing on the base finite element method. Supporting to this algorithm realized code maybe interfaced with finite element codes (Kasimzade and Tuhta 2012). From the finite element model of model aluminum structure, a total of five natural frequencies were attained analytically, which range between 1 and 10 Hz. 3D finite element model of model aluminum structure is constructed with SAP2000 software and dynamic characteristics are determined analytically. From the finite element model of the model aluminum structure, the first five natural frequencies are attained analytically that range between 1 and 10 Hz.

- From the ambient vibration test, the first five natural frequencies are attained experimentally, which range between 1 and 9 Hz.
- When comparing the analytical and experimental results, it is clearly seen that there is best agreement between natural frequencies.
- Analytical and experimental modal frequencies differences between 2.845 and 3.031%.

References

Andersen P, Brincker R, Goursat, Mevel L (2007) Automated modal parameter estimation for operational modal analysis of large systems. In: 2nd International operational modal analysis conference (IOMAC), Copenhagen, Denmark, April

ARTeMIS Extractor, Structural vibration solutions, Aalborg, Denmark (1999)
Bendat JS (1998) Nonlinear systems techniques and applications. Wiley, Hoboken
Brownjohn J, Carden P (2007) Reliability of frequency and damping estimates from free vibration response. In: 2nd International operational modal analysis conference (IOMAC), Copenhagen, Denmark
Caesar B (1986) Update and identification of dynamic mathematical models. In: 4th International modal analysis conference, Los Angeles, California, USA
Chen G (2001) FE model validation for structural dynamics. Ph.D. dissertation, University of London, London, UK
Cunha A, Caetano E, Magalhaes F, Moutinho C (2005) From input-output to output-only modal identification of civil engineering structures. In: 1st International operational modal analysis conference (IOMAC), Copenhagen, Denmark
Dascotte E, Vanhonacker P (1989) Development of an automatic mathematical model updating program. In: 7th International modal analysis conference (IMAC), Las Vegas, Nevada, USA
Gawronski W (2004) Advanced structural dynamics and active control of structures. Springer Science-Business Media
Ge M, Lui EM (2005) "Structural damage identification using system dynamic properties. Comput Struct 83(27):2185–2196
Ibrahim SR, Miculcik EC (1977) A method for the direct identification of vibration parameters from the free response. Shock Vib Bull 47(4):183–194
Johansson R (1993) System modeling and identification. Prentice Hall, Upper Saddle River
Juang JN (1994) Applied system identification. Prentice Hall, Upper Saddle River
Kasimzade AA (2006) Coupling of the control system and the system identification toolboxes with application in structural dynamics. In: International control conference (ICC2006), Glasgow, Scotland, UK
Kasimzade AA, Tuhta S (2005) Ambient vibration test on detection of dynamic properties of rein forced concrete building. In: 1st International operational analyses conference, Copenhagen, Denmark, pp 345–353
Kasimzade AA, Tuhta S (2007a) Ambient vibration analysis of steel structure. Experimental vibration analysis of civil engineering structures (EVACES'07), Porto, Portugal
Kasimzade AA, Tuhta S (2007b) Particularities of monitoring, identification, model update ing hierarchy in experimental vibration analysis of structures. Experimental vibration analysis of civil engineering structures (EVACES'07), Porto, Portugal
Kasimzade AA, Tuhta S (2009) Optimal estimation the building system characteristics for modal identification. In: 3rd International operational modal analysis conference (IOMAC), Porto Novo, Ancona, Italy, May
Kasimzade AA, Tuhta S (2012) Stochastic parametrical system identification approach for validation of finite elements models. TWMS J Pure Appl Math 3(1):11–32
Kasimzade AA, Tuhta S, Uslu A (2006) Finite element model updating of reinforce concrete building use of ambient vibration measurements. In: First European conference on earthquake engineering, Geneva, Switzerland
Kowalczuk Z, Kozlowski J (2000) Continuous-time approaches to identification of continuous-time systems. Automatica 36(8):1229–1236
Labarre D, Grivel E, Najim M, Todini E (2004) Two-kalman filters approach for unbiased ar parameter estimation from noisy observations, application to speech enhancement. In: Signal processing conference, Vienna, Austria
Link M (1993) Updating of analytical models-procedures and experience. In: Modern practice in stress and vibration analysis conference, Sheffield, UK
Ljung L (1999) System identification: theory for the user. Prentice Hall, Upper Saddle River
Nelson A (2000) Non-linear estimation and modeling of noisy time-series by dual kalman filtering methods. Ph.D. dissertation, Oregon Graduate Institute of Science, Oregon
Kalman RE (1960) A new approach to linear filtering and prediction problems. J Basic Eng 82(1): 35–45

Peeters B (2000) System identification and damage detection in civil engineering. Ph.D. dissertation, Katholieke Universiteit Leuven, Leuven, Belgium

Phan MQ, Longman RW, Lee SC, Lee JW (2003) System identification from multiple-trial data corrupted by non-repeating periodic disturbances. Int J Appl Math Comput Sci 13(2):185–192

Raol JR, Madhuranath H (1996) Neural network architectures for parameter estimation of dynamical systems. IEE Proc: Control Theory Appl 143(4):387–394

SAP2000, Integrated finite element analysis and design of structures, computers and structures Inc, Berkeley, California, USA (1997)

Van Overschee P, De Moor BL (1996) Subspace identification for linear systems: theory-implementation-applications. Springer Science-Business Media

Wei FS (1990) Structural dynamic model improvement using vibration test data. AIAA J 28(1): 175–177

Part III
Performance Based Seismic Design

Chapter 11
Performance-Based Seismic Design of Tall Buildings—A USA Perspective

Farzad Naeim

Abstract An overview of current performance-based methodologies utilized for design of tall buildings is presented. The reasons why common prescriptive code provisions are incapable of addressing the needs of tall building design engineers are explained. The performance objectives commonly associated with tall building design are identified and the evolution of current component-based performance objectives to a more rigorous and fully probabilistic approach to performance-based design is discussed. Modeling and acceptance criteria associated with various performance-based design guidelines are explained and special issues such as selection and scaling of ground motion records, soil–foundation–structure interaction issues, and seismic instrumentation and peer review needs are discussed.

Keywords Performance-based method · Tall buildings

11.1 Introduction

This paper is an updated version of the paper accompanying the author's keynote lecture at the 14th European Conference on Earthquake Engineering (Naeim 2010). It provides an overview of standard-of-practice for performance-based seismic design of tall buildings. The paper has a United States based tilt simply because that is where the author is based and most of his practice and research is focused on. However, the concepts and methods reviewed in this paper are equally applicable to any seismic region in the world if proper modifications are made to incorporate regional or local seismicity, engineering practices, and construction quality issues.

We begin by attempting to answer a number of frequently asked questions with respect to tall buildings. Then we proceed with an overview of various performance

---

F. Naeim (✉)
Farzad Naeim, Inc., Irvine, CA, USA
e-mail: farzad@fnaeim.com

A. A. Kasimzade et al. (eds.), *Seismic Isolation, Structural Health Monitoring, and Performance Based Seismic Design in Earthquake Engineering*,
https://doi.org/10.1007/978-3-319-93157-9_11

criteria, modeling procedures, acceptance criteria, ground motion issues, peer review requirements and finish with a brief discussion of seismic instrumentation needs.

11.2 What Is a Tall Building?

Tall buildings are easy to identify but difficult to define. You know one when you see it. There is no universally accepted definition for a tall building.

One way to classify a tall building is to use the overall height, or the height of the highest occupied floor measured from the ground surface. The taller the building, the larger gravity load the vertical members of the structural system have to carry, the more shortening the columns may experience, and for a given cross-section, the members will have less reserve capacity to carry the additional forces and moments imposed by lateral load (i.e., wind or seismic). There can be, however, no clear cut-off point to designate a building as "tall" using this definition. Using the number of floors as an indicator of "tallness" suffers from the same limitations. Council on Tall Buildings and Urban Habitat (CTBUH) maintains an up to date database of "tall buildings" according to the height criteria and number of floors (Council on Tall Buildings and Urban Habitat 2010). Many building codes or standards limit application of certain structural systems to certain height. For example, ASCE 7-10 (American Society of Civil Engineers 2010) prohibits the use of reinforced concrete shear wall only systems (and several other bracing systems) in regions of high seismicity for buildings taller than 160 feet (about 50 m) or 240 feet (75 m).

Another way to classify a building as tall is to measure its largest aspect ratio (the ratio of its overall height to its smaller plan dimension at the base. This approach has the engineering advantage of providing a crude insight as to the importance of design to resist overturning moments caused by lateral forces when the aspect ratio becomes large. However, there is no established aspect ratio barrier for defining tall buildings.

Dynamic characteristics of a building such as prevalence of higher modes in seismic response and/or fundamental period longer than a certain value (say 1.0 s) are more useful to engineers; however, they have not resulted in a generally accepted definition of a tall building either.

11.3 Are Tall Buildings Particularly Vulnerable to Earthquake Ground Motions?

The debate over whether tall and flexible building structures are particularly vulnerable to earthquake ground motions has persisted for a long time (Heaton et al. 1995; Naeim and Garves 2005). What works in favor of tall (flexible) structures is that earthquakes generally release significantly less energy at the longer periods (i.e., 3 s and longer) associated with the fundamental periods of taller, more flexible structures, than they do in the short period range (i.e., 0.2–1.0 s) associated with the fundamental period of stiffer, shorter buildings (Naeim and Garves 2005). What works against them is that earthquakes with larger magnitude release more energy in the long period range than smaller earthquakes and if these earthquakes occur in close proximity of tall buildings then significant directivity and near-source and basin effects could amplify the impact of the earthquake on taller, more flexible, structures. Both sides of the argument probably agree that, generally speaking, if the earthquake problem is viewed from a probabilistic perspective, tall buildings will perform better than similarly designed short (stiff) buildings while if the earthquake problem is looked at from a worst-case scenario, deterministic, perspective for large earthquakes occurring nearby or with very long duration, tall buildings may not be in such a good position and some of them could suffer partial or full collapse.

11.4 Should Tall Buildings Be Treated like Other Buildings?

A tall building represents a significant investment of human and material resources and may be occupied by hundreds, if not thousands, of occupants. Building codes' reaction to this fact, at least in the United States, has been twofold. First, application of certain structural systems has been limited to certain heights. For example, ASCE 7-10 (American Society of Civil Engineers 2010) does not permit the use of certain lateral load resisting systems for buildings taller than code prescribed heights in regions of high seismicity. Second, buildings with high occupancy are required to be designed for larger lateral forces via the use of an importance factor (I) which is taken as unity for ordinary buildings. For example, ASCE 7-10 requires that buildings where more than 300 people congregate in one area or the overall occupancy is more than 5000 persons to be designed using $I = 1.25$. This provision, however, is commonly interpreted in a way that indicates an $I = 1.0$ for most tall buildings. Therefore, you could have a tall building occupied by 4900 people yet designed with $I = 1.0$.

Many, if not most, code imposed height limits on lateral systems are difficult, if not impossible, to rationally justify. Numerous studies and evaluations (Los Angeles Tall buildings structural design Council 2007, 2008, 2009) have shown that it is possible and economical to design tall buildings as safe or safer than code designed buildings while ignoring code imposed height limits. Furthermore, even a strict imposition of an arbitrary 25% or higher premium on elastic design forces does not do much to address the issues of damage and potential collapse which are inherently inelastic and nonlinear.

Prescriptive codes by in large contain a collection of empirical rules and experimental results that have evolved over many years of practice. While these rules and procedures, when followed, have resulted in buildings that have been generally safe and have exhibited a margin of safety larger than that indicated by design analysis calculations, there is no way to quantify the margin of safety provided by following code procedures. By following prescriptive rules that are not tied to a particular performance level we are closing are eyes to detailed building and region specific demand and capacity concerns. Codes provide a "one size fits all" approach to seismic design. Tall buildings as a small class of specialized structures will perform better during earthquakes if special attention is afforded to their individual seismic behavior and engineers are provided with ample opportunities to explore new frontiers, utilize state of the art technologies and the latest research results in order to improve the performance, feasibility, and constructability of such buildings. Prescriptive building codes are simply incapable of offering such attributes.

11.5 Why Performance-Based Design Is a Necessity for Tall Buildings?

For over half a century the implicit objective of prescriptive building codes, at least in the United States, has been to produce buildings which resist minor earthquakes with little or no structural damage; moderate earthquakes with repairable structural damage; and major earthquakes with severe structural damage but no loss of life or limb. While statement of this objective can be found in either preamble or commentary of most prescriptive codes, there is hardly any provisions in the body of the codes that is in one way or another tied to the stated performance objectives. Engineers have been brought up to believe that if they follow the code prescriptive rules, the objectives will be automatically achieved. After many major earthquakes, this collective engineering psyche has suffered a number of blows and has resulted in changes in the code prescriptive rules. For example, in the aftermath of the 1971 San Fernando earthquake massive changes were made to code provisions for design of reinforced concrete members and the 1994 Northridge earthquake caused a rethinking of code's approach to design or steel moment-resisting beam-column connections.

There are other reasons why performance-based design of tall buildings has gathered momentum. The overwhelming majority of construction in United States and worldwide consists of low-rise buildings. According to Portland Cement Association (2000), buildings with one to three floors represent 93% of floor area of construction in United States while buildings with 14 floors or more represent only 1% of floor area of construction. With so much of the construction effort concentrated on low-rise construction it is not surprising that the code writers have these buildings in mind when crafting code provisions. As a result, many of the provisions included in a typical building code either do not have relevance to tall building design, or even worse, do not make much sense for design of tall

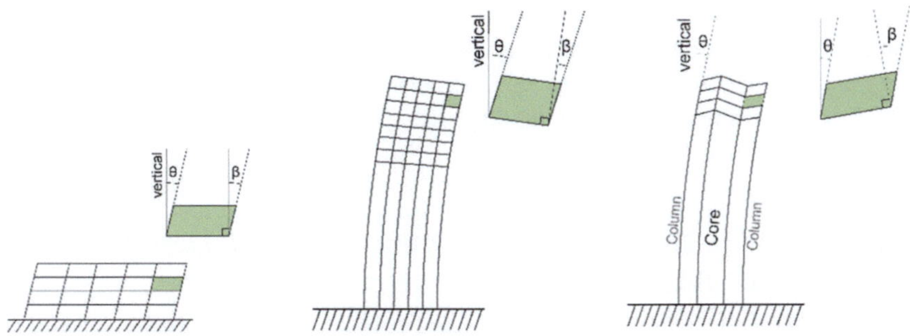

(a) Low rise building: racking deformation angle equals storey drift ratio

(b) 'Tube' high-rise building: racking deformation angle is smaller than the storey drift ratio

(c) Wall-frame high-rise building: racking deformation angle can exceed storey drift ratio

Fig. 11.1 Differences in overall story drift and racking story drift in a tall building (Council on Tall Buildings and Urban Habitat 2008) (reproduced with permission from the Council on Tall Buildings and Urban Habitat)

buildings. For example, until a few years ago, the Los Angeles Building Code had a very peculiar drift design provision (International Conference of Building Officials 2002) requiring story drift not to exceed $0.020/T^{1/3}$ where T is the fundamental vibration period of the building. This provision, which was later retracted, probably did not have a serious effect on design of low-rise buildings but was a huge straightjacket for design of tall buildings with long vibration periods.

Another shortcoming of current prescriptive codes is that they do not distinguish between the racking component of interstorey drift which can lead to significant damage and/or collapse and the rigid body displacement associated with the "rotation" of a tall building as a whole at upper levels caused by axial deformation of columns and walls which generally induces little damage if walls and columns are proportioned to resist the corresponding tensions and compressions (see Fig. 11.1).

Tall buildings present significant monetary investments and command higher engineering fees. Their owners and developers are kin to maximize the benefit of their investment by making the structural system of their building as cost-effective as possible. Larger engineering fees and usually more sophisticated design engineers open the window for critical evaluation of prescriptive code provisions and use of advanced systems and technologies which are either not permitted or not rewarded by a typical prescriptive building code approach.

Last but not least, damage or collapse of a tall building has far more adverse consequences to life and well-being of communities compared to damage or collapse of a small low-rise building. Therefore, owners, developers, potential occupants, and building officials are increasingly more receptive to the idea of requiring a better performance from tall buildings than that expected from ordinary construction.

It was the culmination of all of the above factors, reliance on validated experimental data, and recent advances in our analytical and computational capabilities that made the performance-based seismic design of tall buildings a reality in the first two decades of the twenty-first century.

11.6 What Is Involved in Performance-Based Design of Tall Buildings?

Several procedures need to be established to achieve a meaningful performance-based seismic design of a tall building, namely

1. A set of reasonable performance objectives must be defined
2. A set of rational design procedures must be contemplated
3. A set of sound performance evaluation procedures must be in place, and
4. A set of earthquake ground motion records, consistent with the hazard levels considered, must be selected and processed so that they can be applied towards building performance evaluations.

We will discuss each if the above requirements in the subsections that follow.

11.7 Establishment of Performance Objectives

In 1999, Structural Engineers Association of California (SEAOC) developed perhaps the first conceptual frame work for establishment of performance objectives for design of new buildings (Structural Engineers Association of California (SEAOC) 1999). According to this document, the earthquake levels identified in Table 11.1 were considered proper for performance-based design and evaluation.

Efforts spearheaded by the Federal Emergency Management Agency (FEMA) and carried out by the Applied Technology Council (ATC) culminated in publication of a seminal pre-standard for performance-based seismic rehabilitation of existing structures (Federal Emergency Management Agency (FEMA) 2000). This

Table 11.1 Earthquake levels and associated performance objectives suggested by the 1999 SEAOC document

Event	Recurrence interval (years)	Probability of exceedance	Performance objective
Frequent	43	50% in 30 years	Fully operational
Occasional	72	50% in 50 years	operational
Rare	475	10% in 50 years	Life safe
Very rare	975	10% in 100 years	Near collapse

document was later modified and republished by the American Society of Civil Engineers (ASCE) as a performance-based standard, the latest edition of which is the ASCE 41-13 (American Society of Civil Engineers 2013). ASCE 41-13 defines performance objectives in terms of acceptable performance of structural systems and components as well as nonstructural systems, attachments, and contents. It utilizes three basic performance levels termed Immediate Occupancy (IO), Life Safety (LS) and Collapse Prevention (CP) and ranges of performance bridging the identified performance levels. For each performance level, the anticipated behavior of structural and nonstructural components and contents are identified. As far as seismic hazard is concerned, ASCE-41 places its emphasis on two probabilistic earthquake levels: (1) a 10% in 50 years (475 year mean recurrence interval) event usually associated with the LS performance objective, and (2) a 2% in 50 years (2475 year mean recurrence interval) event usually associated with the CP performance objective.

ASCE 41-13 recognizes four distinct analytical procedures: (1) the linear elastic static analysis procedure (LSP), (2) linear dynamic procedure (LDP) commonly carried out in terms of response spectrum analysis, (3) the nonlinear static procedure (NSP) commonly referred to as the push-over analysis, and (4) the dynamic nonlinear response analysis (NSP). Due to limitations of applicability of LSP and NSP to tall buildings, current performance-based guidelines for design of tall buildings, which we will discuss later, generally permit application of LDP for evaluation of performance where building is anticipated to remain essentially elastic such as the IO performance level and NSP for evaluation of performance for nonlinear stages of building behavior (LS and CP).

11.8 The Component-Based Approach

The current practice of performance-based design of tall buildings in the United States relies on a component-based evaluation as delineated in the ASCE 41-13 document. In the component-based approach, each component of the building (beam, column, wall segment, etc.) is assigned a normalized force/moment—deformation/rotation relation such as the one shown in Fig. 11.2 where segment AB indicates elastic behavior, point C identifies the onset of loss of capacity, segment DE identifies the residual capacity of the component, and point E identifies the ultimate inelastic deformation/rotation capacity of the component. Components are classified as primary (P) or secondary (S) and assigned with different deformation limits corresponding to various performance objectives. The vertical axis in this figure represents the ratio of actual force or moment to the yield force or moment. Primary components are those whose failure results in loss of vertical load carrying capacity and endanger the safety of occupants (i.e., columns). Secondary elements are those whose failure can be sustained by the system via redistribution of forces to adjacent components without endangering the occupants (i.e., link beams in a coupled wall system).

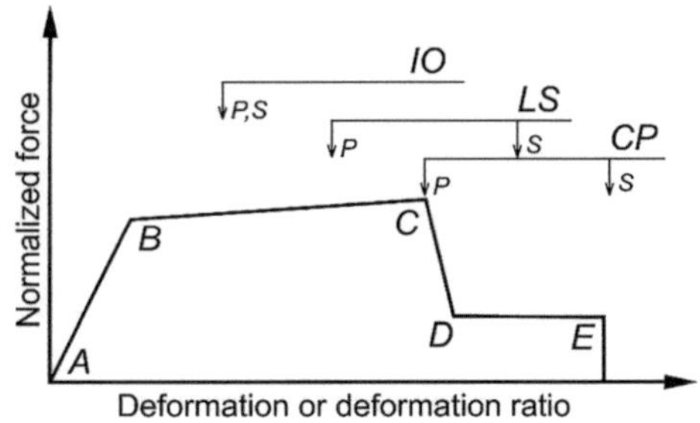

Fig. 11.2 Generalized component force-deformation relations for depicting modeling and acceptance criteria in ASCE 41-13 (Federal Emergency Management Agency (FEMA) 2000)

Although ASCE 41-13 is officially intended for seismic rehabilitation of existing structures, its component-based performance limits for NDP are routinely referenced by guidelines for performance-based design of tall buildings. Engineers, who believe that ASCE 41-13 tabulated limits are not applicable or too conservative for their intended component, perform laboratory testing and obtain confirmation of behavior for their component subject to approval by peer reviewers and approval agencies.

The Los Angeles Tall Buildings Structural Design Council (LATBSDC) was the first professional group in the United States to publish a performance-based alternative seismic analysis and design criteria specifically intended for tall buildings (Los Angeles Tall Buildings Structural Design Council (LATBSDC) 2005) and to obtain approval by the city's building officials in 2010 for its use in lieu of using prescriptive code provisions for buildings of all types and heights. The 2015 edition of LATBSDC criteria (Los Angeles Tall Buildings Structural Design Council (LATBSDC) 2015) sets two performance objectives: (1) serviceable behavior when subjected to frequent earthquakes defined as events having a 50% probability of being exceeded in 30 years (43-year return period); and (2) a very low probability of collapse under extremely rare earthquakes defined as events having a 2% probability of being exceeded in 50 years (2475 year return period) with a deterministic cap. This earthquake is the Risk-Targeted Maximum Considered Earthquake (MCE_R) as defined by ASCE 7-10.

The intent of LATBSDC's serviceability performance objective is to make sure that the building structural and nonstructural components retain their general functionality during and after frequent events. Repairs, if necessary, are expected to be minor and could be performed without substantially affecting the normal use and functionality of the building. Under frequent earthquakes the building structure and nonstructural components associated with the building are expected to remain essentially elastic. Essentially elastic response may be assumed for elements when force demands generally do not exceed provided strength. When demands exceed provided strength, this exceedance shall not be so large as to affect the residual strength or stability of the structure.

The intent of LATBSDC's collapse prevention objective is to validate that collapse does not occur when the building is subjected to MCE_R ground motions and the structure can be repaired following such events. Demands are checked against both structural members of the lateral force resisting system and other structural members. Claddings and their connections to the structure must accommodate MCE_R displacements without failure.

Performance-based tall building design guidelines published by other entities such as CTBUH (2008) and PEER (2010) have followed performance objectives similar to those expressed in the 2015 LATBSDC document. Effort is underway to develop a second edition of the PEER document and a 2017 edition of the LATBSDC document and it is expected that the similarities between these two documents will increase in these upcoming editions.

While component-based approaches are convenient and widely used, they are incapable of assessing the performance of the building as a whole. There is a real difference between a building with hundreds of columns where one column exceeds the CP limit and where all columns in a particular level exceed the CP limit and create a potential for collapse. Component-based methods are incapable of making that distinction and leave such a crucial distinction to the so-called "engineering judgment."

11.9 The System-Based Approach

The Pacific Earthquake Engineering Research Center (PEER) performance-based design framework has made it possible to evaluate the seismic performance of buildings and their attachments and consents in a system-wide rigorous and probabilistic approach. Cornell and Krawinkler (2000) first suggested this approach by stating that "the basis for assessing adequacy of the structure or its design will be a vector of certain key *Decision Variables, DV*, such as the annual earthquake loss and/or the exceedance of one or more limit states (e.g., collapse). These can only be predicted probabilistically. Therefore, the specific objectives of engineering assessment analyses are in effect quantities such as $\lambda_{\$(x)}$, the mean annual frequency (MAF) of the loss exceeding x dollars, or such as $\lambda_{I\mathrm{coll}}$, the MAF of collapse."

Later, this definition was expanded (Krawinkler and Miranda 2004; Moehle and Deierlein 2004) to include the vector of engineering demand parameters, *EDP*, resulting in the now famous triple integral of Eq. 11.1 with the process shown conceptually in Fig. 11.3.

$$\lambda(DV) = \iiint G\langle DV|DM\rangle dG\langle DM|EDP\rangle dG\langle EDP|IM\rangle d\lambda(IM) \tag{11.1}$$

A comprehensive implementation of this approach is contained in the FEMA P-58 reports (Federal Emergency Management Agency (FEMA) 2012a, b) and a companion software tool (Performance Assessment Calculation Tool—PACT 2.0 (Naeim and Hagie 2012). Using this tool and FEMA-P58 methodology it is possible to assess probabilities of various outcomes for a given scenario, intensity of ground motion, or on an annualized time-based basis. Using this tool, once the

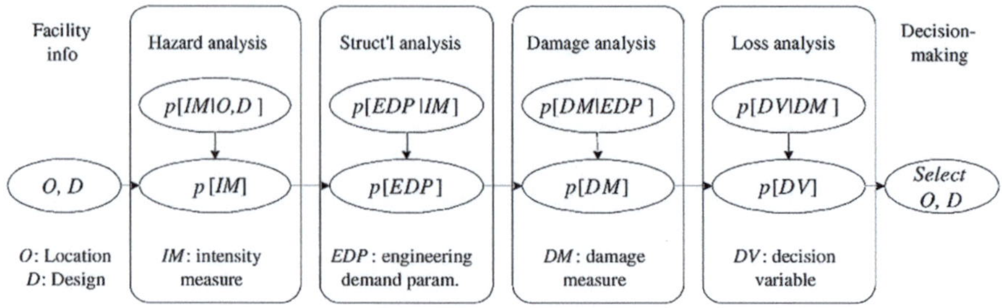

Fig. 11.3 PEER's probabilistic frame work for performance-based design (after Cornell and Porter, Courtesy of Moehle and Deierlein)

fragility specifications for the building core, shell, improvements, and contents (including fragility curves and repair cost, repair time, and casualty consequences) are defined, seismic hazard curve, and earthquake demand parameters in terms of a series of linear and nonlinear structural analyses are introduced, the program carries out the necessary computations and provides a set of probabilistic outcomes.

Unfortunately, as currently envisioned by FEMA-P58, a time-based evaluation requires consideration of eight distinct hazard level and 11 sets of dynamic nonlinear response analyses per hazard level. That totals to 88 sets of nonlinear analyses! It is hard to imagine a design firm having the necessary time, budget, and resources to perform these many nonlinear analyses for performance-based design of a tall building. Particularly given the fact that with today's software and computer speeds each of these analyses could take somewhere between a two to five of days of computer time for a typical tall building. It is expected, however, that the use of parallel computing software and hardware will drastically reduce the run time associated with such evaluations in the near future (see Fig. 11.4).

Fig. 11.4 Run time for NDP analysis of a 52 story building for 48 ground motions can be reduced 52 times (from 78 to 1.5 h) by parallel processing (Courtesy of Armin Bebamzadeh, University of British Columbia)

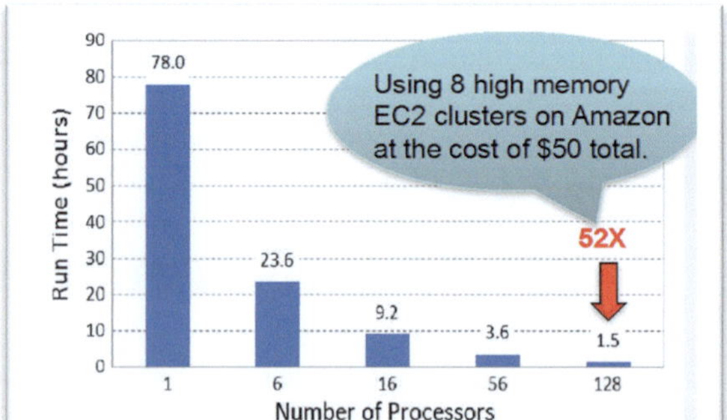

11.10 Design Procedures

Unfortunately, none of the documents guiding the practice of performance-based design of tall buildings contain specific instructions on how to design a tall building so that it will satisfy the delineated performance objectives. For example, the 2015 LATBSDC criteria requires that the building be designed based on capacity design principles, described in project-specific seismic design criteria, clearly explaining how the structural system will achieve a well-defined inelastic behavior where nonlinear actions and members are clearly defined and all other members are stronger than the elements designed to experience nonlinear behavior. Nonlinear action should be limited to the clearly defined members and regions. Yielding due to compression and bending at the base of columns (top of foundation or basement podiums) for steel structures is usually permitted. Other commonly designated zones of nonlinear behavior are listed in Table 11.2.

The document, however, does not explain how the engineer is supposed to achieve this capacity design. Since design by application of nonlinear response history analyses is neither feasible or advisable in a design office environment, most engineers select either a set of lateral forces or a push-over profile, based on their experience, to perform a preliminary capacity design of the building and then subject the building to the evaluation criteria contained in these documents and refine their design as necessary to achieve the specified performance objectives. Fortunately, methodologies for direct design of buildings which will likely satisfy

Table 11.2 Zones and actions commonly designated for nonlinear behavior

Structural system	Zones and actions
Special moment-resisting frames (steel, concrete, or composite)	• Flexural yielding of Beam ends (except for transfer girders) • Shear in Beam-Column Panel Zones
Special concentric braced frames	• Braces (yielding in tension and buckling in compression)
Eccentric braced frames	• Shear Link portion of the beams (shear yielding preferred but combined shear and flexural yielding permitted)
Unbonded braced frames	• Unbonded brace cores (yielding in tension and compression)
Special steel-plate Shear walls	• Shear yielding of web plates • Flexural yielding of Beam ends
R/C shear walls	• P-M-M yielding at the base of the walls (top of foundation or basement podiums) or other clearly defined locations with plastic hinge region permitted to extend to a reasonable height above the lowest plane of nonlinear action as necessary • Flexural yielding and/or shear yielding of link beams
Foundations	• Controlled rocking • Controlled settlement

performance criteria based on nonlinear dynamic response evaluations have emerged (Priestley et al. 2006; Goel and Chao 2008).

11.11 Evaluation Procedures

In this section we provide an overview of evaluation procedures recommended by current performance-based seismic design guidelines for tall buildings. Since the 2015 LATBSDC document is the most widely referenced guideline for design of tall buildings in the United States we will concentrate on the methodologies contained in this document.

11.12 Analysis Methods

A three-dimensional mathematical model of the physical structure is used that represents the spatial distribution of the mass and stiffness of the structure to an extent that is adequate for the calculation of the significant features of the building's dynamic response. Structural models are required to incorporate realistic estimates of stiffness and damping considering the anticipated levels of excitation and damage. Generally, expected material properties (see Table 11.3) are used throughout except when calculating the capacity of brittle elements where specified strength values are used. For serviceability analyses, realistic values of stiffness should be used such as those listed in Table 11.4. Given the current state of

Table 11.3 Suggested expected material strengths

Material		Expected strength
Structural steel		Strength
	Hot-rolled structural shapes and bars	
	ASTM A36/A36 M	$1.5F_y$
	ASTM A572/A572 M Grade 42 (290)	$1.3F_y$
	ASTM A992/A992 M	$1.1F_y$
	All other grades	$1.1F_y$
	Hollow structural sections	
	ASTM A500, A501, A618 and A847	$1.3F_y$
	Steel pipe	
	ASTM A53/A53 M	$1.4F_y$
	Plates	$1.1F_y$
	All other products	$1.1F_y$
Reinforcing steel		1.17 times specified f_y
Concrete		1.3 times specified f'_c

Table 11.4 Suggested effective component stiffness values

Element	Service ability	MCE-level nonlinear models
Structural walls	Flexural—0.75 Ig Shear—1.0 Ag	Flexural—1.0 Ec * Shear—0.5 Ag
Basement walls	Flexural—1.0 Ig Shear—1.0 Ag	Flexural—0.8 Ig Shear—0.5 Ag
Coupling beams	Flexural—0.3 Ig Shear—1.0 Ag	Flexural—0.2 Ig Shear—1.0 Ag
Diaphragms (in-plane only)	Flexural—0.5 Ig Shear—0.8 Ag	Flexural—0.25 Ig Shear—0.25 Ag
Moment frame beams	Flexural—0.7 Ig Shear—1.0 Ag	Flexural—0.35 Ig Shear—1.0 Ag
Moment frame columns	Flexural—0.9 Ig Shear—1.0 Ag	Flexural—0.7 Ig Shear—1.0 Ag

* Nonlinear fiber elements automatically account for cracking of concrete because the concrete fibers have zero tension stiffness.

modeling capabilities and available software systems, there is no reason to estimate the actual three-dimensional behavior of tall buildings by relying on approximate two-dimensional models.

For evaluation of performance under service level earthquakes either linear elastic response spectrum analyses or nonlinear analyses may be used. For evaluation of performance under MCE_R event, nonlinear dynamic response history analysis is required. In both types of analyses, P-Δ effects should be explicitly included as inclusion of P-Δ effects is crucial for establishing the onset of collapse because collapse of tall buildings is ultimately P-Δ related.

In elastic response spectrum analyses at least 90% of the participating mass of the structure is to be included in the calculation of response for each principal horizontal direction and modal responses are combined using the Complete Quadratic Combination (CQC) method.

There are various ways to model the behavior of nonlinear elements (Fig. 11.5). Most commonly, concentrated plasticity models are used for beams and columns and distributed plasticity or fiber models are used for modeling walls and floor diaphragms.

Inherent torsional properties of the structural system should always be considered. The 2015 LATBSDC Guideline requires assessment of accidental eccentricities during the service level building evaluation and if the impact of such eccentricities is proven to be significant, then it requires addressing them in one way or another during the MCE-level evaluation. The PEER Guideline does not require consideration of accidental eccentricities.

In addition to the designated elements and components of the lateral force resisting system, all other elements and components that in combination significantly contribute to or affect the total or local stiffness of the building should be included in the mathematical model. Axial deformation of gravity columns in a core wall system is one example of effects that should be considered in the structural model of the building (Applied Technology Council 2008; Wallace 2010).

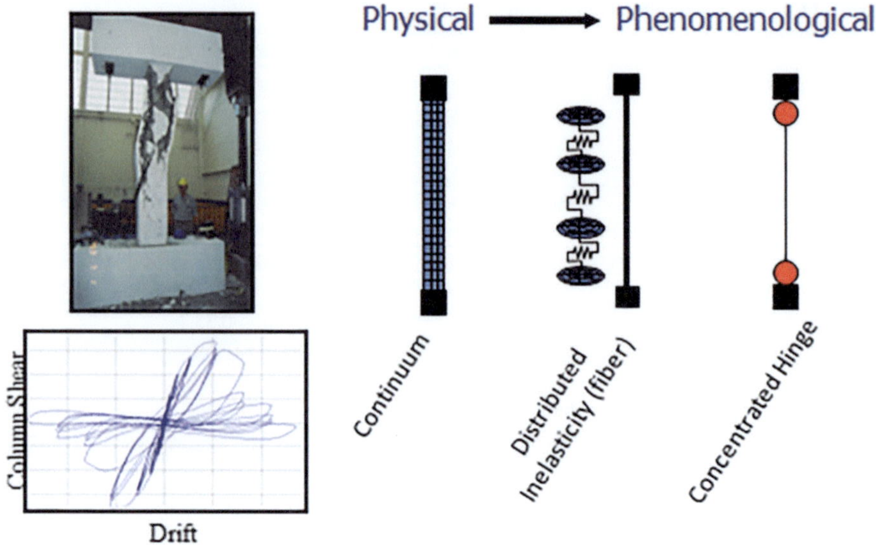

Fig. 11.5 Nonlinear component model types (*Courtesy* of Prof. Greg Deierlein)

11.13 Modeling Criteria

There are four possible ways for modeling nonlinear components. Deformation capacities may be taken equal to the corresponding CP values for primary elements published in ASCE 41-13 for nonlinear response procedures, or may be based on analytical models validated by experimental evidence. When applicable, the ASCE 41-13 component force versus deformation curves may be used as modified backbone curves, with the exception that the drop in resistance following the point of peak strength need not be as rapid as indicated in the ASCE 41-13 curves. This is because the rapid post-peak drop in resistance indicated in the ASCE 41-13 curves is not realistic (unless fracture occurs) and is likely to cause numerical instabilities in the analysis process. The four possible ways for modeling nonlinear components are:

1. Explicitly incorporate cyclic deterioration in the analytical model.
2. Use cyclic envelope curve as a modified backbone curve if cyclic deterioration is not considered explicitly.
3. Use factors for modification of backbone curve if cyclic deterioration is not considered explicitly.
4. Limit deformation capacities so that no deterioration occurs in the analytical model.

For steel moment-resisting frame systems, the contribution of panel zone (beam-column joint) deformations is to be included. If linear models are used for

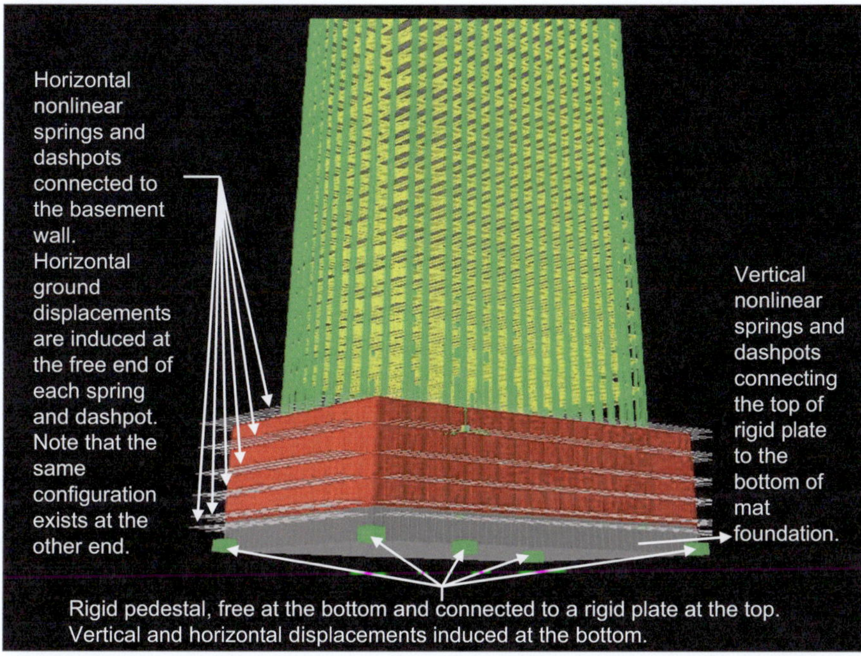

Fig. 11.6 A soil-foundation-structure interaction modeling technique for tall buildings with subterranean floors

service level evaluations, then in lieu of explicit modeling of beam-column panel zone behavior, center-to-center beam dimensions may be used for such evaluations.

Large compressive forces on columns reduce their ductility. To address this issue 2015 LATBSDC limits the MCE compressive force demand on reinforced concrete columns to $0.4f'_cA_g$, where f'_c is the compressive strength of the concrete and A_g is the gross cross sectional area of the column. PEER Guidelines limit MCE compressive force demand to the balanced load which may be taken as $0.3f'_cA_g$.

Naeim et al. (2008) have identified the difficulties involved in accurate modeling soil-foundation-structure interaction (SFSI) and subterranean floors for tall buildings with software tools used in a typical design office (Fig. 11.6). Despite these difficulties a number of tall building projects have utilized sophisticated and detailed SFSI models in their design (Fig. 11.7).

Based on Naeim et al. findings, PEER Guidelines recommends the use of the first approximation suggested by Naeim et al. for service level modeling and the use of a "bathtub" model (in its commentary) for MCE evaluations (Fig. 11.8). 2015 LATBSDC requires only the option (b) of Fig. 11.8. In this author's opinion, the suggested "bathtub" model is not practically implementable in a design office environment given the limitations of existing software tools.

Damping is a particularly thorny issue. Some guidance for tall building design is provided in the ATC-72 report (Applied Technology Council 2008). While 5%

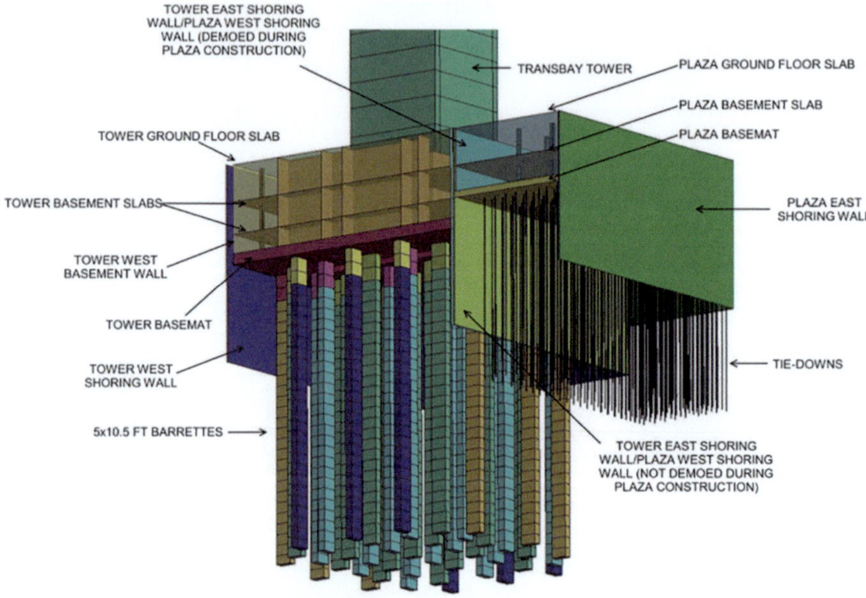

Fig. 11.7 An example of sophisticated soil-foundation-structure interaction model utilized for a tall building in San Francisco (*Courtesy* MKA and ARUP)

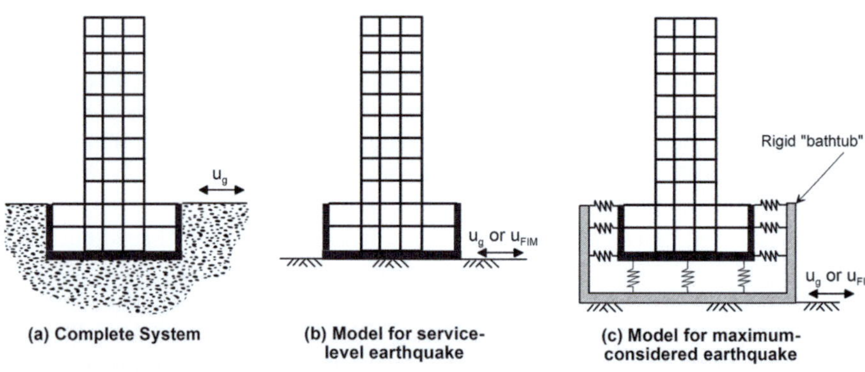

Fig. 11.8 Schematic illustration of tall building with subterranean levels and simple models for analysis in which soil-foundation interaction effects are neglected (Part b) and included in an approximate manner (Part c). Part (c) only shows springs but parallel dashpots are generally also used (Courtesy of Prof. Jonathan Stewart)

critical damping is universally used, right or wrong, in linear analysis of structures for design according to prescriptive codes, Guidelines for performance-based design of tall buildings currently cap the viscous damping to be used in conjunction with nonlinear analysis to 2.5% critical. For nonlinear analyses, most of energy dissipation occurring in the structural components is directly modeled through definition of hysteretic force/moment—displacement/rotation formulations. Some viscous damping is usually included in nonlinear analyses to account for energy dissipation occurring in the nonstructural components and parts of the structural system which is not included in the nonlinear model. Furthermore, a small amount of viscous damping usually alleviates numerical stability issues that are once in a while encountered during such analyses. With these considerations in mind, ATC-72 recommends using a equivalent viscous damping value (D) equal to α/N, where D is the maximum percent critical damping, N is the number of stories in a tall building (>30), and α is a coefficient with a recommended range of $\alpha = 70$–150 for nonlinear analyses with interstory drift amplitudes of 0.005–0.03. System specific recommended values for α are as follows:

- Dual Systems (RC core wall plus RC or steel frame): $\alpha = 130$
- RC Moment Frame Systems: $\alpha = 100$
- RC core wall systems: $\alpha = 80$
- Steel Moment Frame Systems: $\alpha = 80$
- Steel Braced Frame Systems: $\alpha = 70$

Finally, for use with the mass and stiffness proportioned damping (Rayleigh damping), ATC-72 recommends using $\alpha_M = 9\zeta/T_1$ and $\alpha_K = \zeta T_1/15$ where T_1 is the fundamental period of vibration of the building and ζ is percentage of critical damping. A recent comprehensive study of natural periods and damping by Xiang et al. (2016) reports evaluation of more than 100 instrumented building in California and has recommended updated damping values for various structural systems.

11.14 Acceptability Criteria

Acceptability criteria for both serviceability and MCE_R usually contain an absolute ceiling on the permitted drift. For serviceability earthquakes this is intended to minimize damage. For MCE_R event this is to limit the P-Δ effects. Both 2015 LATBSDC and Peer Guidelines limit overall drift ratio to one half of one percent (0.005) for serviceability and three percent (0.030) for MCE-level earthquakes. In addition, the maximum interstory drift at any story is limited to four and half percent (0.045) for any MCE_R level event.

Important component level serviceability criteria in 2015 LATBSDC may be briefly summarized as follows:

- If elastic response spectrum analysis is used the structure is deemed to have satisfied the acceptability criteria if none of the elastic demand to capacity ratios (ratio of the elastic respond parameter to the applicable LRFD limits for steel members or USD limits for concrete members using $\phi = 1.0$) exceed:

(a) 1.50 for deformation-controlled actions for Risk Category I and II Buildings (ASCE 7-10 Table 11.5-1); 1.20 for deformation-controlled actions for Risk Category III Buildings; and a factor smaller than 1.20 as determined by the peer review panel for Risk Category IV Buildings.
(b) 0.70 for force-controlled actions.

- If nonlinear analysis is used, deformation demands shall not exceed a value at which sustained damage requires repair, for reasons of strength deterioration or permanent deformation, as demonstrated by appropriate laboratory testing. Repair, if required, generally should not require removal and replacement of structural concrete, other than cover, or removal or replacement of reinforcing steel or structural steel. In lieu of the use of laboratory test data, the acceptance criteria for Immediate Occupancy performance as contained in ASCE 41 may be utilized.

For MCE_R level response, actions in all lateral load resisting elements must be categorized as either force-controlled or deformation-controlled and if classified as force-controlled, as either critical, or noncritical actions. Table 11.5 shows a representative and acceptable classification of such actions. The acceptance criteria with respect to various actions are determined as:

Table 11.5 Typical classification of component actions

Component	Seismic action	Classification	Criticality
Below grade perimeter walls	Flexure	Force-controlled	Noncritical
	Shear	Force-controlled	Critical
Parking ramp walls	Flexure	Deformation-controlled	N/A
	Shear	Force-controlled	Critical
Podium walls	Flexure	Deformation-controlled	N/A
	Shear	Force-controlled	Critical
Tower core walls (over their entire height)	Flexure	Deformation-controlled	N/A
	Shear	Force-controlled	Critical
Core wall coupling beams	Shear/flexure	Deformation-controlled	N/A
Floor slabs	Out of plane flexure around supports	Deformation-controlled	N/A
Diaphragms with major shear transfer	Flexure	Force-controlled	Noncritical
	Shear	Force-controlled	Critical
Gravity columns	Axial	Force-controlled	Critical
Foundations	Flexure	Force-controlled	Noncritical
	Shear	Force-controlled	Critical

- *Force-Controlled Actions*

(a) Critical Actions
Force-controlled critical actions are those force-controlled actions in which the failure mode poses severe consequences to structural stability under gravity and/or lateral loads. Force-controlled critical actions shall satisfy

$$F_{uc} \leq \kappa_i \phi F_{n,e},$$

where

F_{uc} 1.5 times the mean value of demand.
$F_{n,e}$ nominal strength as computed from applicable material codes but based on expected material properties.
ϕ 1.0.
κ_i Risk reduction factor given in Table 11.5

(b) Noncritical Actions
Noncritical actions are those force-controlled actions for which failure does not result in structural instability or potentially life-threatening damage such as diaphragm shear and axial forces in diaphragm chords and drag members as well as foundation forces. Force-controlled noncritical actions shall satisfy

$$F_u \leq \kappa_i \phi F_{n,e},$$

where

F_u the mean demand obtained from the suite of analyses,
$F_{n,e}$ nominal strength as computed from applicable material codes but based on expected material properties.
ϕ 1.0.
κ_i Risk reduction factor given in Table 11.6

Table 11.6 Risk category reduction factor

Risk category from ASCE 7-10 Table 1.5-1	Risk reduction factor, κ_i
I	1.00
II	1.00
III	0.80
IV	Value to be established by peer review

- *Deformation-Controlled Actions*
  The demand values (member total deformations) shall be permitted to be taken respectively as the average of the values determined from the seven or more pairs of records used in the analyses. Collector elements shall be provided and must be capable of transferring the seismic forces originating in other portions of the structure to the element providing the resistance to those forces. Every structural component not included in the seismic force–resisting system shall be able to resist the gravity load effects, seismic forces, and seismic deformation demands identified in this section.
  Acceptance criterion may be assumed to be equal to κ_i times the corresponding Primary Collapse Prevention values published in ASCE 41 for nonlinear response procedures.
- *Global Requirements*

  (a) Peak Transient Drift
  In each story, the mean of the absolute values of the peak transient drift ratios from the suite of analyses shall not exceed $0.03\kappa_i$. In each story, the absolute value of the maximum story drift ratio from the suite of analyses shall not exceed $0.045\kappa_i$.
  (b) Residual Drift
  In each story, the mean of the absolute values of residual drift ratios from the suite of analyses shall not exceed $0.01\kappa_i$. In each story, the maximum residual story drift ratio in any analysis shall not exceed $0.015\kappa_i$ unless proper justification is provided.
  (c) Loss in Story Strength
  In any nonlinear dynamic response analysis, deformation imposed at any story shall not result in a loss of total story strength that exceeds 20% of the initial strength.

- *Backstay Effects*
  Where applicable (see Fig. 11.9), two sets of analyses shall be conducted to evaluate backstay effects:

  1. A model which uses upper bound (UB) stiffness assumptions for floor diaphragms at the podium and below.
  2. A model which uses lower bound (LB) stiffness assumptions for floor diaphragms at the podium and below.

  Table 11.7 contains recommendations for UB and LB Stiffness parameters for backstay sensitivity analyses. The sensitivity analyses, where applicable, shall be performed in addition to the analyses performed using stiffness properties provided in Table 11.4.
- *Multiple Towers on a Common Podium or Basement*
  Where multiple towers on a common podium or base create a situation in which the number of occupants at or below the podium or ground level may exceed 5000 persons, then

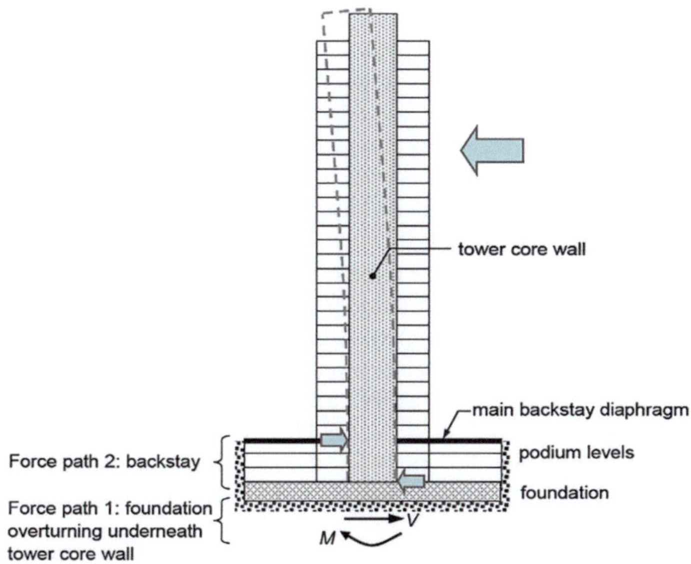

Fig. 11.9 Schematic illustration of the backstay effect (from ATC-72)

Table 11.7 Stiffness parameters for Upper Bound and Lower Bound Models

Stiffness parameters	UB	LB
$E_c\ I_{eff}$	0.5	0.20–0.25
$G_c\ A$	0.5	0.20–0.25

1. The κ_i factor should also be applied to all force-controlled actions including those of the podium diaphragm and below, including the foundations placed under the Risk Category III portion of the structure. The same κ_i factor should be applied to all deformation-controlled and force-controlled elements of the tower passing through the Risk Category III portion of the project.
2. The same κ_i factor shall be also applied to all deformation-controlled actions of the first level of each tower immediately above the common podium. As this level is most likely a location of formation of plastic hinges, special ductile detailing and confinement shall also be provided.

11.15 Ground Motion Record Selection and Scaling

A suite of seven or more pairs of appropriate ground motion time series shall be used in the analysis. This is likely to increase to a minimum of 11 pairs in the upcoming edition of the Guidelines. Ground motion time series and their selection shall comply with the requirements of ASCE 7-10. Either amplitude-scaling procedures or spectrum-matching procedures may be used. In addition, where applicable, an appropriate number of the ground motion time series shall include near fault and directivity effects such as velocity pulses producing relatively large spectral ordinates at relatively long periods.

The use of conditional mean spectrum (CMS) approach is permitted as long as Conditional Spectra that capture the building's response in each significant mode is captured. A minimum of two CMS should be used one to capture the building's first mode translational response in each direction and the other second mode response. In structures where first or second mode periods in the two directions are widely separated, additional CMS are required. A minimum of seven pairs of site-specific ground motion time histories are selected and scaled, or matched to each CMS and used for the nonlinear response history analysis. The envelope of the two suites shall address periods ranging from $0.1T$ to $1.5T$[1] s to the satisfaction of the project's Seismic Peer Review Panel. For purposes of evaluating acceptability of response, the mean response of each suite of motions should be separately evaluated.

11.16 Peer Review Requirements

Stringent peer review conducted during the entire process of structural design and not limited to a review of the end product is an essential quality control and assurance necessity for performance-based seismic design of tall buildings. The generally accepted peer review requirements are as follows:

- Each project needs a Seismic Peer Review Panel (SPRP). The SPRP is to provide an independent, objective, technical review of those aspects of the structural design of the building that relate to seismic performance and to advise the Building Official whether the design generally conforms to the intent of the design criteria established for the project.
- The SPRP participation is not intended to replace quality assurance measures ordinarily exercised by the engineer of record (EOR) in the structural design of a building. Responsibility for the structural design remains solely with the EOR, and the burden to demonstrate conformance of the structural design to the intent of this document and other requirements set forth by the Building Official resides with the EOR. SPRP is not a plan checking entity and the responsibility for conducting structural plan checking resides with the Building Official.
- The SPRP should include a minimum of three members with recognized expertise in relevant fields, such as structural engineering, earthquake engineering research, performance-based earthquake engineering, nonlinear response history analysis, tall building design, earthquake ground motion,

[1] T is the calculated fundamental period of the building.

geotechnical engineering, geological engineering, and other such areas of knowledge and experience relevant to the issues the project poses.

- The SPRP members shall be selected by the Building Official based on their qualifications applicable to the Seismic Peer Review of the project. The Building Official may request the opinion of the Project Sponsor and EOR on proposed SPRP members, with the Building Official making the final decision on the SPRP membership.
- SPRP members shall bear no conflict of interest with respect to the project and shall not be part of the design team for the project.
- The SPRP provides their professional opinion to and acts under the instructions of the Building Official.

11.17 Instrumentation and Structural Health Monitoring

Performance-based design of tall buildings is in its early stages of application and development. The funding necessary to experimentally validate performance of various components and systems utilized in tall buildings probably will not be available for a long time. Analytical simulations, as detailed and elaborate as they may be, cannot replace the need for experimental results and observed performances. It is imperative that we maximize every opportunity at our disposal to learn as much we can and as quickly as possible about performance of tall buildings designed according to these procedures during major earthquakes so that we can improve our design practices and produce more efficient and safe buildings.

Seismic instrumentation can provide valuable insight into performance of structures and help us assess the validity, or lack thereof, of assumptions used and methods applied. It is precisely for this reason that 2015 LATBSDC mandates extensive seismic instrumentation of tall buildings designed according to its provisions. According to 2015 LATBSDC, a tall building must be instrumented with the minimum number of sensors shown in Table 11.8. Each sensor records a single response quantity of interest (e.g., unidirectional floor acceleration, interstory displacement, etc.).

Instrumentation is inexpensive and nonintrusive if planned during the design process and implemented during the construction of a tall building. Given the recent advances in sensor technology, now it is possible to install sensors not only to

Table 11.8 Minimum tall building instrumentation levels

Number of stories above ground	Minimum number of sensors
10–20	15
20–30	21
30–50	24
>50	30

measure accelerations but to measure and record relative or overall displacements (building tilt), and various stresses and strains throughout the structure. Modern information technology has made real-time or near real-time measurements and remote transmission of sensor data and engineering interpretation of them to remote locations via the Internet not only possible, but feasible. Integration of seismic mentoring with broad building health monitoring which includes monitoring buildings during more frequent events and malfunctions (such as wind storms, fires, floor vibrations, flooding, elevator functions, HVAC problems), may finally produce enough tangible benefits for tall building owners and developers to cause them to willingly and enthusiastically embrace the modes cost of building instrumentation and health monitoring.

11.18 Conclusion

An overview of current state of practice for application of performance-based design methodologies to tall building design in the United States was presented. The contents of leading guidelines currently used for performance-based design of tall buildings were introduced and compared. Performance-based design has already established itself as the methodology of choice for design of tall buildings located in seismic regions as it has shown its capability to produce safer, more cost-effective, tall building structures compared to conventional design techniques. Extensive seismic instrumentation and structural health monitoring have the potential of significantly accelerating our path forward and transforming performance-based design of tall buildings from its current state to a fully matured seismic design methodology.

Acknowledgements The author is indebted to his colleagues at LATBSDC and PEER. Special gratitude is due to Professors Jack Moehle, Greg Deierlein, Jonathan Stewart, Farzin Zareian, Mr. Tony Ghodsi and Mr. Ronald Hamburger for graciously providing the author with access to their research results and permission to reproduce their artwork as needed.

References

American Society of Civil Engineers (2010) Minimum design loads for buildings and other structures, ASCE 7-10

American Society of Civil Engineers (2013) ASCE/SEI Standard 41-13. Seismic Rehabilitation of Existing Buildings, Reston, VA

Applied Technology Council (2008) Interim guidelines on modeling and acceptance criteria for seismic design and analysis of tall buildings, PEER tall buildings initiative, ATC-72-1, 95% Draft, September 2008

Cornell CA and Krawinkler H (2000) Progress and challenges in seismic performance assessment, PEER Center News, 3(2), Spring 2000

Council on Tall Buildings and Urban Habitat (2008) Recommendations for the seismic design of high-rise buildings—a consensus document—ctbuh seismic working group. Illinois, Chicago

Council on Tall Buildings and Urban Habitat (2010) CTBUH tall building database, http://www.ctbuh.org/HighRiseInfo/TallestDatabase/tabid/123/language/en-US/Default.aspx

Federal Emergency Management Agency (FEMA) (2000a) Prestandard and commentary for the seismic rehabilitation of buildings, FEMA 356, November 2000, Washington, D.C

Federal Emergency Management Agency (FEMA) (2012a) Seismic performance assessment of buildings volume 1—methodology, FEMA P-58-1, September 2012

Federal Emergency Management Agency (FEMA) (2012b) Seismic performance assessment of buildings volume 2—implementation guide, FEMA P-58-2, September 2012

Goel SC, Chao S-H (2008) Performance-based plastic design: earthquake-resistant steel structures, ICC Press

Heaton TH, Hall JF, Wald DJ, Halling MW (1995) Response of high-rise and base-isolated buildings to a hypothetical Mw 7 0 blind thrust earthquake. Science 267:206–211

International Conference of Building Officials (2002) City of Los Angeles building code, Sec. 1630.10.2, Whittier, California

Krawinkler H, Miranda E (2004) Performance based earthquake engineering. In: Bozorgnia Y, Bertero VV (eds) Earthquake engineering: from engineering seismology to performance-based engineering. CRC Press, Florida

Los Angeles Tall buildings structural design Council (2007) Proceedings of the 2007 annual meeting, Los Angeles, California. May 2007

Los Angeles Tall Buildings Structural Design Council (2008) Proceedings of the 2008 annual meeting, Los Angeles, California. May 2008

Los Angeles Tall Buildings Structural Design Council (2009) Proceedings of the 2009 annual meeting, Los Angeles, California. May 2009

Los Angeles Tall Buildings Structural Design Council (LATBSDC) (2005) An alternative procedure for seismic analysis and design of tall buildings located in the Los Angeles Region, Edition, May 2005

Los Angeles Tall Buildings Structural Design Council (LATBSDC) (2015) An alternative procedure for seismic analysis and design of tall buildings located in the Los Angeles, 2014 Edition with 2015 Supplements, August 2015

Moehle J and Deierlein G (2004) A framework methodology for performance-based earthquake engineering. In: Proceedings 13th world conference on earthquake engineering, Vancouver, British Columbia, paper no. 679

Naeim F (2010) Performance based seismic design of tall buildings. In: Garevsky, Ansal (eds) Earthquake engineering in Europe, pp 147–170A Springer

Naeim F, Garves R (2005) The Case for seismic superiority of well-engineered tall buildings. In: The structural design of tall and special buildings, vol 14, no 5, pp 401–416, Wiley InterScience, London

Naeim F, Hagie S (2012) PACT user manual, seismic performance assessment of buildings volume 2—implementation guide (Appendix C) FEMA P-58-2, September 2012

Naeim F, Tileylioglu S, Alimoradi A, Stewart JP (2008) Impact of foundation modeling on the accuracy of response history analysis of a tall building. In: Proceedings. SMIP2008 Seminar on utilization of strong motion data, California strong motion instrumentation program, Sacramento, CA, pp 19–55

Pacific Earthquake Engineering Research Center (PEER) (2010) Seismic design guidelines for tall buildings, Final Draft, February 2010

Portland Cement Association (2000) Concrete structural floor systems and more, CD013

Priestley MJN, Calvi GM, Kowalsky MJ (2006) Displacement-based seismic design of structures. IUSS Press, Italy

Structural Engineers Association of California (SEAOC) (1999) Recommended lateral force requirements and commentary, 7th Edition, Appendices G and I

Wallace J (2010) Performance-based design of tall reinforced concrete core wall buildings, Theme Lecture. In: 14th european conference on earthquake engineering, contained in this book

Xiang Y, Harris A, Naeim F, Zareian F (2016) Identification and validation of natural periods and modal damping rations for steel and reinforced concrete buildings in California. In: A report to the California strong motion instrumentation program. University of California, Irvine, October

Chapter 12
Performance-Based Evaluation of Hydrocarbon Steel Pipes Under Internal Pressure

Ferit Cakir, Eren Uckan, Bulent Akbas, Bilge Siyahi, Nuri Kanli, Ali Rıza Cakit and Sibel Guven Kaskan

Abstract Hydrocarbon steel pipes have undertaken a very important and fundamental role in the oil and gas industries in the world. Although these pipes are simple structural forms, their structural behaviors are very complex and challenging for the structural engineering community. One of the fundamental problems for pipelines is plastic deformations under internal pressure. This study focuses on to develop a methodology for the performance evaluation of hydrocarbon steel pipes under internal pressure. In this perspective, a hydrocarbon pipe, which is used in natural gas pipelines, is modeled by nonlinear finite element model (FEM) and investigated in terms of structural behavior under different internal pressure conditions; operating pressure (100 bar), design pressure (150 bar) and high pressure (250 bar). In order to obtain an accurate solution, the finite element model is calibrated with a hydrostatic test, which is conducted on a pipe under 133 bar.

F. Cakir (✉)
Department of Civil Engineering, Istanbul Aydin University, Istanbul, Turkey
e-mail: feritcakir@aydin.edu.tr

E. Uckan
Earthquake Engineering Department, Kandilli Observatory and Earthquake Research Institute, Bogazici University, Istanbul, Turkey
e-mail: eren.uckan@boun.edu.tr

B. Akbas · B. Siyahi
Department of Civil Engineering, Gebze Technical University, Kocaeli, Turkey
e-mail: akbasb@gtu.edu.tr

B. Siyahi
e-mail: bilge.siyahi@gtu.edu.tr

N. Kanli
Umran Steel Pipe, Salipazari, Istanbul, Turkey
e-mail: nkanli@umran.com

A. R. Cakit · S. G. Kaskan
Cakit Engineering, Umraniye, Istanbul, Turkey
e-mail: alirizacakit@gmail.com

S. G. Kaskan
e-mail: sgkaskan@cakit.com

A. A. Kasimzade et al. (eds.), *Seismic Isolation, Structural Health Monitoring, and Performance Based Seismic Design in Earthquake Engineering*,
https://doi.org/10.1007/978-3-319-93157-9_12

Keywords Hydrocarbon steel pipes · Internal pressure · Structural behavior of steel pipes · Finite element analysis

12.1 Introduction

Hydrocarbon steel pipes are used to transfer oil and gas from the reservoir to the end user. These pipes are subjected to internal pressure due to transported product along the pipeline during operation. Pipeline integrity is usually defined in terms of pressure-carrying capacity. The hydrocarbon steel pipes might experience failure due to internal pressure. Therefore, structural design of these pipes requires advanced engineering knowledge and in the design stage, internal pressure requirements must be recognized in order to provide desired operational life (McAllister 2009).

Design and evaluation of hydrocarbon pipelines have been described in many standards. The American Society of Mechanical Engineering (ASME) specifies some standards for this purpose. While ASME B31.8 defines gas transmission and distribution piping systems, ASME B31.4 describes design requirements of the liquid transportation systems for hydrocarbons, liquid petroleum gas, anhydrous ammonia, and alcohols. These standards define different failure modes for hydrocarbon pipelines. According to ASME B31.4 and B31.8, the allowable stresses stemming from the operating pressure are limited to a fraction of the specified minimum yield strength of the steel material.

A considerable number of research has implied the critical role of internal pressure in the steel pipes in recent years. Wasmer et al. (2003) focused on creep crack initiation and growth in thick section steel pipes under internal pressure. The study presented that the main effect of the slow pressure cycling was to cause the early cracking. Kamaya et al. (2008) also investigated the failure pressure of straight pipe with wall thinning under internal pressure. Alashti et al. (2015) reported the ductile damage effect on load bearing capacity of a dented API XB pipe subjected to internal pressure. Although performance-based evaluation of hydrocarbon steel pipes has been considered as a challenging issue in engineering research, researchers have rarely focused on performance-based evaluation of hydrocarbon steel pipes under internal pressure. This paper mainly focuses on the methodology for the performance-based evaluation of hydrocarbon steel pipes under internal pressure. The main objective of this research is to develop realistic stress-based fragility curves for hydrocarbon steel pipes.

12.2 Performance-Based Evaluation

Performance-based evaluation of hydrocarbon steel pipes has been quite popular for the last few decades. Although these pipes have simple structural forms, their structural behavior is quite complicated and is challenging for the structural engineering community. Basic principles to be considered in performance-based evaluation of hydrocarbon steel pipes under internal pressure are needed to be described. In addition to that, a performance-based evaluation of existing hydrocarbon pipes is strongly needed.

Performance-based evaluation of hydrocarbon pipes under internal pressure includes five main steps stated below:

1. Description of limit states under internal pressure,
2. Development of fragility curves for the corresponding limit states using material test results,
3. Determination of the stresses and strains due to internal pressure by using 3D finite element models (fems),
4. Comparison of the stresses and strains by using fragility curves developed in step 2,
5. Performance evaluation.

12.2.1 Describing Limit States Under Internal Pressure

The performance levels of a structure can be defined through a threshold of damage limits known as limit states. Different damage conditions can be described for these limit states. For internal pressure, three performance levels are generally defined for pipelines, namely; normal operability, pressure integrity, and rupture. Normal operability performance level is defined as the one where no significant damage occurs to the pipeline and the pipeline is fully operational under internal pressure. Pressure integrity performance level corresponds to the deformed shape of the pipeline where plastic deformations occur in the pipe. Rupture performance level is defined as the damage level where 90% of the rupture capacity of the pipe has been reached. Suggested acceptability criteria corresponding to these three performance levels; normal operability, pressure integrity, and rupture are 2, 10 and 25% strain levels, respectively (Fig. 12.1).

12.2.2 Developing Fragility Curves

Vulnerability of structures can be expressed in terms of fragility curve. The fragility curve, which provides a graphical representation of the vulnerability of a structure, is developed from the structure's capacity and behavior. The curve is a key element for determining the probability of various damage states exceeding specified limit states. There are two main approaches for generating fragility curves. The first one is based on damage data obtained from field observations or from experiments and the second approach is based on the numerical analysis (Zamaro 2013; Shinozuka et al. 2000).

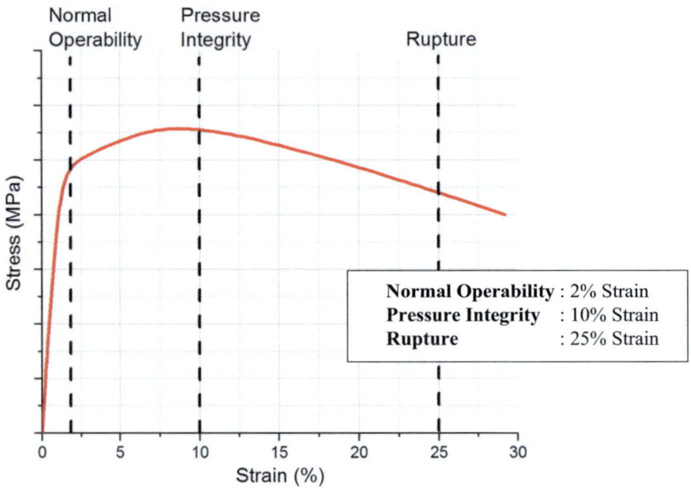

Fig. 12.1 Suggested acceptability criteria for pipelines corresponding to different performance levels

12.2.3 Finite Element Analysis

Determination of structural behavior of a pipe with finite element method is one of the most preferred analysis methods. Finite element models for this purpose are created for structural assessment of hydrocarbon pipes. The accuracy of finite element model depends upon the choice of the accurate model. An updating study on the numerical models is generally needed and the experimental test results for model calibration should be used for updating process.

12.3 Case Study

12.3.1 Material Properties and Limit States

API 5L-X70M grade pipeline steel was used in this study. The mechanical properties of the API5L-X70M steel were determined by mechanical tests (Umran Steel Pipe Corp 2016). The tensile strengths of the steel samples were assessed on the steel specimens. ASTM E8/E8M standard was based on the tensile strength of the samples. The tests were carried out by using a tensile testing machine (UH 1000 kN—Hydraulic Universal Testing Machine), while recording load and deformation data on a computer (Fig. 12.2). The stress–strain curve, which was derived from the average values of the mechanical tests, was presented in Fig. 12.2.

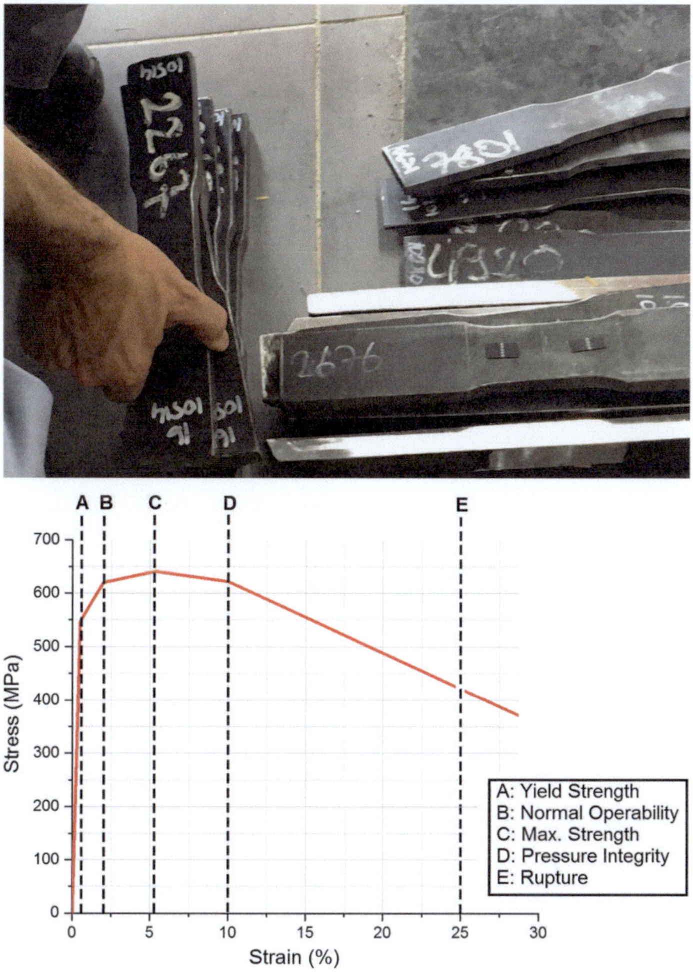

Fig. 12.2 Mechanical test and stress–strain relationships of the material

12.3.2 Fragility Curves

Experimental mechanical test data was collected and fragility curves for the limit states were developed by using the collected data. In this study, 1085 samples were considered. Figure 12.3 shows the fragility curves developed for three limit states.

12.3.3 3D Nonlinear Finite Element Analysis

In this study, finite element analysis software, ANSYS Workbench V15, was used for the analyses. In the numerical model, the pipes were constituted with Solid 186

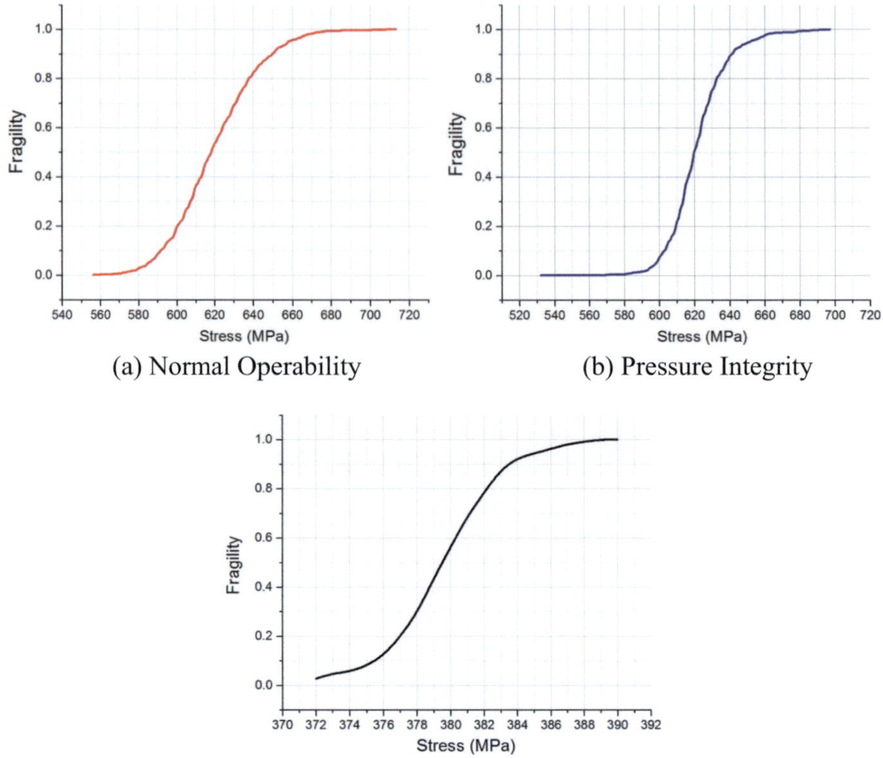

(a) Normal Operability

(b) Pressure Integrity

(c) Rupture

Fig. 12.3 Fragility curves developed by using the experimental data

elements, which have 20 nodes and three degrees of freedom per node; translations in the nodal x, y, and z directions. This element provides plasticity, hyper-elasticity, creep, stress stiffening, large deflection, and large strain capabilities. In the numerical analyses, pipe wall thicknesses, outer diameter, and pipe length were taken as 20, 1400, and 12,000 mm, respectively.

12.3.3.1 Calibration Test

To obtain the most accurate results from the FEM, hydrostatic test results conducted on the pipes were used for calibration of the numerical model. An updating study on the numerical model was then performed by using the experimental results. In cases where the experimental results are significantly different from the numerical ones, the numerical model is updated to minimize the differences.

The hydrostatic test is used for checking the possible defects of welded steel pipes during the forming and welding process. The test system contains an oil–water balance control system. The hydrostatic test on a new pipe was conducted with water. In the calibration test, pipe wall thicknesses, outer diameter of the pipe, and pipe length were measured to be 19.65, 1422.40 and 12,000 mm, respectively. The pipe was subjected to an internal pressure of 133 bars during the testing. The internal pressure was incrementally increased and the pipe was exposed to 133 bars for 20 s (Fig. 12.4). During hydrostatic testing, the pipe was investigated by using visual examination systems. Hydrostatic test was successfully completed without any failure and no visible defect, however, plastic deformation was observed on the pipe. The test results indicated that elastic and plastic deformations were observed during the testing.

Fig. 12.4 Hydrostatic test setup

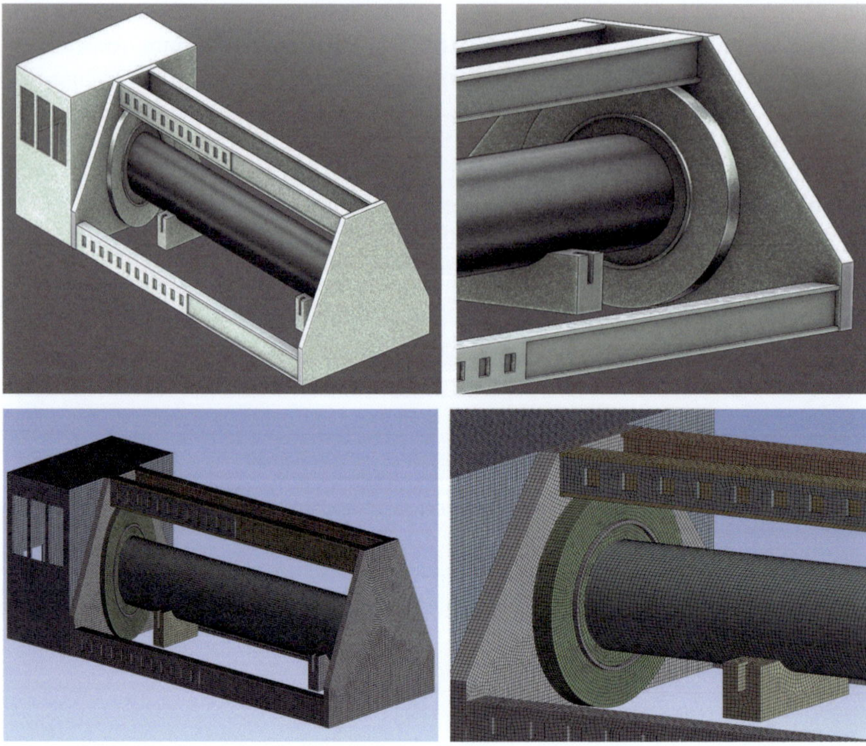

Fig. 12.5 Finite element model of hydrostatic test system

Table 12.1 Calibration of numerical model

Circumferential deformations (mm)	Experimental test	Numerical test	Change (%)	
First model	Max. elastic deformation	9.00	9.57	5.96
	Max. plastic deformation	4.50	5.01	10.1
Calibrated model	Max. elastic deformation	9.00	9.12	1.13
	Max. plastic deformation	4.50	4.62	2.59

Following the hydrostatic tests, a 3D nonlinear FEM was developed and analyzed (Fig. 12.5). The numerical analysis indicated that the results were deviating from the hydrostatic test results for more than 5%. To eliminate this difference, a calibration study on the mesh size was conducted without changing boundary conditions. This amendment enabled consistency between the numerical analyses and testing (Table 12.1). Thus, the new mesh sizes used in the analysis, boundary conditions, and material type were specifically determined after the calibration process.

Table 12.2 Results of the nonlinear numerical analyses

		Operating pressure (100 bar)	Design pressure (150 bar)	High pressure (250 bar)
Normal (Hoop) stress (MPa)		345.11	521.22	601.59
Maximum strain (%)		0.32	0.67	2.36
Circumferential deformations (mm)	Elastic deformation	8.52	8.52	8.52
	Plastic deformation	–	3.34	5.19

12.3.3.2 Nonlinear Finite Element Analyses

A series of 3D nonlinear analyses was carried out. Three different internal pressure conditions, namely, operating pressure (100 bar), design pressure (150 bar), and high pressure (250 bar) were considered in the analyses. The pipe responses obtained from the analyses were compared with the specified stress-based performance criteria and corresponding limit states. The material properties obtained from the mechanical tests were used in the numerical analyses. In this study, hoop stresses, maximum elastic, and maximum plastic deformations were considered and the results obtained from the finite element analyses were compared in terms of the three internal pressure conditions (Table 12.2).

12.4 Conclusions

Steel pipes are usually used in hydrocarbon pipelines. During operation, hydrocarbon steel pipes may be exposed to significant internal pressure. This pressure might cause significant damage to the pipes. This paper briefly investigated the basic principles to be considered in performance-based evaluation of hydrocarbon steel pipes under three different internal pressures. As mentioned before, the analytical fragility curves for the different limit states of the pipe are developed and the numerical results are compared using the fragility curves.

The main findings of the study are summarized as follows;

- Under 133 bar, the stress level within the pipe wall (hoop stress) slightly exceeds the yield strength, but remains below maximum average tensile strength. As the stress level exceeds the yield level, plastic deformations are also observed causing total increase of 13 mm in circumference (9 mm elastic + 4 mm plastic). This numerical result is also verified by the hydrostatic test results

- Under operating pressure, the stress level within the pipe wall (hoop stress) remains below the yield strength, i.e., normal operability performance level is satisfied. Only elastic deformations are observed under operating pressure causing total increase of 8.52 mm elastic deformation in circumference.
- Under design pressure, the stress level within the pipe wall (hoop stress) remains also below the maximum average tensile strength, i.e., pressure integrity performance level is satisfied. As the stress level exceeds the yield level, plastic deformations are observed causing total increase of 11.86 mm in circumference (8.52 mm elastic + 3.34 mm plastic).
- Under high pressure, the stress level within the pipe wall (hoop stress) slightly exceeds the normal operability stress level, but remains below the pressure integrity stress level, i.e., rupture performance level is satisfied. Plastic deformations are measured as a total increase of 13.71 mm in circumference (8.52 mm elastic + 5.19 mm plastic).
- The results indicate that the pipe under high pressure (250 Bar) is the most critical pipe with the highest vulnerability. This is because the pipe under high pressure exhibits higher hoop demands in terms of stress and plastic strain than the pipes under operating and design pressures.

References

Alashti RA, Jafari S, Hosseinipoyr SJ (2015) Experimental and numerical investigation of ductile damage effect on load bearing capacity of a dented API XB pipe subjected to internal pressure. Eng Fail Anal 47(2015):208–228

Kamaya M, Suzuki T, Meshii T (2008) Failure pressure of straight pipe with wall thinning under internal pressure. Int J Press Vessels Pip 85(2008):628–634

McAllister EW (2009) Pipeline rules of thumb handbook: a manual of quick, accurate solutions to everyday pipeline engineering problems. Gulf Professional Publishing, Elsevier, Oxford, UK

Shinozuka M, Grigoriu M, Ingraffea AR, Billington SL, Feenstra P, Soong TT, Reinhorn AM, Maragakis E (2000) Development of fragility information for structures and nonstructural components. In: Fragility Information for Structures and Nonstructural Components, MCCER2000, pp 15–33

Umran Steel Pipe Corp (2016) Mechanical testing laboratory

Wasmer K, Nikbin KM, Webster GA (2003) Creep crack initiation and growth in thick section steel pipes under internal pressure. Int J Press Vessels Pip 80(2003):489–498

Zamora JEM (2013) Seismic fragility analysis of reinforced masonry buildings. Master's Thesis, Department of Civil and Environmental Engineering, Colorado State University, Fort Collins, Colorado

Chapter 13
Seismic Energy Demands of Inverted V-Braced Frames

Selçuk Doğru, Bora Akşar, Bülent Akbaş, Jay Shen and Bilge Doran

Abstract It is known that structures act a nonlinear movement when strong ground motion cycle begins. Energy concept has been progressive tool to evaluate the structural system associated with performance-based design. Energy-based design can be expressed as the balance of energy input and the energy dissipation capacity of the structure. Researches that have been usually done for single degree of freedom system are needed for multi degree of freedom systems (MDOFs) in framework of the energy-based design methodology. In this paper, energy parameters in term of total energy input and hysteretic energy are evaluated and observed changes in height of the buildings for the energy concept. Structures are selected to demonstrate low and medium-rise steel inverted V-braced frames examined in linear and nonlinear dynamic time history analysis. The results are developed to obtain seismic energy demands.

Keywords Special concentrically braced frame · Energy spectrum
Energy demands · Energy-based design · Energy dissipation

S. Doğru (✉) · B. Akşar
Department of Earthquake and Structural Engineering,
Gebze Technical University, Kocaeli, Turkey
e-mail: seltrue@hotmail.com

B. Akşar
e-mail: baksar@gtu.edu.tr

B. Akbaş · B. Doran
Department of Civil Engineering, Yildiz Technical University,
Istanbul, Turkey
e-mail: akbasb@gtu.edu.tr

B. Doran
e-mail: doran@yildiz.edu.tr

J. Shen
Department of Civil, Construction and Environmental Engineering,
Iowa State University, Ames, USA
e-mail: shen@iit.edu.tr

A. A. Kasimzade et al. (eds.), *Seismic Isolation, Structural Health Monitoring, and Performance Based Seismic Design in Earthquake Engineering*,
https://doi.org/10.1007/978-3-319-93157-9_13

13.1 Introduction

Forced-based design has been widely used to consider the seismic effect on structures. The structural members are checked based on the equivalent static lateral load with gravity loads according to seismic design codes. It is known that structures act a nonlinear movement when earthquake cycle begins. To understand the nonlinear behavior of the structure, engineers have to evaluate structural system in the reliable nonlinear evaluation method. Energy-based design parameters ara useful to determine the estimated damage. Energy parameters such as total energy input and hysteretic energy is very important to understand the energy concept. According to elasto-plastic behavior of the structure, part of the total input energy, E_I, is distributed through hysteretic energy, $E_{H,}$ and viscous damping energy, E_D. Dissipated energy can be formed with elastic part and inelastic part.

Viscous damping energy can be defined as elastic dissipated energy and hysteretic energy can be defined as inelastic dissipated energy. The hysteretic energy is related to yielding of the structural members. On the other hand, permanent damage in the structure is occurred as a result of yielding and it will be formed the hysteretic energy.

The minimum required strength (or capacity for lateral force) is based on nonlinear response of the structure. The energy response of a nonlinear system can be written as

$$E_K(t) + E_D(t) + E_e(t) + E_H(t) = E_I(t) \tag{13.1}$$

The right side of Eq. (13.1) represents the total (or seismic) energy input, $E_I(t)$. The first term on the left side of Eq. (13.1) is the kinetic energy and the second term on the left side of Eq. (13.1) is the energy dissipated by damping, $E_D(t)$. The third and fourth terms on the left side of Eq. (13.1) are the sum of the hysteretic energy, $E_H(t)$, and the elastic strain energy, $E_e(t)$. $E_H(t)$ is a cumulative quantity over the plastic deformation throughout the entire duration of the vibration and will be zero if the structure remains elastic. If Eq. (13.1) were considered as a design equation (demand $\leq$ capacity), the four terms on the left-hand side of Eq. (13.1), could be considered as energy response of the structure (capacity) and the term on the right-hand side as energy input (demand). The instant kinetic energy and elastic strain energy consist of relatively small portion of the E_I at any time during the vibration and vanish at the end of the vibration. The E_D and E_H, therefore, are major contributors for dissipating the E_I. Thus, E_k and E_e are negligible in an inelastic response and Eq. (13.1) can be practically written as

$$E_I(t) = E_D(t) + E_H(t) \tag{13.2}$$

An energy-based parameter was proposed by Housner (1956). As mentioned before, despite high irregularity of earthquake ground motions (EQGMs), the energy input to a structure is a stable quantity and mainly based on the mass (m) and fundamental natural period of the structure (T) and affected by the strength of the structure (Akiyama 1985).

Fajfar et al. (1991) studied the energy input into SDOF systems and proposed a formula to predict the maximum energy input. Fajfar and Vidic (1994) proposed inelastic design spectra considering hysteretic energy and input energy. Shen and Akbas (1999) predicted the energy input in low-, medium-, and high-rise steel moment resisting frames (SMRFs) and proposed an energy-based earthquake resisting design method for SMRFs. In their study, they made comparisons of the energy input between the formulas proposed for mainly SDOF systems and concluded that there is a clear difference between the SDOF systems and MDOF systems in terms of energy input. Akbas and Shen (2002) proposed an Energy Spectrum based on the nonlinear time history analysis on SDOF systems within the period range of 0–3.0 s subject to seven different EQGMs. They also defined a damage parameter and studied how to use this parameter in performance-based earthquake resistant design. Shen et al. (2000) investigated the distribution of energy terms in MDOF systems. They provided the distribution of hysteretic energy in regular steel moment frames and emphasized the difference in energy terms between SDOF and MDOF systems. In another study by the same authors (Akbas et al. 2001), a practical design approach based on hysteretic energy was proposed. Akbas et al. (2006) studied on identifying the hysteretic energy demand and distribution throughout the height of regular steel moment resisting frames (SMRFs) subject to severe EQGMs. Benavent-Climent and Zahran (2010) studied the energy input into existing reinforced concrete frames. Habib et al. (2013) proposed a multi-mode energy-based design method for seismic retrofitting with passive energy dissipation systems.

In this study, low and medium-rise steel moment frames and will be studied in linear and nonlinear time history analysis. Steel Inverted V-braced frames in four- and eight-storey with same span length representing typical low and medium-rise steel buildings are designed based on the seismic design requirement in ASCE 7-10 and AISC 341-10. An ensemble of ground motions are selected so that the seismic response of each of the three frames would range from moderate to severe and the seismic energy demand would be evaluated based on the response of the frames. Two sets of ground motions corresponding to 10% and 2% probability of exceedance are used in nonlinear first mode incremental dynamic time history analyses. The results obtained from these analyses are reviewed for inelastic hysteretic energy and energy input demands.

13.2 Evaluation of the Seismic Energy Demands in Parametric Study

13.2.1 Design of the Structures

Inverted V-type steel-braced frames with four-, eight-storey, representing typical low- and medium-rise steel buildings were designed based on the seismic design requirements for SCBFs in accordance with ASCE 7-10, AISC 360-10, AISC 341-10. The plan dimensions of buildings are (width) 45 m and (depth) 45 m with span length of 9.0 m (five equal spans) for all stories in two orthogonal directions. The planes of the buildings are symmetrical. The columns are assumed to be pinned to the ground. All connections of the braces and beams to each other and columns are pinned. The structural system for each building consists of steel perimeter inverted V-braced frames and interior simply connected framing for gravity, i.e., lateral loads are carried by perimeter frames and interior frames are not explicitly designed to resist seismic loads in the direction of the earthquake. The typical storey height for the four- and eight-stories frame is 4.0 m and the height of ground for the eight-storey frame is 5.0 m.

For the eight-storey building, concrete foundation walls and surrounding soil are assumed to prevent any significant horizontal displacement of the structure at the ground level, i.e., the seismic base is assumed to be at the ground level. The linear analysis of buildings are applied by using Load and Resistance Factor Design (LRFD) specification in accordance with AISC 360-10 standard. Dead loads including self-weight of the members and live load used in the study are 5.0 and 2.4 kN/m^2 in except at the roof level, where it is 4.0 and 1.4 kN/m^2, respectively. European wide flange profiles are preferred with quality of S355. Hollow structural sections are used for braces. Member sections of the braced frames are given in Fig. 13.1 Fundamental periods of the four- and eight-storey braced frames are 0.563 and 1.520, respectively. Redundancy factor, $\rho = 1.3$, is assigned to the seismic force-resisting system for all structures in accordance with ASCE 7-10. Shear wave velocity (V_s) of site class is between value of 300 and 770 m/s. These ground motions corresponding to 10% (Design Level) and 2% (Maximum Considered Earthquake Level-MCE) probability of exceedance in 50-year time period were scaled to comply with ASCE 7-10 code spectrum model. Two sets of ground motion records were selected from PEER Strong Ground Motion Database corresponding to 10% and 2% probability of exceedance in 50-year time. Each set

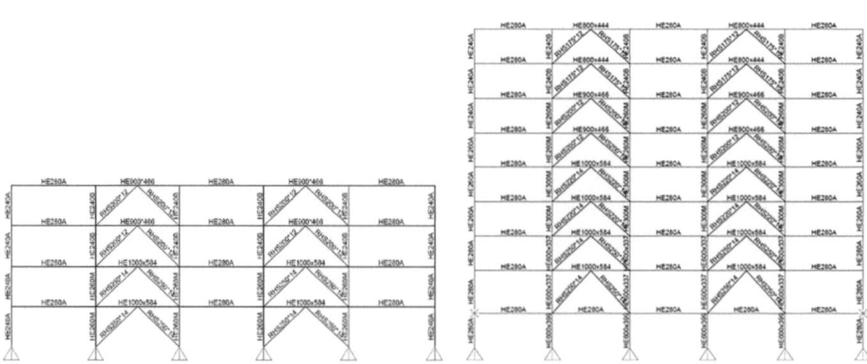

(a) 4-storey inverted V-braced system **(b)** 8-storey inverted V-braced system

Fig. 13.1 Elevations of the four-, eight-storey-braced frames

Table 13.1 Earthquake ground motion characteristics from PEER database

	NGA[a]	Record	Scale factor	Duration (s)	PGA (g)
GM 1	265	Victoria	1.45	24.54	0.93
GM 2	289	İrpinia	2.09	35.22	0.26
GM 3	587	New Zealand	1.56	49.38	0.37
GM 4	881	Landers	1.89	56.07	0.42
GM 5	1119	Kobe	0.60	40.99	0.41
GM 6	4132	Park Field	2.08	21.20	0.76
GM 7	8166	Düzce	1.76	35.02	0.70
GM 8	265	Victoria	2.17	24.54	1.40
GM 9	289	İrpinia	3.14	35.22	0.39
GM 10	587	New Zealand	2.34	49.38	0.56
GM 11	881	Landers	2.84	56.07	0.63
GM 12	1119	Kobe	3.13	40.99	0.62
GM 13	4132	Park Field	2.65	21.20	1.15
GM 14	8166	Düzce	0.90	35.02	1.05

[a]NGA : Sequential number in PEER Strong Ground Motion Database

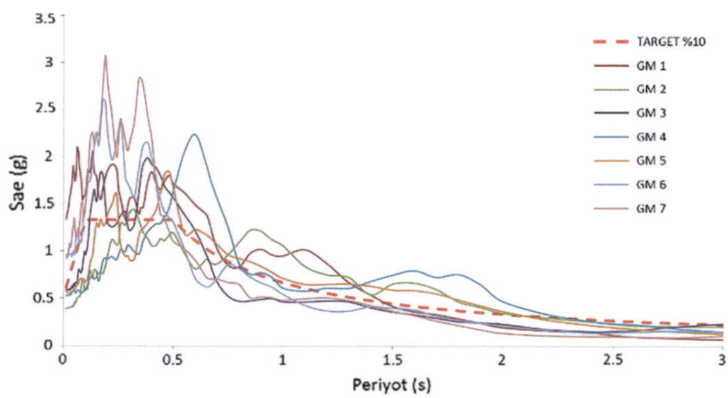

Fig. 13.2 Response spectra of the ground motions selected from PEER database for Design Level

consists of 7 records. Table 13.1 lists the detailed information of these ground motions. Figures 13.2 and 13.3 summarize the elastic response spectra of the ground motions.

In design of steel-braced frames, the appropriate response modification coefficient ($R = 6$), overstrength factor ($\Omega_o = 2$), and the deflection amplification factor ($C_d = 5$), are used in determining the base shear, element design forces, and design storey drift where redundancy factor ρ was taken as 1.3. The buildings were designed for a site where MCE spectral response acceleration parameters S_S is 2.0 g and S_1 is 1.0 g. Design earthquake spectral response acceleration parameters are $S_{DS} = 1.333$ (g), $S_{D1} = 0.666$ (g) and long-period transition period, $T_L = 12.0$ s.

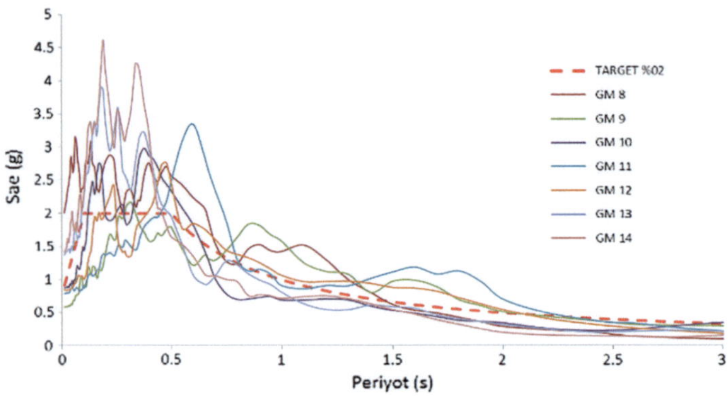

Fig. 13.3 Response spectra of the ground motions selected from PEER database for MCE Level

The frames were designed by selecting an assumed 2% target drift in ASCE 7-10. According to AISC 341-10, Columns and braces are designed as highly ductile members, and beams are satisfied the requirements for moderately ductile members.

13.2.2 Seismic Response of the SCBFs

Inelastic dynamic time history analyses were carried out to study the energy response parameters on the four- and eight-storey braced frames. Column and beams are modeled by PERFORM 3D which is nonlinear time history analysis software. Braces were modeled as an "Inelastic Steel Bar". The interaction between the axial force and bending moment was considered in columns. P-Δ effects were always included in the time history analyses. Ground motions corresponding to 10% (Design Level) probability of exceedance in 50-year time period are presented as GM 1–GM 7 that cause moderate structural damage for life safety performance level (LS). and ground motions corresponding to 2% (Maximum Considered Earthquake Level-MCE) probability of exceedance in 50-year time period are presented as GM 8–GM 14 that cause heavy structural damage for collapse prevention performance level (CP). The seismic energy response parameters; total energy input (E_I/m) and hysteretic energy (E_H/m) normalized with respect to mass in each bracing type are given in Figs. 13.4 and 13.5.

13.3 Conclusion

In this paper, total energy input and inelastic hysteretic energy demands are presented to understand how to change for low- and medium-rise steel special concentrically braced frames (SCBFs) with inverted V type. All results are obtained

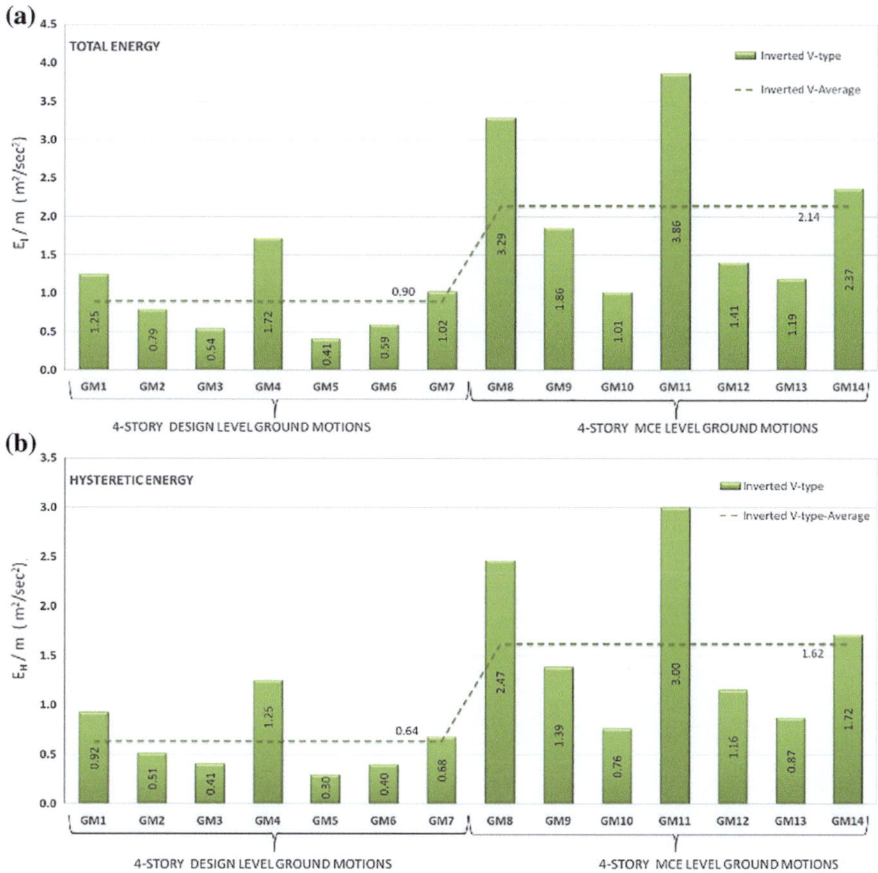

Fig. 13.4 Seismic energy parameters (E_I/m, E_H/m) for the four-storey frame

from the inverted V-type steel braced frame systems designed in accordance with AISC 360-10, AISC 341-10, and ASCE 7-10 specifications. Four- and eight-storey SCBFs are selected to represent low- and medium-rise buildings. Main results obtained in this study are reviewed below.

- For the four-storey frame with inverted V-bracing the average total energy parameter, E_I/m, when subjected to GM1 to GM7 was 0.90 m^2/s^2 while it was 2.14 m^2/s^2 when subjected to GM8 to GM14 in Fig. 13.4a.
- For the four-storey frame with inverted V-bracing, the average hysteretic energy parameter, E_H/m, when subjected to GM1 to GM7 was 0.64 m^2/s^2 while it was 1.62 m^2/s^2 when subjected to GM8 to GM14 in Fig. 13.4b.
- For the eight-storey frame with inverted V-bracing the average total energy parameter, E_I/m, when subjected to GM1 to GM7 was 0.90 m^2/s^2 while it was 1.64 m^2/s^2 when subjected to GM8 to GM14 in Fig. 13.5a.

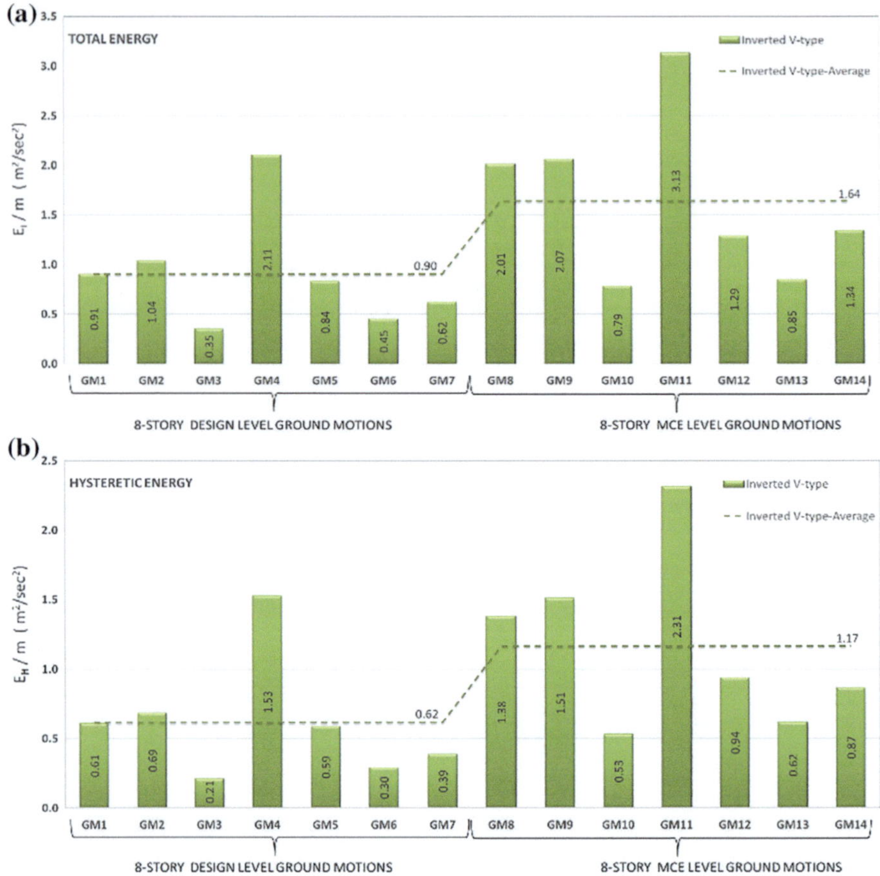

Fig. 13.5 Seismic energy parameters (E_I/m, E_H/m) for the eight-storey frame

- For the eight-storey frame with inverted V-bracing, the average hysteretic energy parameter, E_H/m, when subjected to GM1 to GM7 was 0.62 m^2/s^2 while it was 1.17 m^2/s^2 when subjected to GM8 to GM14 in Fig. 13.5b.
- The study is carried out by considering only regular structure models. It would be studying to define how different the effects would be for non-regular structures.

Acknowledgements The authors gratefully acknowledge The Scientific and Technological Research Council of Turkey (TUBITAK) for supporting this research (Grant #114R044). However, the views expressed in this paper belong to the authors alone and do not necessarily represent the position of any other organization or person.

References

AISC 341-10 (2010) Seismic provisions for steel structural buildings. American Institute of Steel Construction, Chicago, IL

AISC 360-10 (2010) Specification for structural steel buildings. American Institute of Steel Construction, Chicago, IL

Akbas B, J, Hao H (2001) Energy approach in performance-based seismic design of steel moment resisting frames for basic safety objective. Struct Des Tall Buildings 10:193-217

Akbas B, Shen J (2002) Energy approach in performance-based earthquake-resistant design. In: Twelfth European conference on earthquake engineering, London, England

Akbas B, Shen J, Temiz H (2006) Identifying the hysteretic energy demand and distribution in regular steel frames. Steel Compos Struct 6(6):479–491

Akiyama H (1985) Earthquake-resistant limit-state design for buildings. University of Tokyo Press

ASCE 7-10 (2010) Minimum design loads for buildings and other structures. American Society of Civil Engineers, Reston, VA

Benavent-Climent A, Zahran R (2010) Seismic evaluation of existing RC frames with wide beams using an energy-based approach. Earthq Struct 1(1):93–108

Fajfar P, Vidic T (1994) Consistent inelastic design spectra: hysteretic and input energy. J Earthq Eng Struct Dyn 23:523–537

Fajfar P, Vidic T, Fischinger M (1991) On the energy input into structures. In: Proceedings of the Pacific conference on earthquake engineering, Auckland, New Zealand, vol 1, pp 81–92

Habibi A, Chan RWK, Albemani F (2013) Energy-based design method for seismic retrofitting with passive energy dissipation systems. Eng Struct 46:77–86

Housner GW (1956) Limit design of structures to resist earthquakes. In: Proceedings of the first world conference on earthquake engineering, Berkeley, CA, 5-1-5-13

PEER Database, Peer.berkeley.edu/peer_ground_motion_database. Pacific Earthquake Engineering Research Center, 325 Davis Hall, University of California, Berkeley, CA 94720

PERFORM-3D (2011) Nonlinear analysis and performance assessment for 3D structures, version 5

Shen J, Akbas B (1999) Seismic energy demand in steel moment frames. J Earthquake Eng 3(4): 519–559

Shen J, Hao H, Akbas B (2000) Hysteresis energy in moment frames. In: Shen J (ed) The engineering science of structures, a special volume honoring Sidney A. Guralnick. Illinois Institute of Technology, April 112–138

Chapter 14
A New Approach to Improving Seismic Safety Based on the Energy Theory of Reinforced Concrete Resistance

Valeriy M. Mitasov, Vladimir V. Adischev and Mariya A. Logunova

Abstract The article presents the main principles of a new approach to the calculation of reinforced concrete structures. A new chart-energy method for the calculation is developed with an additional stress–strain state stage. It helps to solve the problem of a stochastic crack. The absence of transition from a state of continuity to the state of "section with a crack" does not allow to make uniform the calculation of strength, stiffness, and fracture toughness. According to the energy theory of reinforced concrete resistance at the time of crack formation stretched concrete energy redistributes to armature, which acts as a brake element and restrains crack propagation in the cross section. The sudden nature of a crack appearance is accompanied by an instantaneous change of cross-section stress state. The change is dynamic in nature. The problem of a reinforced concrete section transition from "solid" state to state "with a crack" in a stretched zone and determining the depth of cracks was solved. The proposed approach allows to create a uniform methodology for calculating strength, stiffness, and fracture toughness, as well as offer a number of constructive measures to improve the earthquake resistance of buildings and structures. A series of experiments was conducted in relation to key provisions of the energy theory of reinforced concrete resistance. The proposed variant of a seismic load damper will significantly dissipate the earthquake energy and protect the structural system from damage. The use of internal clamps in vertical elements significantly increases a resistance to shear, while enhancing the junction of column and slab.

V. M. Mitasov (✉) · V. V. Adischev · M. A. Logunova
Reinforced Concrete Structures Department, Faculty of Civil Engineering,
Novosibirsk State University of Architecture and Civil Engineering (Sibstrin),
Novosibirsk, Russia
e-mail: mitassovv@mail.ru

V. V. Adischev
e-mail: adishch@mail.ru

M. A. Logunova
e-mail: m-nblack@yandex.ru

A. A. Kasimzade et al. (eds.), *Seismic Isolation, Structural Health Monitoring, and Performance Based Seismic Design in Earthquake Engineering*,
https://doi.org/10.1007/978-3-319-93157-9_14

14.1 Introduction

One of the main tasks of designing buildings and structures operated in earthquake zones is to ensure the ability to perceive impacts excluding the infliction of harm to health and lives of people. When assessing the risks they need to monitor a condition of structural systems from background influences and to assume the admissibility of structures injuries, sites, to preclude collapse of the building. Analysis of different cases of damage under seismic loading allows to identify the three most typical case of the destruction:

- General loss of a building stability;
- the destruction of the vertical frame elements;
- the destruction of horizontal and vertical elements junction.

14.2 General Statements

From the energy point of view, the destruction of nodes or elements takes place due to a concentrated pulse energy impact of the external environment on the structure. At the Reinforced Concrete Structures Department of The Novosibirsk State University of Architecture and Civil Engineering (Sibstrin) the energy theory of reinforced concrete resistance was established (Mitasov and Adishchev 2010a, b). In the framework of this theory the problem of "smearing" of the pulse energy to the whole structural system, as well as the dissipation of energy through special events can be posed and solved.

With increasing loads in the bending a cross section of the reinforced concrete goes through several stages of the stress–strain state, which are qualitatively different from each other. Herein the so-called limit states differ, and calculation methods on two groups of limit states are based on contradictory hypotheses and assumptions. Each stage is described by the equations of equilibrium and the relevant hypotheses, and the change of cross-section parameters in the transition processes from one state to another are calculated by means of empirical coefficients. In fact, during a crack formation there is not only a "catastrophic" change in the stress–strain state in the cross section with a crack, but also an actual change in the structure itself. Calculation of the two groups of limit states virtually ignores the real physical properties of the concrete, and also additionally does not take into account the peculiarities of the stress–strain state and physically adequate conditions for the transition from one state to another. The transition from stage I (before cracking) to stage II (operational) is conditioned by the limit value of deformations of the extreme stretched fiber of the concrete. Conditions of the transition from the operational stage to the fracture stage (hypothetical) in the theory of reinforced concrete resistance are not determined. Thus, the "pass-through" calculation from the beginning of loading to the loss of bearing capacity of the reinforced concrete element is impossible in the frame of this theory.

For the evaluation of the bearing capacity, hardness, and crack resistance of the reinforced concrete element only the operational stage presents an interest, namely, the beginning of this stage (crack initiation) and the end (loss of load-carrying ability). Consequently, the problems of a physically adequate description of the stress–strain state of reinforced concrete structures at the main stages of loading, as well as the conditions of the transition from one stage to another come to the foreground. Stage III cannot be realized, and it is of interest to assess the nature of the fracture—plastic or brittle.

At the heart of the calculation method of limit states there are three stages of the stress state of the cross section, which are qualitatively different from each other. Stage I is characterized with the displacements (deflections) of the construction, stage II—the moment of cracking, stage II—displacements and crack width. Stage IIa is a pre-destruction stage at which the formation of a plastic hinge takes place. The calculation of the bearing capacity of the first group of limit states assumes implementation of a hypothetical stage III. Analyzing the second group of limit states, it should be noted that the applied methods resemble "allowable stresses" but instead of stresses the control values of deflections and crack width are used. The absence of conditions of transition from a "continuity' state to a "section with a crack" state does not allow us to make a consistent calculation of strength, hardness, and crack resistance.

Taking into account the nonlinear deformation of concrete and discreteness of reinforcement of concrete out of the known integral calculation methods, suitable for the quite rigorous description of the deformation process of reinforced concrete structures, we find the method, based on the use of energy relations and criteria, to be the most acceptable. As the area of a subgraph of a diagram of the deformation of the material corresponds to the energy density (or specific energy) of deformation, using diagrams of deformation of materials in the construction of the computational model organically fits into the scope of the problem under consideration.

Since before cracking the qualitative state of cross sections is presented through stages Ia, and the stabilized condition—through stage II, to preserve the existing gradation, we introduce stage Ib—the stage of formation and stabilization of the cracks.

14.3 The Basic Hypotheses

Let us formulate the basic stages of changing the stress–strain state of stage Ib:

(1) at the time of the cracking the energy of concrete in tension is redistributed to the reinforcement, which acting as a braking element, restrains crack propagation along the cross section of the element:

$$\sum_j A_{\mathrm{sj}} \int\limits_{\varepsilon_{\mathrm{sj}}}^{\varepsilon_{\mathrm{sj}} + \Delta\varepsilon_{\mathrm{sj}}^{\mathrm{st}}} \sigma(\varepsilon)\mathrm{d}\varepsilon - \Delta W_{\mathrm{cr}}^{\mathrm{bt}} = 0, \tag{14.1}$$

where $\Delta\varepsilon_{\mathrm{sj}}^{\mathrm{st}}$—increment of deformation in the *j*-layer of reinforcement after the redistribution of energy of the concrete in tension $\Delta W_{\mathrm{cr}}^{\mathrm{bt}}$ on the reinforcement;

(2) the sudden nature of the crack appearance is accompanied by an instantaneous change in the stress state of the cross section, which is dynamic in nature; as there are no changes along the length of the reinforced concrete element (all redistribution occurs in the vicinity of one cross section), the voltage $\sigma_{\mathrm{s}}^{\mathrm{d}}$ («dynamic» value) in the reinforcement at the maximum disclosure of a newly formed crack is determined from the condition of equality of the increment of the specific energy of deformation and work of stresses $\sigma_{\mathrm{s}}^{\mathrm{d}}$ («static» value) on the total increment of deformations $\varepsilon_{\mathrm{s}}^{\mathrm{d}} - \varepsilon_{\mathrm{s}}^{*}$:

$$\int\limits_{\varepsilon_{\mathrm{s}}^{*}}^{\varepsilon_{\mathrm{s}}^{\mathrm{d}}} \sigma(\varepsilon)\mathrm{d}\varepsilon = \sigma_{\mathrm{s}}^{\mathrm{st}}\left[\varepsilon_{\mathrm{s}}^{\mathrm{d}} - \varepsilon_{\mathrm{s}}^{*}\right], \tag{14.2}$$

where σ_{s}^{*}, $\varepsilon_{\mathrm{s}}^{*}$—stresses and deformations in the reinforcement, respectively before cracking, "fictitious static" stresses and deformations σ_{s}^{*}, $\varepsilon_{\mathrm{s}}^{\mathrm{st}}$ after redistribution of the energy of the fractured zone of the concrete cross section are determined by the Eq. (14.1); "dynamic" stresses and deformations $\sigma_{\mathrm{s}}^{\mathrm{d}}$, $\varepsilon_{\mathrm{s}}^{\mathrm{d}}$ for each layer of the reinforcement are determined by the Eq. (14.2) and correspond to the maximum (unbalanced by external loads) crack opening Fig. 14.2;

(3) at the moment of the realization of the "dynamic" stresses $\sigma_{\mathrm{sj}}^{\mathrm{d}}$ in the reinforcing layers the statistical equilibrium in a section with cracks is absent; then the stresses in the reinforced metal decrease, the process of vibrations is of a fading character, stresses in the reinforced metal are stabilized and become equal to $\sigma_{\mathrm{sj}}^{\mathrm{cr}} < \sigma_{\mathrm{sj}}^{\mathrm{d}}$. This condition is characterized by a stable equilibrium. In the general case for the two-valued stress distribution in the cross section the equilibrium equation after the crack stabilization the can be written as follows:

$$\int\limits_{-x}^{h_{\mathrm{cr}}-x} \sigma(y)b(y)\mathrm{d}y - \sum_j \sigma_{\mathrm{sj}}^{\mathrm{cr}} A_{\mathrm{sj}} = \pm N, \tag{14.3}$$

where h_{cr}—the distance from the most compressed fiber to the tip of the crack; by—a variable width of the cross section; *N*—the resultant of all the external longitudinal forces (the sign is taken according to the nature of the impact: compression-tension).

The scheme of changing stresses in the reinforcement after the crack formation and its stabilization is shown in Figs. 14.1 and 14.2.

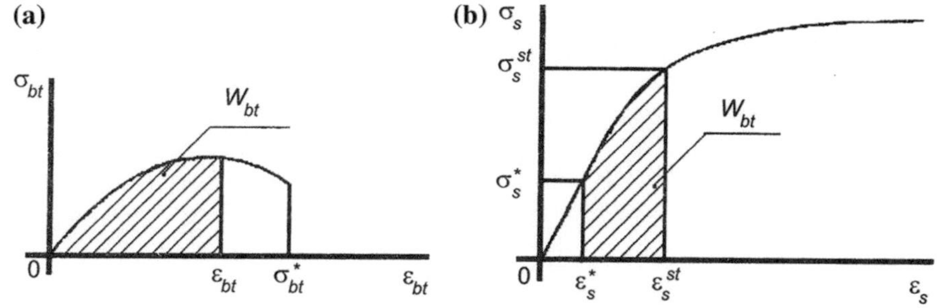

Fig. 14.1 The scheme of changes of stresses and deformations in the reinforcement after cracking **a**—The specific energy of deformation of concrete in tension before cracking; **b**—The scheme of determination of stresses in the reinforcement in a cross section with a crack

Fig. 14.2 The scheme determining the maximum stress in the reinforcement at the time of the formation of cracks

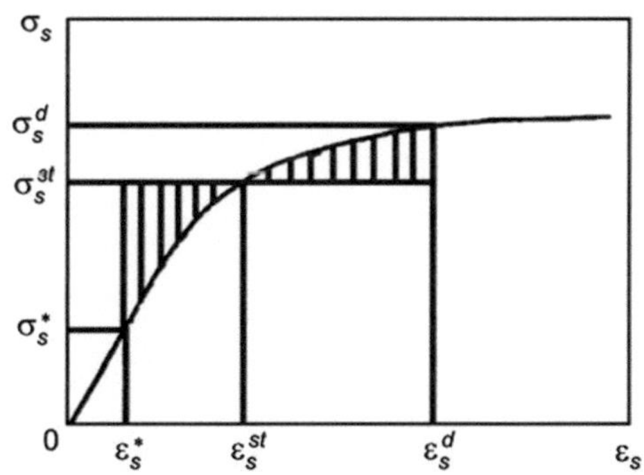

14.4 Deformation of Reinforced Concrete Beam with a Crack

Stage II occurs after the cracks formation and their stabilization in the tension zone. It is characterized by cracks growth along the height of the cross section and the formation of new ones with increasing of the external load.

The study of this stage is required for the determination of displacements, crack width and a possibility of their closure at lower loads.

Stage IIa is characterized by the formation of a "plastic hinge" in a cross section. The development of this stage can occur in two scenarios.

(a) The stresses in the reinforcement have reached the yield point (physical or conditional). The determination of stress–strain state is estimated according to the scheme of Stage Ib with the replacement of the unknown quantity, instead of a certain moment of the external forces the limit stress value in the reinforcement is given. For a closed solution in this case one equilibrium equation is enough (14.1). By (14.2) the load-carrying capacity of the cross section is determined;

(b) The compressive zone of the concrete has entered the stage of pre-fracture. This state should be evaluated in accordance with the well-known postulate: the specific energy in a pre-fracture state does not depend on the method of the load application and the deformation rate, i.e.,

$$\int_0^{\varepsilon_b^*} \sigma_b(\varepsilon)d\varepsilon = \text{const.} \tag{14.4}$$

Studying this stage allows us to describe the pre-fracture state from energy positions.

Let us consider the process of cracks formation and their stabilization.

The moment of cracking corresponds to achieving the critical value by the specific energy of deformation of concrete in tension:

$$\Delta W_{cr}^{bt} = \int_0^{h-x} b(y)\left[\int_0^{\varepsilon(y)} \psi(\varepsilon)d\varepsilon\right]dy, \tag{14.5}$$

where $\varepsilon(y)$—fiber deformation, located at a distance y from the neutral layer of the element.

After crack formation in the cross section there is a redistribution of forces. The energy of concrete in tension (14.5) is transmitted to the reinforcement, a crack "grows" in width and height, stresses in the reinforcement in the crack increase. The instantaneous nature of the cracking process generates the dynamics of the process of the redistribution forces in the cross section—on the tip of a dynamically grown crack there are the highest deformations of concrete in tension ε_{bt}^*. Then under the action of elastic forces in the reinforcement a crack is partially closed, its height decreases, the oscillations are faded. This state corresponds to the balancing of the work of the external load on the appropriate movements by the work of the internal forces.

The specific energy of deformation of the concrete in tension before the crack formation at the two-valued stress distribution in the cross section is represented by the expression (14.5). In accordance with the accepted hypotheses the dependence of any fiber deformations in the cross section from the vertical coordinate is as follows:

$$\varepsilon(y) = \left(\frac{\varepsilon_{bt}}{h-x}\right)y. \tag{14.6}$$

Then the energy of concrete in tension is determined from the formula

$$\Delta W_{\mathrm{cr}}^{\mathrm{bt}} = \int_{0}^{h-x} b(y) \int_{0}^{(\varepsilon_{\mathrm{bt}}/h-x)y} \psi(\varepsilon)\mathrm{d}\varepsilon\mathrm{d}y. \tag{14.7}$$

After the crack formation there is an unloading of the concrete in tension and energy $\Delta W_{\mathrm{cr}}^{\mathrm{bt}}$ passes completely into the energy of the tensile reinforcement. When this happens, the stresses in the reinforcement have increased to a value σ_{s}^{*}, a and the corresponding increase in the energy of reinforcement deformation is equal to σ_{s}^{*}:

$$\Delta W_{\mathrm{s}} = \sum_{j} A_{\mathrm{sj}} \int_{\varepsilon_{\mathrm{sj}}^{*}}^{\varepsilon_{\mathrm{sj}}^{\mathrm{st}}} \sigma_{\mathrm{s}}(\varepsilon)\mathrm{d}\varepsilon = \sum_{j} A_{\mathrm{sj}} \int_{\sigma_{\mathrm{sj}}^{*}}^{\sigma_{\mathrm{sj}}^{\mathrm{st}}} \sigma\varphi'(\sigma)\mathrm{d}\sigma, \tag{14.8}$$

where $\varepsilon_{\mathrm{s}} = \varphi(\sigma_{\mathrm{s}})$—the analytical expression of the diagram of reinforcement deformation.

Then $\sigma_{\mathrm{sj}}^{\mathrm{st}}$ can be found from the equation

$$\Delta W_{\mathrm{cr}}^{\mathrm{bt}} = \sum_{j} A_{\mathrm{sj}} \int_{\sigma_{\mathrm{s}}^{*}j}^{\sigma_{\mathrm{sj}}^{\mathrm{st}}} \sigma\varphi'(\sigma)\mathrm{d}\sigma. \tag{14.9}$$

Cracking occurs so quickly that a change of stress in the reinforcement can be considered as instantaneous. Stresses in the reinforcement thus increase to some dynamic value $\sigma_{\mathrm{s}}^{\mathrm{d}} > \sigma_{\mathrm{s}}^{\mathrm{st}}$. As already noted, this quantity is calculated from the Eq. (14.1), which takes the following form:

$$\int_{\sigma_{\mathrm{sj}}^{*}}^{\sigma_{\mathrm{sj}}^{\mathrm{d}}} \sigma\varphi'(\sigma)\mathrm{d}\sigma = \sigma_{\mathrm{sj}}^{\mathrm{st}}\left[\varphi\left(\sigma_{\mathrm{sj}}^{\mathrm{d}}\right) - \varphi\left(\sigma_{\mathrm{sj}}^{*}\right)\right]. \tag{14.10}$$

The most "dynamic" deformation of the extreme fiber from the compressive zone of the concrete $\varepsilon_{\mathrm{b}}^{\mathrm{d}}$ is determined from the equation

$$\frac{x_{\mathrm{cr}}}{\varepsilon_{\mathrm{b}}^{\mathrm{d}}} \int_{\varepsilon_{\mathrm{b}}^{\mathrm{d}}}^{\varepsilon_{\mathrm{bt}}^{*}} \psi(\varepsilon) b\left(\frac{x_{\mathrm{cr}}}{\varepsilon_{\mathrm{b}}^{\mathrm{d}}}\varepsilon\right)\mathrm{d}\varepsilon = \sum_{j} \sigma_{\mathrm{sj}}^{\mathrm{d}} A_{\mathrm{sj}} \tag{14.11}$$

After reaching the highest values of stresses in reinforcement and concrete the unloading take place. The unloading for the concrete can be considered in accordance with the recommendations of the work (Zhunusov and Bespaev 1972), for the reinforcement—with a linear law

$$\varepsilon_{\mathrm{sj}}^{\mathrm{cr}} = \varepsilon_{\mathrm{sj}}^{\mathrm{d}} - \frac{\left(\sigma_{\mathrm{sj}}^{\mathrm{d}} - \sigma_{\mathrm{sj}}^{\mathrm{st}}\right)}{E_{\mathrm{s}}} \tag{14.12}$$

Thus, the energy relations to solve one of the problems of the theory of resistance of reinforced concrete have been obtained—the problems of determining the stresses in the reinforcement in the cross section with a crack.

Stage III, as noted above, cannot be used in the calculation method and can serve only to distinguish between fracture scenarios—quasi-plastic and brittle.

We obtained fundamental solutions to the problem of transition from continuity state to the "cross-section with a crack" state that will allow us to create a uniform calculation method of strength, hardness, and crack resistance.

The fundamental solution of the problem of transition from the state continuity to the state "section with a crack" was obtained. That will allow creating a uniform method for calculating strength, stiffness, and fracture toughness. Analysis of Eqs. (14.1)–(14.12) allowed us to outline a series of constructive events (Fig. 14.3), which will allow to prevent the collapse of buildings and structures by increasing the stability of vertical elements, improving resistance to the efforts of extension

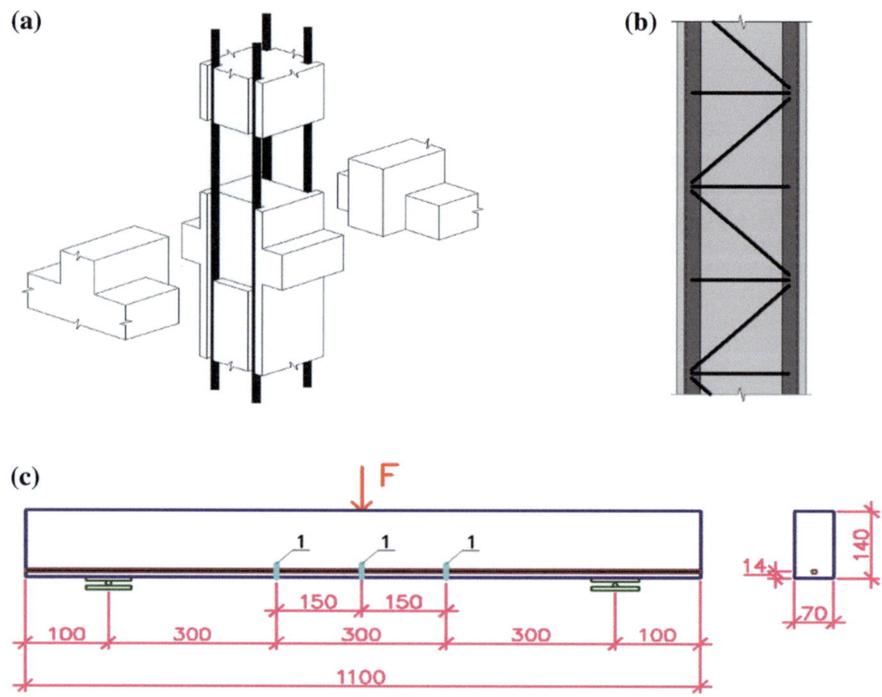

Fig. 14.3 The proposed structural measures to improve the seismic resistance

and shear, as well as due to energy dissipation through pre-organized cracks (special plates).

14.5 Constructive Measures

If the schemas of increasing stability by pre-stressing (Fig. 14.3a) are well known and used in the construction of high-rise buildings and earthquake-resistant constructions (Zhunusov and Bespaev 1972; Zhou and Jiang 1985; Saatcioglu and Yalcin 2003; Moghaddam and Samadi 2009; Dong et al. 2006), the other two (Fig. 14.3b, c) were developed at our University (RF patent No. 67603, No. 112693).

The authors conducted theoretical and experimental studies, the results of which prove that stochastic cracking (the greatest damage is to the first crack) generate shock waves that spread ahead of them, preparing the ground for the destruction of construction, and low stretch of concrete at the impact on compressed elements leads to fault them on the inclined sections.

(a) the increasing resistance of vertical elements;
(b) the increasing resistance to the efforts of extension and shear;
(c) the increase of hardness and fracture toughness by controlling crack formation

1—plates

Figure 14.4 shows diagrams and photographs of the physical experiments in the studies that were conducted by the authors at the Department of reinforced concrete structures (Logunova and Peshkov 2011; Panteleev et al. 2012; Mitasov and Logunova 2011). There were studied the work of compression elements with the interior forming frame to a vertical impact and shear with bending.

(a) the junction of the columns with the internal forming frame and slab;
(b) test compressed elements with various types of internal cages;
(c) tests of columns with inner cage on shear and bending;
(d) tests of beams with pre-organized cracks and graphs of deflections.

Experiments conducted by the authors (Fig. 14.4) allow to draw the following observations:

(1) post-tensioning of compressed elements ensures their stability and together with the internal forming frame will facilitate the work of plate-column junction.
(2) The internal forming frame almost eliminates the shear of compressed element on the inclined section.
(3) Pre-organized cracks retains the reinforced concrete element not only before but also after a shock, dissipating part of the energy of the seismic impact.

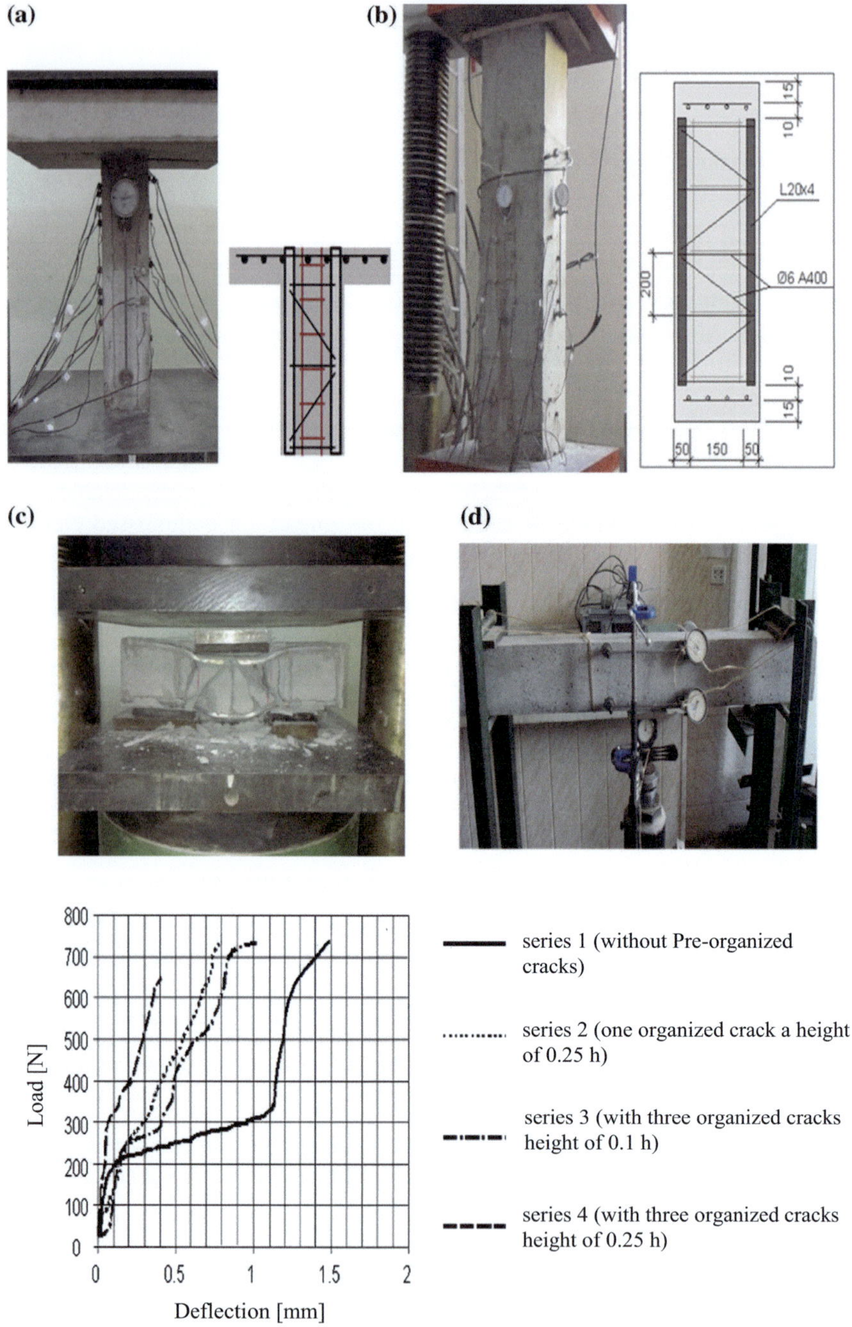

Fig. 14.4 Experimental study of constructive solutions to improve seismic resistance

14.6 Conclusions

The article presents main provisions of energy theory of reinforced concrete resistance and methods to improve the earthquake resistance of structures. The presented information allows to draw the following conclusions:

- The problem of a reinforced concrete section transition from “”solid” state to state “with a crack” in a stretched zone was solved.
- The energy theory of reinforced concrete resistance allows to provide a uniform methodology for calculating strength, stiffness and fracture toughness of structures.
- The use of pre-stressing with internal forming frame in compressed elements and a pre-organized cracks in bending elements allows to increase shear resistance of columns, their stability and to damp part of seismic energy.

References

Dong TF, Li ZB, Zhou XY (2006) Experimental study on seismic performances of precast prestressed concrete beam-to-column connections with unbonded tendons. J Beijing Univ Technol 32(2):144–148

Logunova MA, Peshkov AS (2011) Experimental research of concrete beams without cracks and with specified cracks. News High Educ Inst Constr 625(1):116–120

Mitasov VM, Adishchev VV (2010a) Main preconditions of creating energy theory of reinforced concrete strength. News High Educ Inst Constr 617(5):3–9

Mitasov VM, Adishchev VV (2010b) Main regulations of energy theory of reinforced concrete strength. News High Educ Inst Constr 618(6):3–8

Mitasov VM, Logunova MA (2011) Precast-monolithic reinforced concrete frame for housing and civil construction. In: Problems of modern concrete and reinforced concrete. P.1. concrete and reinforced concrete structures, Minsk, November

Moghaddam HA, Samadi M (2009) Seismic retrofit of large-scale reinforced concrete columns by prestressed high-strength metal strips. In: Proceedings of the structures congress (ASCE’09), Austin, May

Panteleev NN, Mitasov VM, Logunova MA (2012) The determination of bearing capacity of reinforced concrete columns with internal clip. News High Educ Inst Constr 639(3):105–110

Saatcioglu M, Yalcin C (2003) External prestressing concrete columns for improved seismic shear resistance. J Struct Eng 129(8):1057–1070

Zhou XZ, Jiang WS (1985) Experimental study on the seismic behavior of short RC columns under high axial loading. Xi’an Univ Archit Technol 6:103–106

Zhunusov TZh, Bespaev AA (1972) A study of the influence of pre-stress on seismic stability of reinforced concrete columns. Study Seismic Stab Build Struct, Alma-Ata

Chapter 15
Free Flexural Vibrations of Axially Loaded Timoshenko Beams with Internal Viscous Damping Using Dynamic Stiffness Formulation and Differential Transformation

Baran Bozyigit, Yusuf Yesilce and Hikmet Huseyin Catal

Abstract The effects of axial compressive load and internal viscous damping on the free vibration characteristics of Timoshenko beams are carried out using the dynamic stiffness formulation and the differential transformation method. The governing equations of motion are derived using the Hamilton's principle. After the analytical solution of the equation of motion has been obtained, the dynamic stiffness method (DSM) is used and the dynamic stiffness matrix of the axially loaded Timoshenko beam with internal viscous damping is constructed to calculate natural frequencies. Moreover, an efficient mathematical technique called the differential transform method (DTM) is used to solve the governing differential equations of motion. The calculated natural frequencies of Timoshenko beams with various combinations of boundary conditions using the DSM and DTM are presented and compared with the analytical results where a very good agreement is observed.

Keywords Axial load · Timoshenko beam · Internal viscous damping Differential transformation · Dynamic stiffness · Natural frequencies

15.1 Introduction

Axially loaded beams with distributed internal viscous damping are of great importance in a wide class of civil engineering and mechanical engineering structures. The effects of internal viscous damping and the axial load of a beam play an important role on its vibration characteristics and dynamic stability.

B. Bozyigit (✉) · Y. Yesilce · H. H. Catal
Department of Civil Engineering, Dokuz Eylul University,
35160 Buca, Izmir, Turkey
e-mail: baran.bozyigit@deu.edu.tr

A. A. Kasimzade et al. (eds.), *Seismic Isolation, Structural Health Monitoring, and Performance Based Seismic Design in Earthquake Engineering*,
https://doi.org/10.1007/978-3-319-93157-9_15

Several studies on vibrations and stability of axially loaded beams with viscous damping have been reported. Gürgöze and Erol (2004) investigated the eigen-characteristics of multistep Bernoulli–Euler beams carrying a tip mass subjected to nonhomogeneous external viscous damping. Cai et al. (2006) studied on an analytical approach for vibration response analysis of a Bernoulli–Euler beam with a single active constraining layer damping patch. In another study, the modal analysis of nonhomogeneous Timoshenko beams with generalized damping distributions is investigated (Sorrentino et al. 2007). Dohnal et al. (2008) investigated a uniform cantilever beam under the effect of a time-periodic axial force by using a finite-element approach. In the other study, an enhanced beam model for constrained layer damping and a parameter study of damping contribution are studied by Xie and Shepard (2009). Bending-bending vibration equations of a twisted beam with internal damping of Kelvin–Voigt type are studied using Timoshenko beam theory (Chen et al. 2013). The dynamic response of a Timoshenko beam with distributed internal viscous damping is investigated (Capsoni et al. 2013). Lin (2014) studied the forced vibration of beam subjected to a harmonic external force and with the squeezing film and thermos elastic damping in non-Fourier model. In the other study, bending–bending vibrations of an axially loaded twisted Timoshenko beam with locally distributed Kelvin–Voigt damping are investigated by Chen (2014a). In the other study, Chen (2014b) studied the vibration behavior of a cantilevered twisted Timoshenko beam with partially distributed Kelvin–Voigt damping using a finite-element method.

DSM is an effective method for free and forced vibration analyses of structures such as beams, plates, and their assemblies. Banerjee (1997) noted that the Wittrick–Williams algorithm can be used as a nonlinear eigenvalue problem is experienced. Free vibration analysis of laminated composite beams under axial compressive force is performed by Jun et al. (2008) using DSM. The effectiveness of DSM for solving free vibration problem of laminated composite beams is observed. Dynamic stiffness approach is used for free vibration analysis of pipe conveying fluid according to Timoshenko beam theory (Bao-Hui et al. 2011). The first three natural frequencies of a multi-span pipe conveying fluid are calculated. Banerjee (2012) researched free vibrations of beams carrying spring–mass systems using the DSM. The natural frequencies of rotating tapered beams are obtained according to Rayleigh beam theory by using the DSM by Banerjee and Jackson (2013). The results are compared with the first natural frequencies obtained according to Euler–Bernoulli beam theory. Su and Banerjee (2015) calculated nondimensional natural frequencies of functionally graded Timoshenko beams for different boundary conditions. DSM is applied to in-plane free vibration problem and response analysis of isotropic rectangular plates. Different boundary conditions and length/width ratios are considered (Nefovska-Danilovic and Petronijevic 2015). Bozyigit and Yesilce (2016) applied dynamic stiffness approach for free vibration analysis of moving beams according to high order shear deformation theory. Different axial tensile force and axial speed values are used to reflect their effects on natural frequencies.

The concept of DTM was first introduced by Zhou (1986). Ozgumus and Kaya (2006) investigated the out-of-plane free vibration analysis of a double tapered Bernoulli–Euler beam using DTM. Çatal (2006, 2008) used DTM for the free vibration analysis of Timoshenko beams with fixed and simply supported ends. Çatal and Çatal (2006) calculated the critical buckling loads of partially embedded Timoshenko pile in elastic soil by DTM. Free vibration analysis of a rotating, double tapered Timoshenko beam featuring coupling between flapwise bending and torsional vibrations is performed using DTM by Ozgumus and Kaya (2007). In another study, Kaya and Ozgumus (2007) used DTM to analyze the free vibration response of an axially loaded, closed section composite Timoshenko beam. Yesilce (2010, 2013) investigated the free vibration analysis of moving Bernoulli–Euler and Timoshenko beams by using DTM. Yesilce (2015) described the determination of the natural frequencies and mode shapes of the axially loaded Timoshenko multiple-step beam carrying a number of intermediate lumped masses and rotary inertias by using the numerical assembly technique and the DTM. Previous studies have shown that the DTM is an efficient tool and it has been applied to solve boundary value problems in fluid mechanics, viscoelasticity, control theory, acoustics, etc. Besides the variety of the problems that DTM may be applied to, its accuracy and simplicity in calculating the natural frequencies and plotting the mode shapes make this method outstanding among many other methods.

The free vibration analysis of simply supported, one end fixed, the other end simply supported, fixed supported and axially loaded Timoshenko beams with distributed internal viscous damping is performed in this study. At the beginning of the study, the governing equations of motion are derived by applying Hamilton's principle. In the next step, the equations of motion, including the parameters for the damping factor and the nondimensionalized multiplication factor for the axial compressive force, are solved using an efficient mathematical technique, called DTM. Besides DTM, DSM is used for calculating the natural frequencies of the axially loaded Timoshenko beams with distributed internal viscous damping. The first four mode shapes are plotted and the effects of the parameters, mentioned above, are investigated. A suitable example that studies the effects of axial compressive load and internal viscous damping on the free vibration analysis of Timoshenko beam using DSM and DTM has not been investigated by any of the studies in open literature so far.

15.2 The Mathematical Model and Formulation

An axially loaded uniform Timoshenko beam with distributed internal viscous damping is presented in Fig. 15.1.

The total kinetic energy T and the total potential energy V of the axially loaded Timoshenko beam can be written as

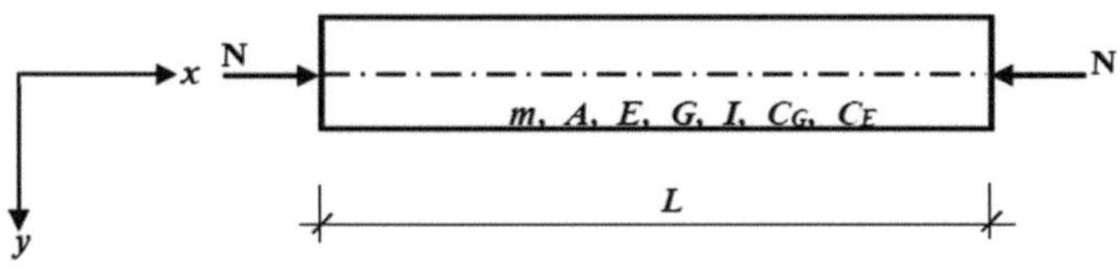

Fig. 15.1 Axially loaded Timoshenko beam with internal viscous damping

$$T = \frac{1}{2}\int_0^L \left\{ m\left(\frac{\partial y(x,t)}{\partial t}\right)^2 + \frac{mI}{A}\left(\frac{\partial \phi(x,t)}{\partial t}\right)^2 \right\} \mathrm{d}x \tag{15.1}$$

$$V = \frac{1}{2}\int_0^L \left\{ EI\left(\frac{\partial \phi(x,t)}{\partial x}\right)^2 + \frac{AG}{\bar{k}}\left(\frac{\partial y(x,t)}{\partial x} - \phi(x,t)\right)^2 - N\left(\frac{\partial y(x,t)}{\partial x}\right)^2 \right\} \mathrm{d}x \tag{15.2}$$

where $y(x,t)$ is the total transverse deflection, $\phi(x,t)$ is the angle of rotation due to bending, m is the mass per unit length of the beam, L is the length of the beam, N is the axial compressive force, A is the cross-section area, $\bar{k}$ is the shape factor due to cross-section geometry, I is the moment of inertia, E and G are the Young's modulus and shear modulus of the beam, respectively, x is the beam position, and t is time variable.

The concept of Rayleigh's dissipation function is utilized to express the dissipation function V_D of Timoshenko beam with distributed internal viscous damping as:

$$V_\mathrm{D} = \frac{1}{2}\int_0^L \left\{ C_E I\left(\frac{\partial^2 \phi(x,t)}{\partial x \partial t}\right)^2 + \frac{C_G A}{\bar{k}}\left(\frac{\partial^2 y(x,t)}{\partial x \partial t} - \frac{\partial \phi(x,t)}{\partial t}\right)^2 \right\} \mathrm{d}x \tag{15.3}$$

In Eq. (15.3), for an isotropic material, the relationship between the coefficients of the internal damping C_E and C_G is assumed to be similar to those between E and G, respectively.

The equations of motion for the axially loaded Timoshenko beam with distributed internal viscous damping are derived by applying Hamilton's principle, which is given by

$$\delta \int_{t_1}^{t_2} L_g \mathrm{d}t = 0 \tag{15.4a}$$

where

$$L_g = T - V - V_D \tag{15.4b}$$

is termed as the Lagrangian density function.

Taking the variation of the Lagrangian density function and integrating Eq. (15.4a) by parts, the equations of motion for the axially loaded Timoshenko beam with distributed internal viscous damping can be derived as $(0 \leq x \leq L)$:

$$\begin{aligned} &\frac{AG}{\bar{k}}\left(\frac{\partial^2 y(x,t)}{\partial x^2} - \frac{\partial \varphi(x,t)}{\partial x}\right) - N\frac{\partial^2 y(x,t)}{\partial x^2} - m\frac{\partial^2 y(x,t)}{\partial t^2} \\ &+ \frac{C_G A}{\bar{k}}\left(\frac{\partial^3 y(x,t)}{\partial x^2 \partial t} - \frac{\partial^2 \varphi(x,t)}{\partial x \partial t}\right) = 0 \end{aligned} \tag{15.5}$$

$$\begin{aligned} &EI\frac{\partial \phi^2(x,t)}{\partial x^2} - \frac{mI}{A}\frac{\partial \phi^2(x,t)}{\partial t^2} + \frac{AG}{\bar{k}}\left(\frac{\partial y(x,t)}{\partial x} - \phi(x,t)\right) \\ &+ C_E I\frac{\partial \phi^3(x,t)}{\partial x^2 \partial t} + \frac{C_G A}{\bar{k}}\left(\frac{\partial^2 y(x,t)}{\partial x \partial t} - \frac{\partial \phi(x,t)}{\partial t}\right) = 0 \end{aligned} \tag{15.6}$$

The bending moment function $M(x, t)$ and the shear force function $T(x, t)$ of the axial-loaded Timoshenko beam with distributed internal viscous damping are written as

$$M(x,t) = EI\frac{\partial \phi(x,t)}{\partial x} + C_E I\frac{\partial^2 \phi(x,t)}{\partial x \partial t} \tag{15.7}$$

$$\begin{aligned} T(x,t) &= \frac{AG}{\bar{k}}\gamma(x,t) - \frac{C_G A}{\bar{k}}\frac{\partial \gamma(x,t)}{\partial t} = \frac{AG}{\bar{k}}\left(\frac{\partial y(x,t)}{\partial x} - \phi(x,t)\right) \\ &- \frac{C_G A}{\bar{k}}\left(\frac{\partial^2 y(x,t)}{\partial x \cdot \partial t} - \frac{\partial \phi(x,t)}{\partial t}\right) \end{aligned} \tag{15.8}$$

where $\gamma(x,t)$ is the corresponding shear deformation.

Assuming that the motion is harmonic we substitute for $y(x, t)$ and $\phi(x,t)$ the following:

$$y(x,t) = y(x)e^{i\omega t} \tag{15.9a}$$

$$\phi(x,t) = \phi(x)e^{i\omega t} \tag{15.9b}$$

where $y(x)$ and $\phi(x)$ are the amplitudes of the total transverse deflection and the angle of rotation due to bending, respectively; ω is the natural circular frequency of the vibrating system and $i = \sqrt{-1}$. By using Eq. (15.9a, b) and introducing the dimensionless coordinate $z = x/L$, Eqs. (15.5) and (15.6) can be converted into the ordinary differential equations:

$$\left(\frac{AG + C_G Ai\omega}{\bar{k}L^2} - \frac{N}{L^2}\right)\frac{d^2y(z)}{dz^2} - \left(\frac{AG + C_G Ai\omega}{L\bar{k}}\right)\frac{d\phi(z)}{dz} + \left(m\omega^2\right)y(z) = 0 \tag{15.10}$$

$$\left(\frac{EI + C_E Ii\omega}{L^2}\right)\frac{d^2\phi(z)}{dz^2} + \left(\frac{AG + C_G Ai\omega}{L\bar{k}}\right)\frac{dy(z)}{dz} + \left(\frac{m\omega^2 I}{A} - \frac{(AG + C_G Ai\omega)}{\bar{k}}\right)\phi(z) = 0 \tag{15.11}$$

$$y(z) = C \cdot e^{isz} \tag{15.12}$$

$$\phi(z) = P \cdot e^{isz} \tag{15.13}$$

Substituting Eqs. (15.12) and (15.13) into Eqs. (15.10) and (15.11) results in

$$\left[m\omega^2 - \left(\frac{AG + C_G Ai\omega}{\bar{k}L^2} - \frac{N}{L^2}\right)s^2\right]C - \left[\left(\frac{AG + C_G Ai\omega}{L \cdot \bar{k}}\right)is\right]P = 0 \tag{15.14}$$

$$\left[\left(\frac{AG + C_G Ai\omega}{L\bar{k}}\right)is\right]C + \left(\frac{m\omega^2 I}{A} - \frac{(AG + C_G Ai\omega)}{\bar{k}} - \left(\frac{EI + C_E Ii\omega}{L^2}\right)s^2\right)P = 0 \tag{15.15}$$

Equations (15.14)–(15.15) can be written in matrix form as

$$\begin{bmatrix} A_{11} & A_{12} \\ A_{21} & A_{22} \end{bmatrix}\begin{Bmatrix} C \\ P \end{Bmatrix} = \begin{Bmatrix} 0 \\ 0 \end{Bmatrix} \tag{15.16}$$

where

$$A_{11} = m\omega^2 - \left(\frac{AG + C_G Ai\omega}{\bar{k}L^2} - \frac{N}{L^2}\right)s^2 \tag{15.17a}$$

$$A_{21} = -A_{12} = \left(\frac{AG + C_G Ai\omega}{L\bar{k}}\right)is \tag{15.17b}$$

$$A_{22} = \frac{m\omega^2 I}{A} - \frac{(AG + C_G Ai\omega)}{\bar{k}} - \left(\frac{EI + C_E Ii\omega}{L^2}\right)s^2 \tag{15.17c}$$

The nontrivial solution is obtained when the determinant of the coefficient matrix is set up to zero. Thus, we have a fourth-order equation with the unknowns, resulting in four values and the general solution functions can be written as:

$$y(z,t) = \left[C_1 e^{is_1 z} + C_2 e^{is_2 z} + C_3 e^{is_3 z} + C_4 e^{is_4 z}\right] e^{i\omega t} \tag{15.18}$$

$$\phi(z,t) = \left[P_1 e^{is_1 z} + P_2 e^{is_2 z} + P_3 e^{is_3 z} + P_4 e^{is_4 z}\right] e^{i\omega t} \tag{15.19}$$

The eight constants, C_1, …, C_4 and P_1, …, P_4 will be found from Eqs. (15.14), (15.15) and boundary conditions.

The bending moment and shear force functions of the axially loaded Timoshenko beam with distributed internal viscous damping can be obtained by using Eqs. (15.7) and (15.8) as:

$$M(z,t) = \left[\left(\frac{EI + C_E I i\omega}{L}\right)\frac{d\phi(z)}{dz}\right] e^{i\omega t} \tag{15.20}$$

$$T(z,t) = \left[\frac{(AG - C_G A i\omega)}{\bar{k}L}\frac{dy(z)}{dz} - \left(\frac{AG - C_G A i\omega}{\bar{k}}\right)\phi(z)\right] e^{i\omega t} \tag{15.21}$$

15.3 The Differential Transform Method (DTM)

DTM is a semi-analytic transformation technique based on Taylor series expansion and is a useful tool to obtain analytical solutions of the differential equations. Certain transformation rules are applied and the governing differential equations and the boundary conditions of the system are transformed into a set of algebraic equations in terms of the differential transforms of the original functions in DTM. The solution of these algebraic equations gives the desired solution of the problem. The DTM differs from Taylor series as Taylor series method requires symbolic computation of the necessary derivatives of the data functions and is expensive for large orders. DTM is an iterative procedure to obtain analytic Taylor series solutions of differential equations (Yesilce 2015).

A function $y(z)$, which is analytic in a domain D, can be represented by a power series with a center at $z = z_0$, any point in D. The differential transform of the function $y(z)$ is given by

$$Y(k) = \frac{1}{k!}\left(\frac{d^k y(z)}{dz^k}\right)_{z=z_0} \tag{15.22}$$

where $y(z)$ is the original function and $Y(k)$ is the transformed function. The inverse transformation is defined as

$$y(z) = \sum_{k=0}^{\infty} (z - z_0)^k Y(k) \tag{15.23}$$

From Eqs. (15.22) and (15.23), we get

$$y(z) = \sum_{k=0}^{\infty} \frac{(z - z_0)^k}{k!} \left(\frac{d^k y(z)}{dz^k} \right)_{z=z_0} \tag{15.24}$$

Equation (15.24) implies that the concept of the differential transformation is derived from Taylor's series expansion, but the method does not evaluate the derivatives symbolically. However, relative derivatives are calculated by iterative procedure that are described by the transformed equations of the original functions. In real applications, the function $y(z)$ in Eq. (15.23) is expressed by a finite series and can be written as:

$$y(z) = \sum_{k=0}^{\bar{N}} (z - z_0)^k Y(k) \tag{15.25}$$

Equation (15.25) implies that $\sum_{k=\bar{N}+1}^{\infty} (z - z_0)^k Y(k)$ is negligibly small. Where $\bar{N}$ is the series size and the value of $\bar{N}$ depends on the convergence of the eigenvalues.

Theorems that are frequently used in differential transformation of the differential equations and the boundary conditions are introduced in Tables 15.1 and 15.2, respectively.

15.3.1 Application of DTM for Solving Equations of Motion

Equations (15.10) and (15.11) can be rewritten as follows:

$$\begin{aligned} \frac{d^2 y(z)}{dz^2} = & \left[\frac{(AG + \eta AGi\omega) \cdot L^3}{(AG + \eta AGi\omega)L^2 - \bar{k}N_r EI} \right] \frac{d\phi(z)}{dz} \\ & - \left[\frac{m\omega^2 \bar{k} L^4}{(AG + \eta AGi\omega)L^2 - \bar{k}N_r EI} \right] y(z) \end{aligned} \tag{15.26}$$

Table 15.1 DTM theorems used for equations of motion

Original function	Transformed function
$y(z) = u(z) \pm v(z)$	$Y(k) = U(k) \pm V(k)$
$y(z) = a \cdot u(z)$	$Y(k) = a \cdot U(k)$
$y(z) = \frac{d^m u(z)}{dz^m}$	$Y(k) = \frac{(k+m)!}{k!} \cdot U(k+m)$
$y(z) = u(z) \cdot v(z)$	$Y(k) = \sum_{r=0}^{k} U(r) \cdot V(k-r)$

Table 15.2 DTM theorems used for boundary conditions

$z=0$		$z=1$	
Original boundary conditions	Transformed boundary conditions	Original boundary conditions	Transformed boundary conditions
$y(0)=0$	$Y(0)=0$	$y(1)=0$	$\sum_{k=0}^{\infty} Y(k)=0$
$\frac{dy}{dz}(0)=0$	$Y(1)=0$	$\frac{dy}{dz}(1)=0$	$\sum_{k=0}^{\infty} kY(k)=0$
$\frac{d^2y}{dz^2}(0)=0$	$Y(2)=0$	$\frac{d^2y}{dz^2}(1)=0$	$\sum_{k=0}^{\infty} k(k-1)Y(k)=0$
$\frac{d^3y}{dz^3}(0)=0$	$Y(3)=0$	$\frac{d^3y}{dz^3}(1)=0$	$\sum_{k=0}^{\infty} k(k-1)(k-2)Y(k)=0$

$$\begin{aligned}\frac{d^2\phi(z)}{dz^2} = & -\left[\frac{(AG+\eta AGi\omega)L}{(EI+\eta EIi\omega)\bar{k}}\right]\frac{dy(z)}{dz} \\ & + \left[\frac{(AG+\eta AGi\omega)L^2}{(EI+\eta EIi\omega)\bar{k}} - \frac{mI\omega^2L^2}{(EI+\eta EIi\omega)A}\right]\phi(z)\end{aligned} \tag{15.27}$$

where

$$N_r = \frac{NL^2}{EI}\,(\text{Nondimensionalized multiplication factor for compressive force}) \tag{15.28a}$$

$$C_E = \eta E \tag{15.28b}$$

$$C_E = \eta G \tag{15.28c}$$

$$\eta = \xi\sqrt{\frac{mI}{EA^2}}(\xi\ \text{is nondimensionalized damping value called as damping factor}) \tag{15.28d}$$

The differential transformation is applied to Eqs. (15.26) and (15.27) by using the theorems introduced in Table 5.1 and the following expressions are obtained:

$$\begin{aligned}Y(k+2) = & \left[\frac{(AG+\eta AGi\omega)L^3}{(AG+\eta AGi\omega)L^2 - \bar{k}N_rEI}\right]\frac{\Phi(k+1)}{(k+2)} \\ & - \left[\frac{m\omega^2\bar{k}L^4}{(AG+\eta AGi\omega)L^2 - \bar{k}N_rEI}\right]\frac{Y(k)}{(k+1)(k+2)}\end{aligned} \tag{15.29}$$

$$\Phi(k+2) = -\left[\frac{(AG+\eta AGi\omega)L}{(EI+\eta EIi\omega)\bar{k}}\right]\frac{Y(k+1)}{(k+2)} + \left[\frac{(AG+\eta AGi\omega)L^2}{(EI+\eta EIi\omega)\bar{k}} - \frac{mI\omega^2L^2}{(EI+\eta EIi\omega)A}\right]\frac{\Phi(k)}{(k+1)(k+2)} \tag{15.30}$$

where $Y(k)$ is the transformed function of y(z) and $\Phi(k)$ is the transformed function of $\phi(z)$.

The boundary conditions of a simply supported Timoshenko beam with distributed internal viscous damping are given below:

$$y(z=0)=0 \tag{15.31a}$$

$$M(z=0)=0 \tag{15.31b}$$

$$y(z=1)=0 \tag{15.31c}$$

$$M(z=1)=0 \tag{15.31d}$$

Applying DTM to Eqs. (15.31a)–(15.31d) and using the theorems introduced in Table 15.2, the transformed boundary conditions of a simply supported beam are obtained as

$$\text{for } z=0; \quad Y(0)=\Phi(1)=0 \tag{15.32a}$$

$$\text{for } z=1; \quad \sum_{k=0}^{\bar{N}} Y(k) = \sum_{k=0}^{\bar{N}} \bar{M}(k) = 0 \tag{15.32b}$$

where $\bar{M}(k)$ is the transformed function of $M(z)$.

The boundary conditions of a fixed-fixed Timoshenko beam with distributed internal viscous damping are given below

$$y(z=0)=0 \tag{15.33a}$$

$$\varphi(z=0)=0 \tag{15.33b}$$

$$y(z=1)=0 \tag{15.33c}$$

$$\varphi(z=1)=0 \tag{15.33d}$$

Applying DTM to Eqs. (15.33a)–(15.33d), the transformed boundary conditions of a fixed-fixed beam are obtained as:

$$\text{for } z=0; \quad Y(0)=\Phi(0)=0 \tag{15.34a}$$

$$\text{for } z = 1; \quad \sum_{k=0}^{\bar{N}} Y(k) = \sum_{k=0}^{\bar{N}} \Phi(k) = 0 \tag{15.34b}$$

The boundary conditions of one end (z = 0) fixed and the other end (z = 1) simply supported Timoshenko beam are given below:

$$y(z = 0) = 0 \tag{15.35a}$$

$$\varphi(z = 0) = 0 \tag{15.35b}$$

$$y(z = 1) = 0 \tag{15.35c}$$

$$M(z = 1) = 0 \tag{15.35d}$$

Applying differential transformation to Eqs. (15.35a)–(15.35d), the transformed boundary conditions of one end fixed and the other end simply supported beam are obtained as

$$\text{for } z = 0; \quad Y(0) = \Phi(0) = 0 \tag{15.36a}$$

$$\text{for } z = 1; \quad \sum_{k=0}^{\bar{N}} Y(k) = \sum_{k=0}^{\bar{N}} \bar{M}(k) = 0 \tag{15.36b}$$

For simply supported beam, substituting the boundary conditions expressed in Eqs. (15.32a) and (15.32b) into Eqs. (15.29) and (15.30), and taking $Y(1) = \mathrm{c}_1$, $\Phi(0) = \mathrm{c}_2$; for fixed-fixed supported beam, substituting the boundary conditions expressed in Eqs. (15.34a) and (15.34b) into Eqs. (15.29) and (15.30), and taking $Y(1) = \mathrm{c}_1$, $\Phi(1) = \mathrm{c}_2$; for one end fixed and the other end simply supported beam, substituting the boundary conditions expressed in Eqs. (15.36a) and (15.36b) into Eqs. (15.29) and (15.30), and taking $Y(1) = \mathrm{c}_1$, $\Phi(1) = \mathrm{c}_2$; the following matrix expression is obtained:

$$\begin{bmatrix} \bar{A}_{11}^{(\bar{N})}(\omega) & \bar{A}_{12}^{(\bar{N})}(\omega) \\ \bar{A}_{21}^{(\bar{N})}(\omega) & \bar{A}_{22}^{(\bar{N})}(\omega) \end{bmatrix} \begin{Bmatrix} c_1 \\ c_2 \end{Bmatrix} = \begin{Bmatrix} 0 \\ 0 \end{Bmatrix} \tag{15.37}$$

where c_1 and c_2 are constants and $\bar{A}_{j1}^{(\bar{N})}(\omega)$, $\bar{A}_{j2}^{(\bar{N})}(\omega)$ (j = 1, 2) are polynomials of ω corresponding $\bar{N}$.

In the last step, for nontrivial solution, equating the coefficient matrix that is given in Eq. (15.37) to zero one determines the natural frequencies of the vibrating system as given in Eq. (15.38).

$$\begin{vmatrix} \bar{A}_{11}^{(\bar{N})}(\omega) & \bar{A}_{12}^{(\bar{N})}(\omega) \\ \bar{A}_{21}^{(\bar{N})}(\omega) & \bar{A}_{22}^{(\bar{N})}(\omega) \end{vmatrix} = 0 \quad (15.38)$$

The jth estimated eigenvalue, $\omega_j^{(\bar{N})}$ corresponds to $\bar{N}$ and the value of $\bar{N}$ is determined as

$$\left|\omega_j^{(\bar{N})} - \omega_j^{(\bar{N}-1)}\right| \leq \varepsilon \quad (15.39)$$

where $\omega_j^{(\bar{N}-1)}$ is the jth estimated eigenvalue corresponding to $(\bar{N}-1)$ and ε is the small tolerance parameter. If Eq. (15.39) is satisfied, the jth estimated eigenvalue, $\omega_j^{(\bar{N})}$ is obtained.

The procedure explained below can be used to plot the mode shapes of the axially loaded Timoshenko beam with distributed internal viscous damping. The following equalities can be written by using Eq. (15.37):

$$\bar{A}_{11}(\omega)c_1 + \bar{A}_{12}(\omega)c_2 = 0 \quad (15.40)$$

Using Eq. (15.40), the constant c_2 can be obtained in terms of c_1 as follows:

$$c_2 = -\frac{\bar{A}_{11}(\omega)}{\bar{A}_{12}(\omega)}c_1 \quad (15.41)$$

All transformed functions can be expressed in terms of ω, c_1 and c_2. Since c_2 has been written in terms of c_1 above, $Y(k)$, $\Phi(k)$ and $\bar{M}(k)$ can be expressed in terms c_1 as follows:

$$Y(k) = Y(\omega, c_1) \quad (15.42a)$$

$$\Phi(k) = \Phi(\omega, c_1) \quad (15.42b)$$

$$\bar{M}(k) = \bar{M}(\omega, c_1) \quad (15.42c)$$

The mode shapes can be plotted for several values of ω by using Eq. (15.42a).

15.4 Dynamic Stiffness Formulation

The dynamic stiffness matrix of a beam relates the amplitudes of end forces to the amplitudes of end displacement of a beam. The vector of end displacements of beam and the vector of coefficients are given in Eqs. (15.43) and (15.44), respectively.

$$\delta = [y_0 \quad y_1 \quad \phi_0 \quad \phi_1]^{\mathrm{T}} \tag{15.43}$$

$$C = [C_1 \quad C_2 \quad C_3 \quad C_4]^{\mathrm{T}} \tag{15.44}$$

where

$$y_0 = y(z=0), y_1 = y(z=1), \phi_0 = \phi(z=0), \phi_1 = \phi(z=1)$$

Equations (15.18) and (15.19) are used to obtain Eq. (15.45):

$$\begin{bmatrix} y_0 \\ y_1 \\ \phi_0 \\ \phi_1 \end{bmatrix} = \begin{bmatrix} 1 & 1 & 1 & 1 \\ e^{is_1} & e^{is_2} & e^{is_3} & e^{is_4} \\ K_1 & K_2 & K_3 & K_4 \\ K_1 e^{is_1} & K_2 e^{is_2} & K_3 e^{is_3} & K_4 e^{is_4} \end{bmatrix} \begin{bmatrix} C_1 \\ C_2 \\ C_3 \\ C_4 \end{bmatrix} \tag{15.45}$$

where

$$K_1 = \frac{-\left(\frac{AG + C_G A i\omega}{kL^2} - \frac{N}{L^2}\right) s_1^2 + m\omega^2}{\left(\frac{AG + C_G A i\omega}{Lk}\right) i s_1}; \quad K_2 = \frac{-\left(\frac{AG + C_G A i\omega}{kL^2} - \frac{N}{L^2}\right) s_2^2 + m\omega^2}{\left(\frac{AG + C_G A i\omega}{L \cdot k}\right) i s_2};$$

$$K_3 = \frac{-\left(\frac{AG + C_G A i\omega}{kL^2} - \frac{N}{L^2}\right) s_3^2 + m\omega^2}{\left(\frac{AG + C_G A \cdot i\omega}{Lk}\right) i s_3}; \quad K_4 = \frac{-\left(\frac{AG + C_G A i\omega}{kL^2} - \frac{N}{L^2}\right) s_4^2 + m\omega^2}{\left(\frac{AG + C_G A i\omega}{Lk}\right) i s_4}$$

The closed form of Eq. (15.45) is presented in Eq. (15.46):

$$\delta = \Delta C \tag{15.46}$$

where

$$\Delta = \begin{bmatrix} 1 & 1 & 1 & 1 \\ e^{is_1} & e^{is_2} & e^{is_3} & e^{is_4} \\ K_1 & K_2 & K_3 & K_4 \\ K_1 e^{is_1} & K_2 e^{is_2} & K_3 e^{is_3} & K_4 e^{is_4} \end{bmatrix}$$

The vector of end forces of a member is given by

$$F = [T_0 \quad T_1 \quad M_0 \quad M_1]^T \tag{15.47}$$

where

$$T_0 = T(z=0), T_1 = T(z=1), M_0 = M(z=0), M_1 = M(z=1)$$

It should be noted that the following sign convention is valid for DSM.

$$T_0 = -T_1, M_0 = -M_1 \tag{15.48}$$

Equations (15.20) and (15.21) are used to construct the matrix form below

$$\begin{bmatrix} T_0 \\ T_1 \\ M_0 \\ M_1 \end{bmatrix} = \kappa \begin{bmatrix} C_1 \\ C_2 \\ C_3 \\ C_4 \end{bmatrix} \tag{15.49}$$

The closed form of Eq. (15.49) is presented in Eq. (15.50):

$$F = \kappa C \tag{15.50}$$

where

$$\kappa = \begin{bmatrix} R_1 is_1 - R_2 K_1 & R_1 is_2 - R_2 K_2 & R_1 is_3 - R_2 K_3 & R_1 is_4 - R_2 K_4 \\ (R_1 is_1 - R_2 K_1)e^{is_1} & (R_1 is_2 - R_2 K_2)e^{is_2} & (R_1 is_3 - R_2 K_3)e^{is_3} & (R_1 is_4 - R_2 K_4)e^{is_4} \\ R_3 is_1 K_1 & R_3 is_2 K_2 & R_3 is_3 K_3 & R_3 is_4 K_4 \\ (R_3 is_1 K_1)e^{is_1} & (R_3 is_2 K_2)e^{is_2} & (R_3 is_3 K_3)e^{is_3} & (R_3 is_4 K_4)e^{is_4} \end{bmatrix}$$

$$R_1 = \frac{(AG - C_G Ai\omega)}{\bar{k}L}; \quad R_2 = \left(\frac{AG - C_G Ai\omega}{\bar{k}}\right); \quad R_3 = \left(\frac{EI + C_E Ii\omega}{L}\right)$$

The dynamic stiffness matrix of the beam can be obtained by using Eqs. (15.46) and (15.50):

$$F = \kappa(\Delta)^{-1}\delta \tag{15.51}$$

$$K^* = \kappa(\Delta)^{-1} \tag{15.52}$$

K^* denotes dynamic stiffness matrix of a Timoshenko beam with internal viscous damping and subjected to axial compression force.

15.5 Numerical Analysis and Discussions

Axially loaded Timoshenko beams with distributed internal viscous damping are considered in the numerical analysis. The first four natural frequencies, f_i ($i = 1, \ldots, 4$ where $f_i = 2\pi\omega_i$) are calculated by using computer programs prepared in Matlab by the authors. The natural frequencies are calculated by equating the determinant of the coefficient matrix to zero for the analytical and differential transformation solutions. In the DSM, the natural frequencies are calculated by applying boundary conditions to K^* and using the equation below:

$$|K^*| = 0 \tag{15.53}$$

The numerical results of this paper are obtained based on uniform, rectangular Timoshenko beams with the following data:

m = 0.31250 kNs^2/m; $EI = 7.28086 \times 10^4$ kNm^2; $AG = 1.344159 \times 10^6$ kN; $\bar{k} = 6/5$; L = 3.0 m; N_r = 0.00, 0.50 and 1.00; ξ = 0.00, 0.10 and 0.20

Using the DTM and DSM, the frequency values of the simply supported Timoshenko beam for the first four modes are presented in Table 15.3. In Table 15.4, the first four frequency values of one end fixed, the other end simply supported Timoshenko beam can be seen. The fixed-fixed Timoshenko beam's first four frequency values are presented in Table 15.5. The first four mode shapes of the axially loaded Timoshenko beam with various boundary conditions for N_r = 1 and ξ = 0.2 are shown in Figs. 15.2, 15.3 and 15.4.

For all boundary conditions, as the axial compressive force acting to beams is increased with constant damping, the natural frequency values are decreased. This result indicates that the increasing axial compressive force leads to the reduction in natural frequencies for all types of boundary conditions. This result is very important for the effect of axial compressive force.

For the fixed supported axially loaded Timoshenko beams, an increase in natural frequency values is observed for the condition of N_r being constant and the values of the damping factor is increased. This result indicates that the increasing damping factor leads to an augmentation in natural frequency values for fixed-fixed boundary condition.

For simply supported and one end fixed, the other end simply supported boundary conditions, when the damping factor is increased with constant axial compressive load, a decrease is observed in natural frequency values of the first mode and the fourth mode and an increase is observed in natural frequency values of the second mode.

For the constant N_r and increasing damping factor, it is observed that the natural frequency values of the third mode of simply supported beam is increased. However, the third mode frequency of fixed-simple supported beam is decreased when the damping factor is increased with constant axial compressive force.

In the application of the DTM, the natural frequency values of the axially loaded Timoshenko beams with internal viscous damping are calculated by increasing series size $\bar{N}$. In Tables 15.3, 15.4 and 15.5, convergences of the first four natural frequencies are introduced. It is seen that the series size varies between 14 and 30 for perfect convergence in the DTM application for the first four modes of axially loaded Timoshenko beams with internal viscous damping. Additionally, it is observed that higher modes appear when more terms are taken into account in the DTM applications. Thus, depending on the order of the required mode, one must try a few values for the term number at the beginning of the calculations in order to find the adequate number of terms.

It is observed that the DSM is a reliable method for free vibration analysis of axially loaded Timoshenko beams with internal viscous damping.

Table 15.3 The first four natural frequencies of the simply supported Timoshenko beam for different values of N_r and ξ

f_i (Hz)	Method	$\xi = 0$			$\xi = 0.1$			$\xi = 0.2$		
		N_r			N_r			N_r		
		0	0.5	1	0	0.5	1	0	0.5	1
f_1	DTM ($\bar{N} = 14$)	80.5949	78.3774	76.0954	80.5949	78.3774	76.0954	80.5948	78.3773	76.0953
	ANM	80.5949	78.3774	76.0954	80.5949	78.3774	76.0954	80.5948	78.3773	76.0953
	DSM	80.5949	78.3774	76.0954	80.5949	78.3774	76.0954	80.5944	78.3773	76.0953
f_2	DTM ($\bar{N} = 20$)	289.2450	286.8833	284.5019	289.2472	286.8854	284.5040	289.2539	286.8919	284.5102
	ANM	289.2450	286.8833	284.5019	289.2472	286.8854	284.5040	289.2539	286.8919	284.5102
	DSM	289.2450	286.8833	284.5019	289.2429	286.8811	284.4997	289.2366	286.8747	284.0629
f_3	DTM ($\bar{N} = 24$)	569.4642	566.8363	564.1960	569.4742	566.8464	564.2062	569.5047	566.8772	564.2373
	ANM	569.4642	566.8363	564.1960	569.4742	566.8464	564.2062	569.5047	566.8772	564.2373
	DSM	569.4642	566.8363	564.1960	569.5492	566.9205	564.2795	569.8036	567.1727	564.5292
f_4	DTM ($\bar{N} = 28$)	883.6495	880.6716	877.6833	883.5194	880.5446	877.5593	883.1457	880.1796	877.2030
	ANM	883.6495	880.6716	877.6833	883.5194	880.5446	877.5593	883.1457	880.1796	877.2030
	DSM	883.6495	880.6716	877.6833	884.7772	881.7946	878.8015	883.4128	880.4127	877.4021

Table 15.4 The first four natural frequencies of one end fixed, the other end simply supported Timoshenko beam for different values of N_r and ξ

		$\xi = 0$			$\xi = 0.1$			$\xi = 0.2$		
f_i (Hz)	Method	N_r			N_r			N_r		
		0	0.5	1	0	0.5	1	0	0.5	1
f_1	DTM ($\bar{N} = 16$)	119.1143	117.4462	115.7516	119.1141	117.4460	115.7514	119.1136	117.4455	115.7509
	ANM	119.1143	117.4462	115.7516	119.1141	117.4460	115.7514	119.1136	117.4455	115.7509
	DSM	119.1143	117.4462	115.7516	119.1150	117.4469	115.7522	119.1171	117.4489	115.7543
f_2	DTM ($\bar{N} = 20$)	338.5120	336.4291	334.3328	338.5129	336.4298	334.3335	338.5156	336.4322	334.3355
	ANM	338.5120	336.4291	334.3328	338.5129	336.4298	334.3335	338.5156	336.4322	334.3355
	DSM	338.5120	336.4291	334.3328	338.3975	336.4734	334.3769	338.6911	336.6071	334.5099
f_3	DTM ($\bar{N} = 24$)	613.9415	611.4768	609.0019	613.9375	611.4729	608.9981	613.9263	611.4621	608.9877
	ANM	613.9415	611.4768	609.0019	613.9375	611.4729	608.9981	613.9263	611.4621	608.9877
	DSM	613.9415	611.4768	609.0019	614.5247	612.0577	609.5806	614.3902	611.9153	609.4303
f_4	DTM ($\bar{N} = 28$)	917.9218	915.0452	912.1592	917.7442	914.8706	911.9876	917.2354	914.3707	911.4965
	ANM	917.9218	915.0452	912.1592	917.7442	914.8706	911.9876	917.2354	914.3707	911.4965
	DSM	917.9218	915.0452	912.1592	918.4556	915.5612	912.6574	917.8426	915.0124	911.9886

Table 15.5 The first four natural frequencies of the fixed-fixed Timoshenko beam for different values of N_r and ξ

		$\xi = 0$			$\xi = 0.1$			$\xi = 0.2$		
f_i (Hz)	Method	N_r			N_r			N_r		
		0	0.5	1	0	0.5	1	0	0.5	1
f_1	DTM ($\bar{N} = 16$)	162.5348	161.2862	160.0261	162.5357	161.2871	160.0270	162.5385	161.2897	160.0295
	ANM	162.5348	161.2862	160.0261	162.5357	161.2871	160.0270	162.5385	161.2897	160.0295
	DSM	162.5348	161.2862	160.0261	162.6000	161.3459	160.0802	162.6746	161.4148	160.1435
f_2	DTM ($\bar{N} = 20$)	385.2480	383.3589	381.4597	385.2666	383.3772	381.4777	385.3221	383.4321	381.5319
	ANM	385.2480	383.3589	381.4597	385.2666	383.3772	381.4777	385.3221	383.4321	381.5319
	DSM	385.2480	383.3589	381.4597	386.2298	384.3244	382.4088	387.3330	385.4102	383.4772
f_3	DTM ($\bar{N} = 24$)	655.4122	653.0751	650.7295	655.4927	653.1554	650.8095	655.7342	653.3963	651.0497
	ANM	655.4122	653.0751	650.7295	655.4927	653.1554	650.8095	655.7342	653.3963	651.0497
	DSM	655.4122	653.0751	650.7295	656.2571	653.8838	651.5017	656.2989	653.8904	651.4732
f_4	DTM ($\bar{N} = 30$)	949.7997	947.0060	944.2036	949.8921	947.1003	944.3001	950.1785	947.3929	944.5986
	ANM	949.7997	947.0060	944.2036	949.8921	947.1003	944.3001	950.1785	947.3929	944.5986
	DSM	949.7997	947.0060	944.2036	950.4602	947.6031	944.7371	950.6756	947.7414	944.9369

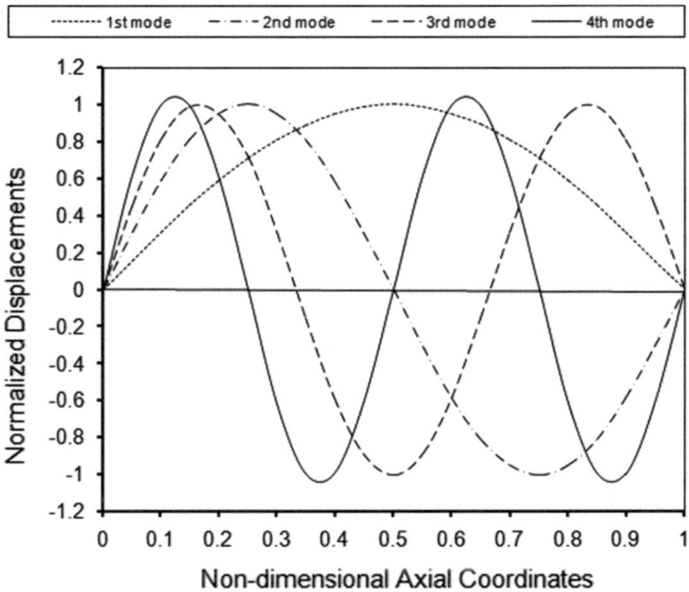

Fig. 15.2 The first four mode shapes of the simply supported and axial-loaded Timoshenko beam with internal viscous damping, $N_r = 1.00$ and $\xi = 0.20$

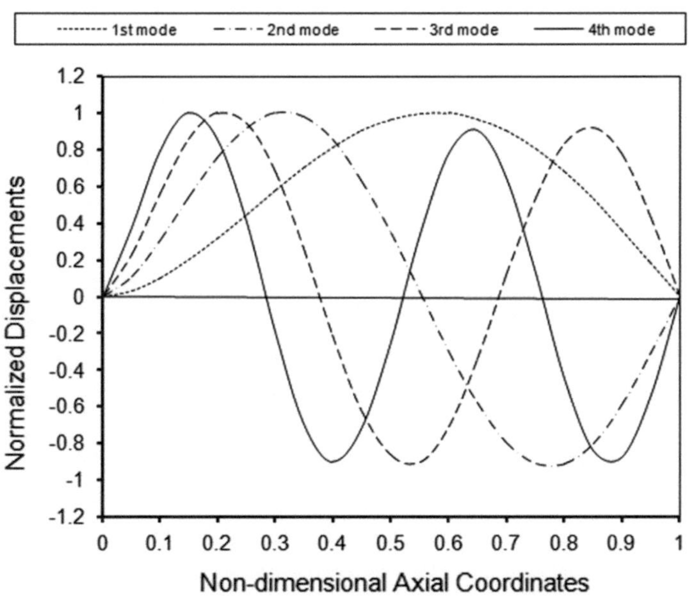

Fig. 15.3 The first four mode shapes of one end fixed, the other end simply supported and axial-loaded Timoshenko beam with internal viscous damping, $N_r = 1.00$ and $\xi = 0.20$

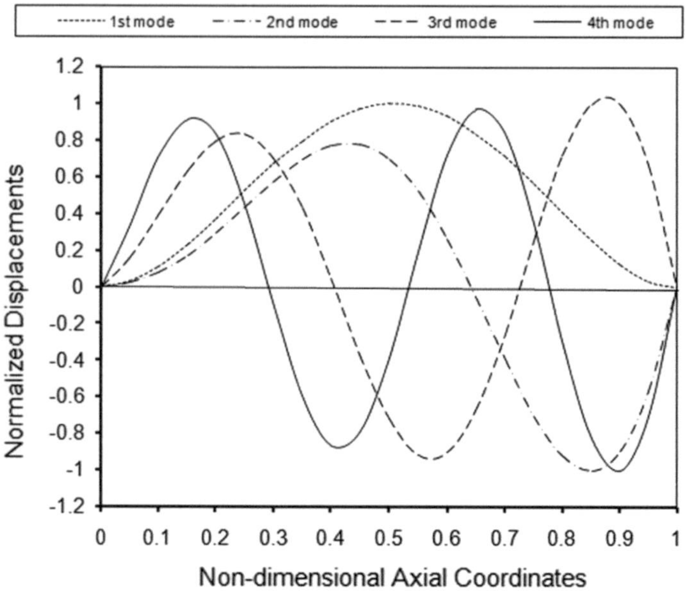

Fig. 15.4 The first four mode shapes of the fixed-fixed and axial-loaded Timoshenko with internal viscous damping, $N_r = 1.00$ and $\xi = 0.20$

15.6 Conclusions

The effects of viscous damping with axial compressive load on natural frequencies of Timoshenko beams are observed for different support conditions. This study reveals that DSM and DTM can be used effectively for free vibration analysis of axially loaded Timoshenko beams including internal viscous damping. The procedure of DSM is simple when compared to DTM. The application of the DTM to both the equations of motion and the boundary conditions seem to be involved computationally. However, all the algebraic calculations are finished quickly using symbolic computational software. Besides all these, the results show that DTM solutions converge fast. When the results of the DTM and DSM are compared with the results of analytical method, very good agreement is observed.

References

Banerjee JR (1997) Dynamic stiffness for structural elements: a general approach. Comput Struct 63:101–103

Banerjee JR (2012) Free vibration of beams carrying spring-mass systems—a dynamic stiffness approach. Comput Struct 104–105:21–26

Banerjee JR, Jackson DR (2013) Free vibration of a rotating tapered Rayleigh beam: a dynamic stiffness method of solution. Comput Struct 124:11–20
Bao-hui L, Hang-shan G, Hong-bo Z et al (2011) Free vibration analysis of multi-span pipe conveying fluid with dynamic stiffness method. Nucl Eng Des 241:666–671
Bozyigit B, Yesilce Y (2016) Dynamic stiffness approach and differential transformation for free vibration analysis of a moving Reddy-Bickford beam. Struct Eng Mech 58(5):847–868
Cai C, Zheng H, Hung KC et al (2006) Vibration analysis of a beam with an active constraining layer damping patch. Smart Mater Struct 15:147–156
Capsoni A, Viganò GM, Hani KB (2013) On damping effects in Timoshenko beams. Int J Mech Sci 73:27–39
Çatal S (2006) Analysis of free vibration of beam on elastic soil using differential transform method. Struct Eng Mech 24(1):51–62
Çatal S (2008) Solution of free vibration equations of beam on elastic soil by using differential transform method. Appl Math Model 32:1744–1757
Çatal S, Çatal HH (2006) Buckling analysis of partially embedded pile in elastic soil using differential transform method. Struct Eng Mech 24(2):247–268
Chen WR (2014a) Parametric studies on bending vibration of axially-loaded twisted Timoshenko beams with locally distributed Kelvin-Voigt damping. Int J Mech Sci 88:61–70
Chen WR (2014b) Effect of local Kelvin-Voigt damping on eigenfrequencies of cantilevered twisted Timoshenko beams. Procedia Eng 79:160–165
Chen WR, Hsin SW, Chu TH (2013) Vibration analysis of twisted Timoshenko beams with internal Kelvin-Voigt damping. Procedia Eng 67:525–532
Dohnal F, Ecker H, Springer H (2008) Enhanced damping of a cantilever beam by axial parametric excitation. Arch Appl Mech 78:935–947
Gürgöze M, Erol H (2004) On the eigencharacteristics of multi-step beams carrying a tip mass subjected to non-homogeneous external viscous damping. J Sound Vib 272:1113–1124
Jun L, Hongxing H, Rongying H (2008) Dynamic stiffness analysis for free vibrations of axially loaded laminated composite beams. Comput Struct 84:87–98
Kaya MO, Ozgumus OO (2007) Flexural-torsional-coupled vibration analysis of axially loaded closed-section composite Timoshenko beam by using DTM. J Sound Vib 306:495–506
Lin SM (2014) Analytical solutions for thermoelastic vibrations of beam resonators with viscous damping in non-Fourier model. Int J Mech Sci 87:26–35
Matlab R2014b (2014) The MathWorks, Inc.
Nefovska-Danilovic M, Petronijevic M (2015) In-plane free vibration and response analysis of isotropic rectangular plates using the dynamic stiffness method. Comput Struct 152:82–95
Ozgumus OO, Kaya MO (2006) Flapwise bending vibration analysis of double tapered rotating Euler-Bernoulli beam by using the differential transform method. Meccanica 41:661–670
Ozgumus OO, Kaya MO (2007) Energy expressions and free vibration analysis of a rotating double tapered Timoshenko beam featuring bending-torsion coupling. Int J Eng Sci 45: 562–586
Sorrentino S, Fasana A, Marchesiello S (2007) Analysis of non-homogeneous Timoshenko beams with generalized damping distributions. J Sound Vib 304:779–792
Su H, Banerjee JR (2015) Development of dynamic stiffness method for free vibration of functionally graded Timoshenko beams. Comput Struct 147:107–116
Xie Z, Shepard WS Jr (2009) An enhanced beam model for constrained layer damping and a parameter study of damping contribution. J Sound Vib 319:1271–1284
Yesilce Y (2010) Differential transform method for free vibration analysis of a moving beam. Struct Eng Mech 35(5):645–658

Yesilce Y (2013) Determination of natural frequencies and mode shapes of axially moving Timoshenko beams with different boundary conditions using differential transform method. Adv Vib Eng 12(1):90–108

Yesilce Y (2015) Differential transform method and numerical assembly technique for free vibration analysis of the axial-loaded Timoshenko multiple-step beam carrying a number of intermediate lumped masses and rotary inertias. Struct Eng Mech 53(3):537–573

Zhou JK (1986) Differential transformation and its applications for electrical circuits. Huazhong University Press, Wuhan, China

Chapter 16
Behavior of RC Square Column Strengthening with CFRP Strips Subjected to Low Velocity Lateral Impact Loading

Ömer Mercimek, Rahim Ghoroubi and Özgür Anil

Abstract Reinforced concrete (RC) square columns are vulnerable to sudden dynamic impact loadings such as vehicle impact to bridge column or air blast shock due to blast effects. In this experimental study, RC square columns strengthened with CFRP strip subjected to sudden low-velocity lateral impact loading were investigated. A free falling weight test setup was used to apply the impact loading to RC square columns. The test specimens were manufactured with square cross sections with one-third geometric scale. In scope of the study, four test specimens were manufactured and tested. The main variables considered in the study were the application point of impact loading and CFRP strip spacing. A 9.0 kg mass was allowed to free fall from a height of 1.0 m to apply the impact loading on the columns. During the impact tests, acceleration, impact force, column midpoint displacement, and CFRP strip strains measurements were taken. The general behavior of test specimens, collapse mechanisms, acceleration, displacement, impact load, and strain–time relationships were interpreted and the load–displacement relationships were obtained. The data from the experimental study was used to investigate the effect of variables on the impact performances of RC columns.

Keywords RC column · Impact load · Free drop test · CFRP strip

16.1 Introduction

Designing structures to resist sudden impact loads is a popular issue due to the risks of vehicle, ice, ship impacts, explosions, etc. On the other hand, behavior of reinforced concrete structures exposed to impact loads is not a completely explained topic (Othman and Marzouk 2016). Accordingly, many studies focused on the impact tests have been conducted to identify the response of reinforced concrete

Ö. Mercimek (✉) · R. Ghoroubi · Ö. Anil
Gazi Üniversitesi, Ankara, Turkey
e-mail: omercimek@ybu.edu.tr

A. A. Kasimzade et al. (eds.), *Seismic Isolation, Structural Health Monitoring, and Performance Based Seismic Design in Earthquake Engineering*,
https://doi.org/10.1007/978-3-319-93157-9_16

elements to sudden impact loads (Remennikov and Kaewunruen 2006; Bao and Li 2010; Thilakarathna et al. 2010; Wu et al. 2011; Hao and Wu 2005; Li et al. 2007; Lu and Chong 2005; Yang and Lok 2007; Wu et al. 2005; Bayraktar et al. 2010). Xu et al. 2016 investigated the behavior of ultra-high performance fiber reinforced concrete columns (UHPFRC) subjected to blast loading. In the present study, a series of field tests were conducted to investigate the behavior of UHPFRC columns subjected to blast loading. In total, four 0.2 m × 0.2 m × 2.5 m UHPFRC columns were tested under different designed explosions but all at a standoff distance of 1.5 m. Blast tests were also performed on four high strength reinforced concrete (HSRC) columns with the same size and reinforcement as UHPFRC columns to evaluate their behavior under the same loading conditions. The post-blast crack patterns, permanent deflections, and different levels of damage observations showed that UHPFRC columns performed superior in blast loading resistance as compared with HSRC columns. Huynh et al. (2015) studied about the high strength and reactive powder concrete columns subjected to lateral impact load. This experimental study investigates the performance of 16 RC columns of three different types, i.e., high strength concrete, high strength concrete core with reactive powder concrete shell, and reactive powder RC under low- to medium-velocity impacts. The effect of axial force, loading eccentricity, and the use of steel fiber reactive powder concrete, as a replacement for conventional concrete, on the impact performance of columns are considered. The experimental results showed that axial force and its eccentricity had significant influence on both the impact resistance and the failure mode of the columns subjected to impact. Also, it was observed that application of reactive powder concrete can enhance the impact performance of the columns.

From the literature review, it is observed that experimental studies focused on the response of RC columns to (i) air pressure waves resulting from impact loads, and (ii) vehicle impact conducted. However, in the literature there is no comprehensive study focused on the impact behavior of CFRP retrofitted RC columns. In relation to that in this study, an experimental procedure is employed to measure the performance of CFRP retrofitted RC columns under impact loading. The bonding spacing of CFRP strips and application point of impact loading are the considered variables of the conducted experimental study. The results of the experimental study are interpreted using the acceleration, impact loading, central displacement, and CFRP strain measurements.

16.2 Experimental Program

The experimental study was conducted using an impact test setup specially designed to test the reduced scale test specimens. The impact tests involved four RC square columns. The bonding spacing of CFRP strips and application point of impact loading are the considered variables in the study. The dimensions of the test specimens and reinforcement details are given in Fig. 16.1. The steel reinforcement used in the test specimens is provided in the same batch to have identical

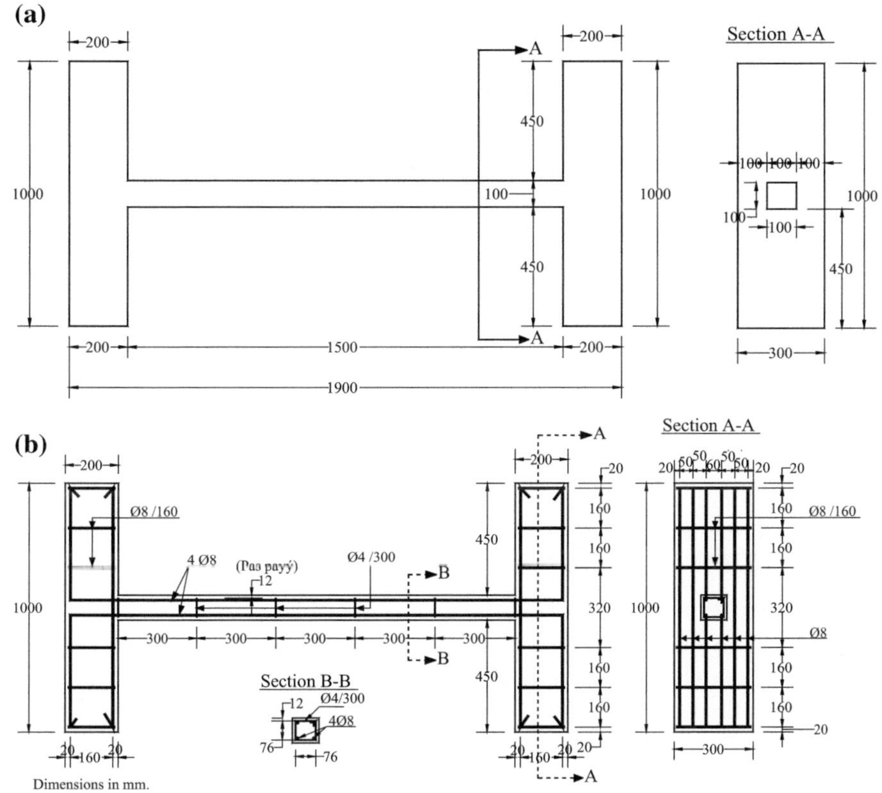

Fig. 16.1 Dimensions and reinforcement details of specimens

mechanical properties. In the test specimens, four longitudinal reinforcements with 8 mm diameter are used. Stirrups with 4 mm diameter and 300 mm spacing are used as the transverse reinforcement of the test specimens. Such a reinforcement design lead to shear deficient columns and these columns with inadequate shear force capacity are retrofitted using externally bonded CFRP strips. Also, it should be noted that the specimens were tested with fixed support conditions. The fixed support conditions were provided by constructing large RC blocks at both ends of the tested columns represent with floor slab and foundation of column. The details of these RC blocks are given in Fig. 16.1. For the determination of compressive strength of concrete at least five standard cube test samples with dimensions of 150 mm × 150 mm were taken from each test element and tested. The cube samples are kept in the same curing conditions with specimens and are tested in the same date in order to determine compressive strength of concrete. Targeted compressive strength of the concrete is 10 MPa and obtained concrete compressive strengths after testing are given in Table 16.1.

Table 16.1 Properties of specimens

Spec. No.	Compression strength of concrete (MPa)	CFRP strip spacing (mm)	Loading point
1	10.2	150	Center point
2	9.8	150	500 mm from column below point
3	10.6	200	Center point
4	9.9	200	500 mm from column below point

Table 16.2 Properties of CFRP sikawrap 160-C (unidirectional) and resin sikadur 330

Properties of CFRP	Remarks of CFRP*	properties of resin	Remarks of resin*
Thickness (mm)	0.12	Tensile strength (MPa)	30
Tensile strength (MPa)	4100	Elastic modulus (MPa)	3800
Elastic modulus (MPa)	231,000		
Ultimate tensile Strain (%)	1.7		

* This information has been obtained from the manufacturer

Spacing of 25 mm wide CFRP strips are provided as 150 mm for test specimens 1 and 2 and 200 mm for test specimens 3 and 4. In the study, unidirectional CFRP strips and a resin composed of two compounds are used. The mechanical properties of the CFRP strips and resin, given by the supplied, are presented in Table 16.2.

Before bonding of CFRP members on to concrete surface, special consideration was given to the RC column surface preparation. CFRP strip locations on to RC column were roughened mechanically by a grinding machine down to aggregate level, and then grinded surface was brushed. Surfaces were vacuum cleaned for removing loose particles and dust. Prepared mixture of epoxy was spread over the grinded surface up to 0.5 mm thickness approximately. CFRP strips were bonded on their predefined places at the RC columns. After bonding of CFRP strips, some pressure was applied on them by hand along the fiber directions to get rid of air bubbles entrapped between CFRP strips and concrete surface, and CFRP strips were soaked with applied epoxy on concrete. The temperature during application was 20 ± 2 °C in all cases. After bonding operations were completed, specimens were cured for 7 days under laboratory conditions before testing. The placement of CFRP strips used in the retrofitting procedure is given in Fig. 16.2.

Two accelerometers mounted on the specimens were used for each test to measure the accelerations during the vibrations arise from impact loading. Two potentiometers were attached to the side surface of the specimen at 50 mm left and 50 mm right of the impact point, to measure vertical displacements due to impact loading. One dynamic load cell was mounted on top of the free-falling hammer to measure the magnitude of the impact force at the instance of impact. Additionally, strain measurements are taken from CFRP strips bonded to RC columns. Strain measurements are taken from the left and right sides of the impact point in each test

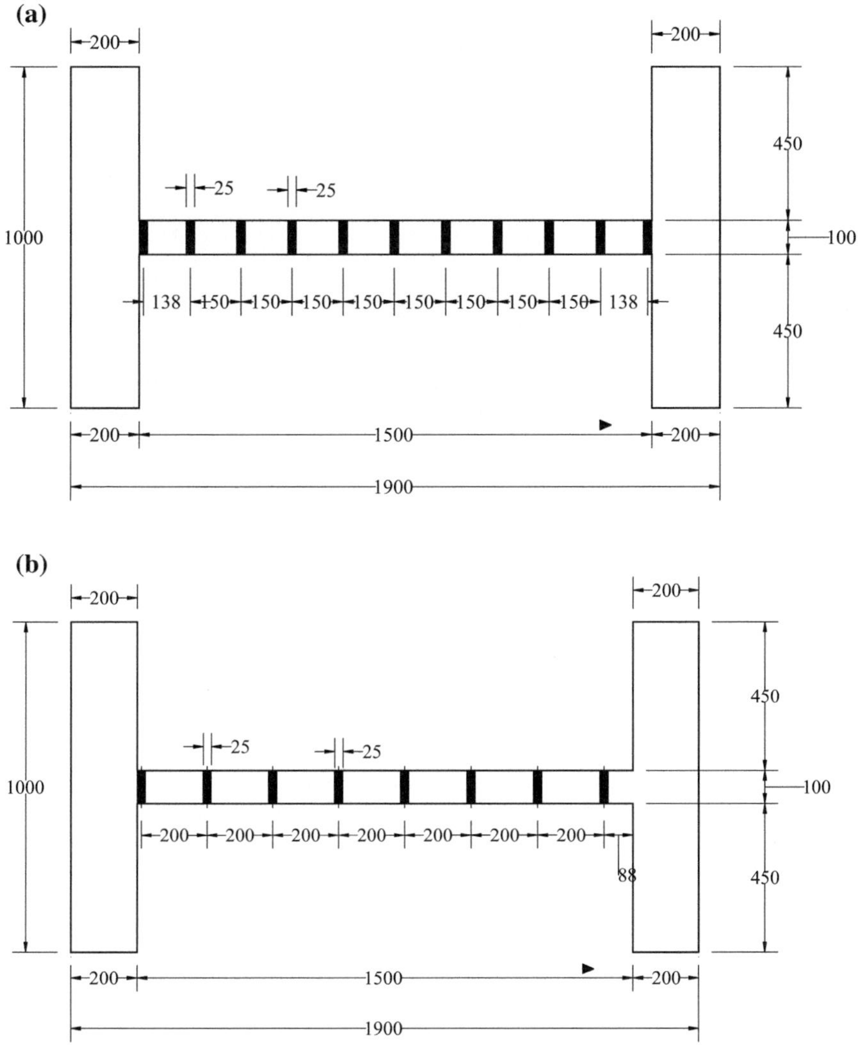

Dimensions in mm.

Fig. 16.2 Strengthening scheme of specimens. **a** Specimen 1 and 2 CFRP strip layout, **b** specimen 3 and 4 CFRP strip layout

specimens in the fiber direction of CFRP strips. The test setup is presented in Fig. 16.3. The data acquisition system was used to capture data from potentiometers, accelerometers, strain gauges, and load cell. The drop weight used in the tests was 9.0 kg and manufactured from aluminum. Impact loads were applied via the free fall of the drop weights impacting the specimens at the mid-span or 500 mm apart from concrete block. The free-falling weights were dropped from a clear height 1.0 m above the specimens. All specimens were subjected to multiple

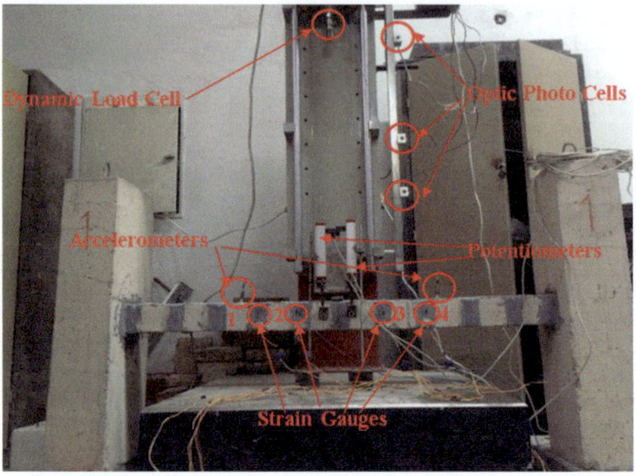

Fig. 16.3 Test setup and instrumentations

(five) impacts. Three photocells, distributed along the falling height, were used to measure the velocity of the free-falling weight.

16.3 Experimental Results

In order to investigate the behavior of CFRP retrofitted RC columns, two acceleration, two displacement and two strain measurements are taken from the test specimens for each drop of impact load. Results of the study are given in Table 16.3. Variation of impact load and strain values with time are given in Fig. 16.4, for test specimen 1. The crack distributions observed from the test specimens, after five drops, are presented in Fig. 16.5. In Fig. 16.6, the comparison of (i) acceleration, (ii) displacement, and (iii) strain distribution of test specimens 1–3 and 2–4 are given to observe the effect of CFRP strip spacing on the response. The effect of application point of impact load is presented in Fig. 16.7, by illustrating the comparative results of test specimens 1–2 and 3–4.

16.4 Conclusions

In this study, response of shear retrofitted RC columns under impact loadings due to rock, ice, and vehicle impacts or pressure waves due to explosions are investigated. In the study, the retrofitting procedure is applied to increase the shear force capacities of RC columns with square cross-sections. The main variables considered in the study were the application point of impact loading, and CFRP strip spacing. The results of the experimental study are interpreted using the acceleration, impact loading, central displacement, and CFRP strain measurements. In the

Table 16.3 Experimental results

Spec. No.	Drop#	Left. accel. (g)		Right. accel. (g)		Left. disp. (mm)		Right. disp. (mm)		Max. disp. (mm)	Impact load (kN)	Max. strain (mm/mm)	
		Max.	Min.	Max.	Min.	Max.	Min.	Max.	Min.			Max.	Min.
1	1	340.23	−136.62	322.74	−475.70	0.21	−0.91	0.20	−0.57	1.12	26.58	0.0334	−0.0245
	2	364.50	−132.60	363.30	−190.10	0.91	−0.93	0.53	−0.89	1.84	25.54	0.0390	−0.0200
	3	417.10	−535.46	419.81	−195.33	1.24	−0.71	0.85	−0.73	1.95	25.49	0.0409	−0.0796
	4	470.65	−690.29	468.16	−648.83	1.52	−0.82	1.31	−0.83	2.34	25.76	0.0532	−0.0216
	5	612.82	−559.45	613.34	−446.11	1.88	−1.22	1.88	−1.17	3.10	25.99	0.0774	−0.0154
2	1	375.81	−265.16	374.02	−215.35	0.21	−1.11	0.20	−1.14	1.34	26.06	0.0376	−0.0003
	2	406.41	−485.41	403.19	−212.68	0.79	−1.37	0.70	−1.41	2.16	25.76	0.0435	−0.0005
	3	473.21	−160.78	477.67	−428.61	1.47	−1.19	1.01	−1.67	2.68	25.87	0.0493	−0.0004
	4	524.93	−495.66	514.98	−376.75	1.61	−1.57	1.41	−1.73	3.18	25.70	0.0534	−0.0003
	5	627.19	−259.70	626.82	−513.94	2.36	−1.46	1.71	−2.10	3.82	25.32	0.0780	0.0000
3	1	301.81	−320.08	298.11	−130.21	0.70	−1.56	0.29	−1.81	2.26	25.91	0.0129	−0.0008
	2	330.25	−370.73	328.87	−539.13	1.50	−2.70	0.43	−3.66	4.20	26.10	0.0255	−0.0057
	3	391.03	−189.58	389.10	−182.80	2.36	−3.71	2.76	−3.20	6.07	26.02	0.0323	−0.0051
	4	443.89	−162.10	439.29	−244.65	3.38	−5.15	4.11	−3.70	8.53	26.17	0.0498	−0.0027
	5	528.02	−156.31	519.02	−294.87	4.26	−6.24	5.65	−4.09	10.50	25.76	0.0551	−0.0050
4	1	345.24	−391.27	340.55	−114.67	0.65	−2.70	0.37	−2.84	3.35	26.09	0.0158	0.0000
	2	387.50	−322.40	383.95	−347.81	1.58	−3.82	1.43	−3.70	5.40	25.90	0.0305	−0.0001
	3	452.61	−408.97	449.82	−322.40	2.20	−4.74	2.73	−4.11	6.94	26.02	0.0397	−0.0003
	4	485.03	−120.26	482.97	−355.56	3.46	−6.08	3.90	−5.55	9.54	26.12	0.0507	−0.0004
	5	548.46	−427.86	543.44	−424.00	3.16	−8.27	3.76	−7.47	11.43	26.08	0.0560	−0.0002

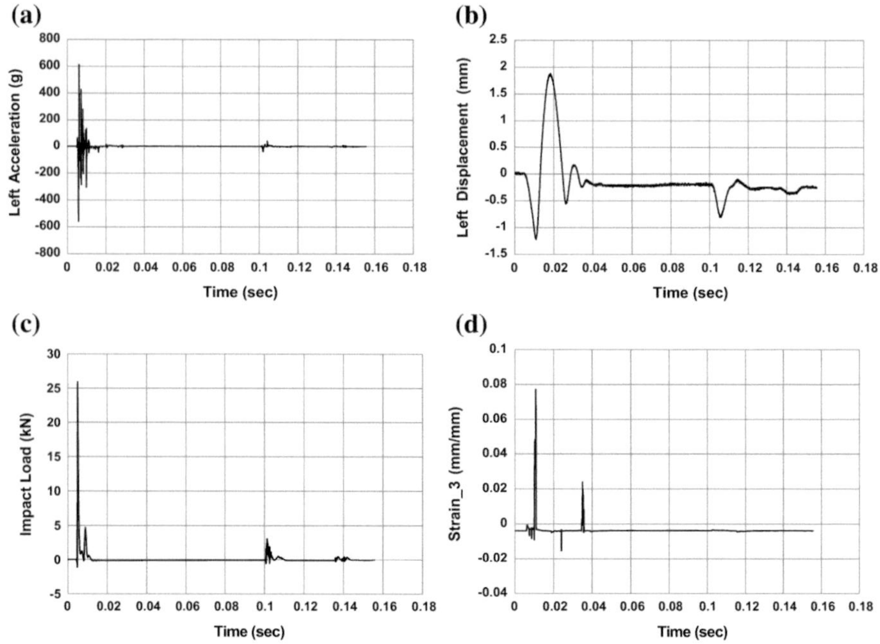

Fig. 16.4 **a** Acceleration, **b** displacement, **c** impact load, **d** strain—time graphs of specimen-1 for last drop

experimental procedure, a constant weight was dropped to test specimens to provide an impact loading with constant energy. In the tests, the free weight was dropped five times to each test specimen The results deduced from the study are given below.

- Increasing the spacing of CFRP strips leads to a reduction in the measured acceleration values of test specimens. From the test results, it is observed that increasing the CFRP strip spacing leads to a 9.5% mean reduction in the measured acceleration values of the test specimens.
- The application point of impact load also affected the results. The shift of application point of impact load from the center of the column to the lower end of the column resulted in 11% mean increase in the measured acceleration values.
- Increase CFRP strip spacing significantly increased the measured central displacement values. The displacement values of specimens 3 and 4 with 200 mm CFRP strip spacing were 2.8 times higher than those of specimens 1 and 2 with 150 mm CFRP strip spacing.
- The application point of impact load was less effective than the CFRP strip spacing in terms of central column displacement values. The central displacement values of specimens 2 and 4, with shifted impact load points, were 25% larger than those of specimens 1 and 3 with central impact load points.

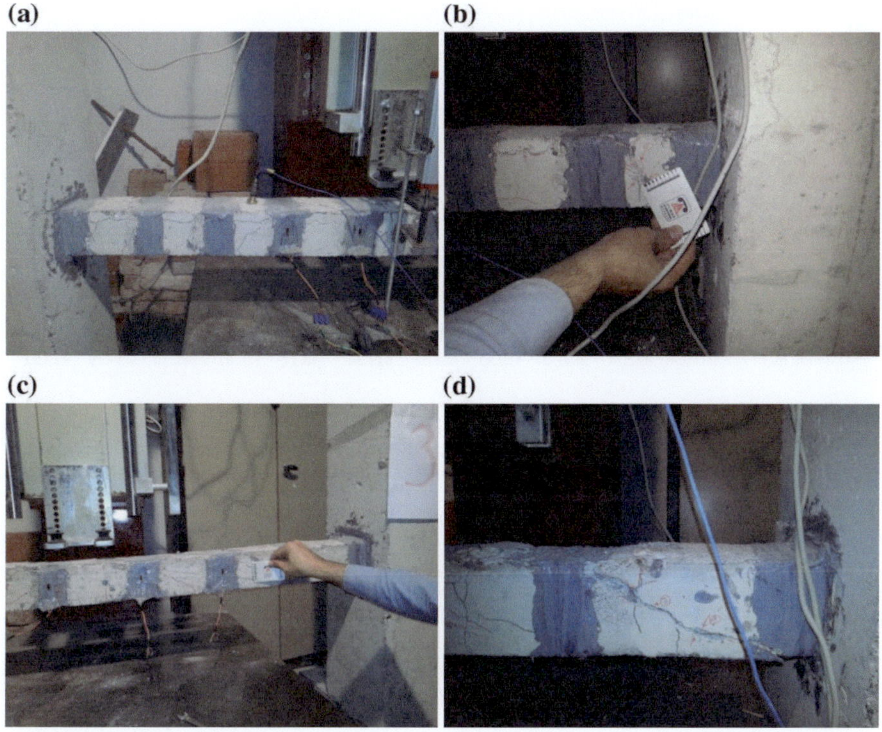

Fig. 16.5 a–d Crack distribution of specimens after 5 drop

- Spacing of CFRP strips was more effective than the application point of impact load on the strain distribution of CFRP strips. The maximum strain values of specimens 1 and 2, with 150 mm CFRP strip spacing, was 53% larger than those of specimens 3 and 4 with 200 mm CFRP strip spacing. The central displacement values of specimens 2 and 4, with shifted impact load points, were 11% larger than those of specimens 1 and 3 with central impact load points.
- From the investigation of crack distributions given in Fig. 16.5, it is observed that the number of shear cracks on the test specimens with 150 mm CFRP strip spacing was significantly smaller than those of test specimens with 200 mm CFRP strip spacing.
- The overall impact behavior of specimens with 150 mm CFRP strip spacing was better than the impact behavior of specimens with 200 mm CFRP strip spacing.

(a)

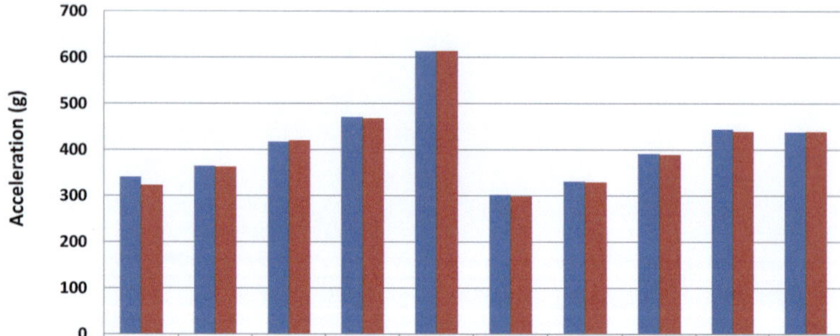

	1	2	3	4	5	1	2	3	4	5
Maximum Left Acceleration	340.23	364.50	417.10	470.65	612.82	301.81	330.25	391.03	443.89	438.02
Maximum Right Acceleration	322.74	363.30	419.81	468.16	613.34	298.11	328.87	389.10	439.29	439.02

(b)

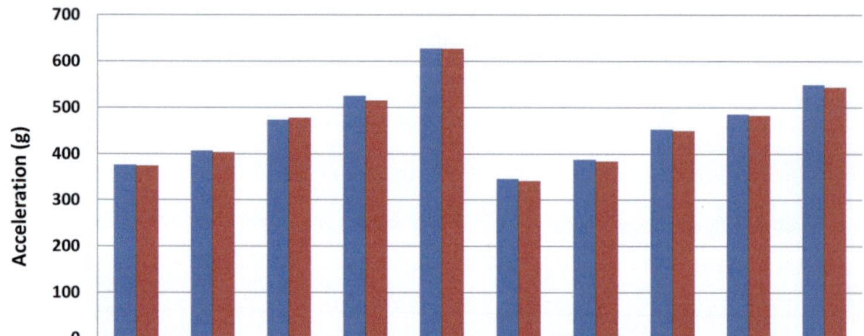

	1	2	3	4	5	1	2	3	4	5
Maximum Left Acceleration	375.81	406.41	473.21	524.93	627.19	345.24	387.50	452.61	485.03	548.46
Maximum Right Acceleration	374.02	403.19	477.67	514.98	626.82	340.55	383.95	449.82	482.97	543.44

(c)

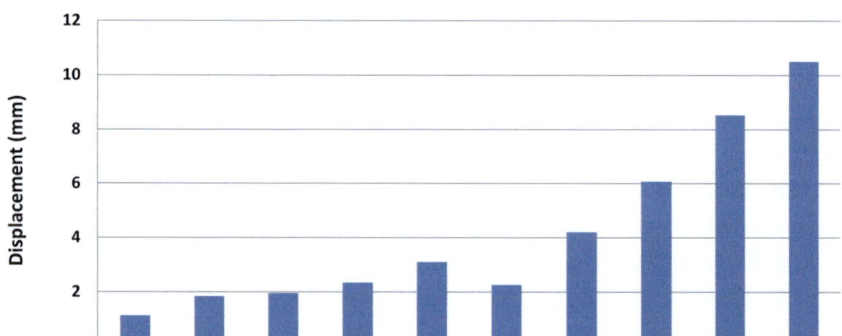

	1	2	3	4	5	1	2	3	4	5
Maximum Displacement	1.12	1.84	1.95	2.34	3.10	2.26	4.20	6.07	8.53	10.50

Fig. 16.6 Comparison of acceleration, displacement and strain values for CFRP strip span. **a** Comparison of acceleration for specimen-1 and specimen-3, **b** comparison of acceleration for specimen-2 and specimen-4, **c** comparison of displacement for specimen-1 and specimen-3, **d** comparison of displacement for specimen-2 and specimen-4, **e** comparison of strain for specimen-1 and specimen-3, **f** comparison of strain for specimen-2 and specimen-4

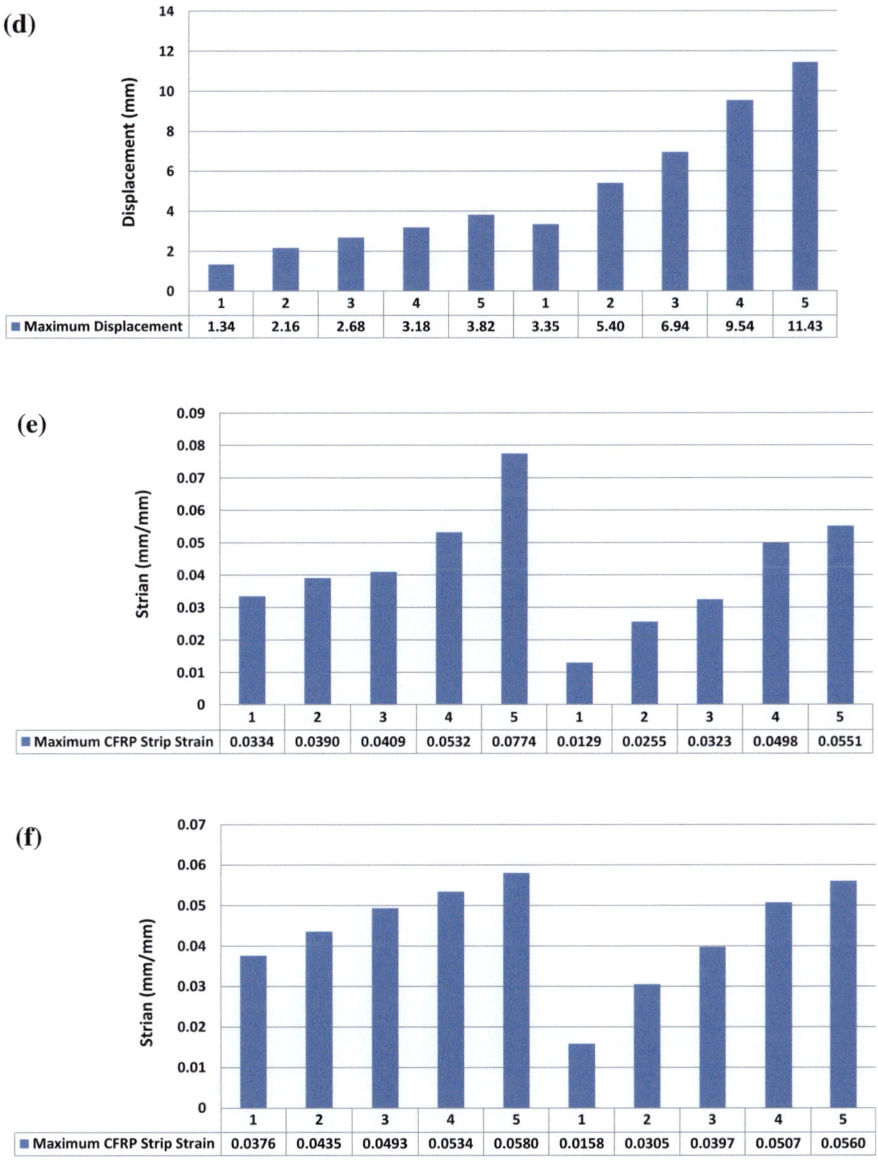

Fig. 16.6 (continued)

(a)

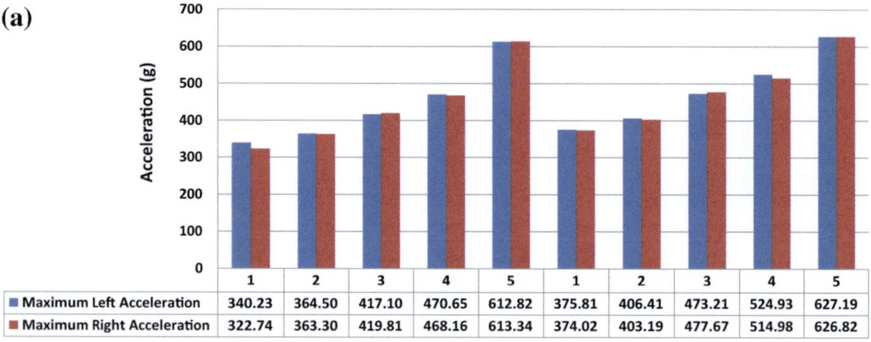

(b)

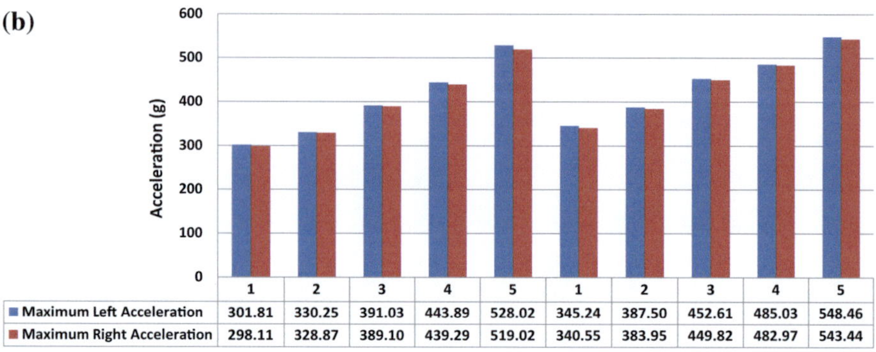

(c)

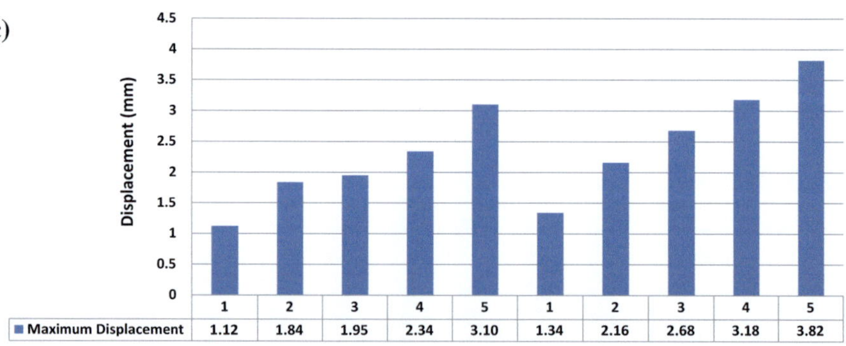

Fig. 16.7 Comparison of acceleration, displacement and strain values for impact loading point. **a** Comparison of acceleration for specimen-1 and specimen-2, **b** comparison of acceleration for specimen-3 and specimen-4, **c** comparison of displacement for specimen-1 and specimen-2, **d** comparison of displacement for specimen-3 and specimen-4, **e** comparison of strain for specimen-1 and specimen-2, **f** comparison of strain for specimen-3 and specimen-4

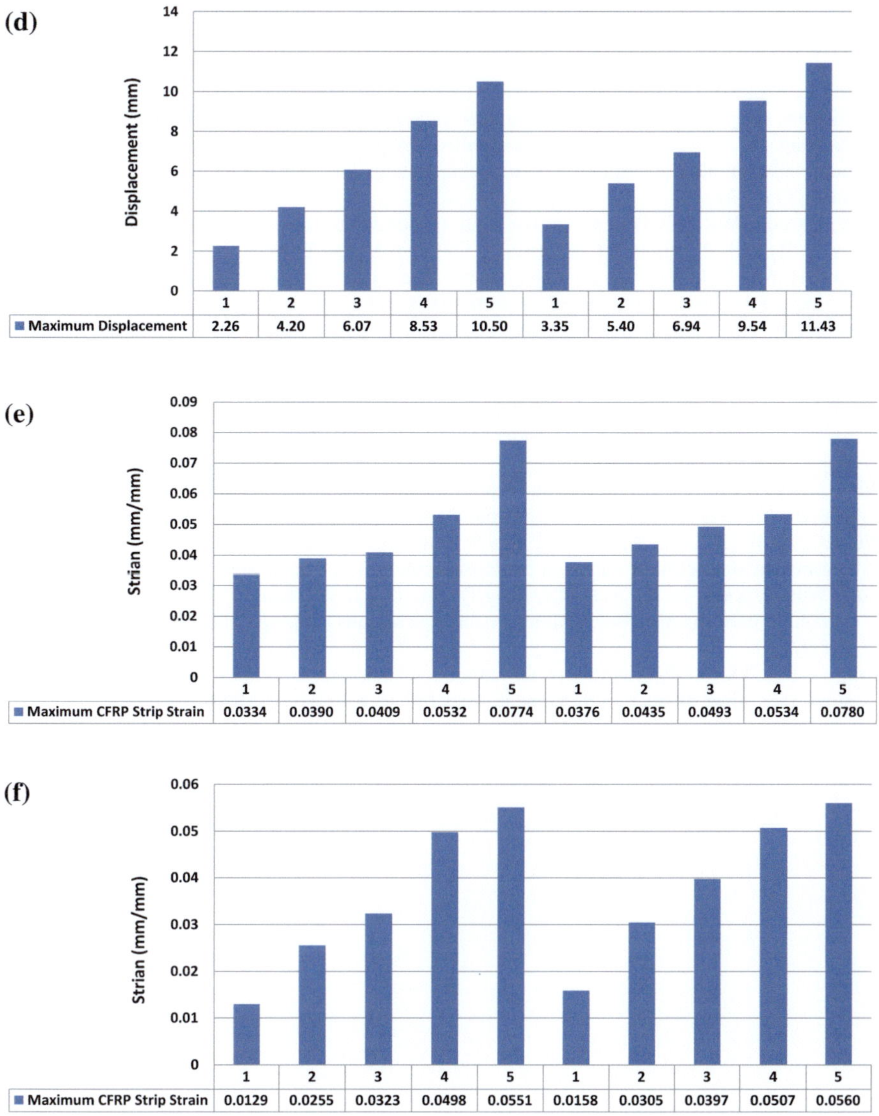

Fig. 16.7 (continued)

References

Bayraktar A, Türker T, Altunişik AC, Sevim B (2010) Evaluation of blast effects on reinforced concrete buildings considering operational modal analysis results. Soil Dyn Earthq Eng 30(5):310–319

Bao X, Li B (2010) Residual strength of blast damaged reinforced concrete columns. Int J Impact Eng 37(3):295–308

Hao H, Wu C (2005) Numerical study of characteristics of underground blast induced surface ground motion and their effect on above-ground structures. Part II: effects on structural responses. Soil Dyn Earthq Eng 25(1):39–53

Huynh L, Foster S, Valipour H, Randall R (2015) High strength and reactive powder concrete columns subjected to impact: Experimental investigation. Constr Build Mater 78:153–171

Li B, Rong HC, Pan TC (2007) Drift-controlled design of reinforced concrete frame structures under distant blast conditions—Part II: implementation and evaluation. Int J Impact Eng 34(4):755–770

Lu Y, Chong ZWK (2005) Comparative study of buried structure in soil subjected to blast load using 2D and 3D numerical simulations. Soil Dyn Earthq Eng 25(4):275–288

Othman H, Marzouk H (2016) An experimental investigation on the effect of steel reinforcement on impact response of reinforced concrete plates. Int J Impact Eng 88:12–21

Remennikov AM, Kaewunruen S (2006) Impact resistance of reinforced concrete columns: experimental studies and design considerations. In: 19th Australasian conference on the mechanics of structures and materials, Nov 29–Dec 1, Christchurch, New Zealand, pp 817–824

Thilakarathna HMI, Thambiratnam DP, Dhanasekar M, Perera N (2010) Numerical simulation of axially loaded concrete columns under transverse impact and vulnerability assessment. Int J Impact Eng 37(11):1100–1112

Wu C, Hao H, Lu Y (2005) Dynamic response and damage analysis of masonry structures and masonry infilled RC frames to blast ground motion. Eng Struct 27(3):323–333

Wu KC, Li B, Tsai KC (2011) The effects of explosive mass ratio on residual compressive capacity of contact blast damaged composite columns. J Constr Steel Res 67(4):602–612

Xu J, Wua C, Xiang H, Su Y, Li ZX, Fang Q, Hao H, Liu Z, Zhang Y, Li J (2016) Behavior of ultra-high performance fibre reinforced concrete columns subjected to blast loading. Eng Struct 118:97–107

Yang G, Lok TS (2007) Analysis of RC structures subjected to air-blast loading accounting for strain rate effect of steel reinforcement. Int J Impact Eng 34(12):1924–1935

Chapter 17
Strain-Based Seismic Performance Evaluation of Prefabricated Structures

Aytug Seckin, Hilal M. Atalay, Hakan Erdogan, Bilge Doran and Bulent Akbas

Abstract Precast concrete frame systems are widely preferred for the single storey industrial buildings in Turkey and Europe due to its rapid and economical construction practice. Field investigations after past major earthquakes revealed that the damage in such structures was primarily due to the improper detailing in beam–column connections and lack of required lateral stiffness. In this study, two different precast industrial buildings, damaged during the 1999 Marmara earthquake, were numerically investigated. The seismic performances of the buildings were evaluated by using Incremental Equivalent Seismic Load Method and strain-based damage definitions as recommended in Turkish Earthquake Code. The analytical results obtained from performance evaluations are compared with the findings from the field investigations.

Keywords Precast concrete structures · Seismic damage · Incremental equivalent seismic load method · Strain-based seismic performance

A. Seckin (✉) · H. M. Atalay · H. Erdogan
Department of Civil Engineering, Kocaeli University, Kocaeli, Turkey
e-mail: aytug.seckin@kocaeli.edu.tr

H. M. Atalay
e-mail: hilal.meydanli@kocaeli.edu.tr

H. Erdogan
e-mail: hakan.erdogan@kocaeli.edu.tr

B. Doran
Department of Civil Engineering, Yildiz Technical University, Istanbul, Turkey
e-mail: doranbilge@gmail.com

B. Akbas
Department of Civil Engineering, Gebze Technical University, Kocaeli, Turkey
e-mail: akbasb@gtu.edu.tr

A. A. Kasimzade et al. (eds.), *Seismic Isolation, Structural Health Monitoring, and Performance Based Seismic Design in Earthquake Engineering*,
https://doi.org/10.1007/978-3-319-93157-9_17

17.1 Introduction

In recent decades, precast concrete frame systems which are fixed at the column base and having pinned beam-to-column connections at the roof level have been mostly preferred for the single storey industrial buildings due to its high construction speed and relatively low cost. Besides those advantages, most of those structures are very susceptible to seismic effects. There are numerous studies in the literature conducted on the damage and failure of reinforced concrete precast buildings after past major earthquakes all over the world (Iverson and Hawkins 1994; Muguruma et al. 1995; Posada and Wood 2002; Arslan et al. 2006; Khare et al. 2011; Ghosh and Cleland 2012; Ozden et al. 2014). The field investigations after the major earthquakes revealed that the main reasons for the damage in precast frame structures were the lack of diaphragm action at the roof level, inadequate lateral stiffness and insufficient connection strength and detailing. Observed damage modes were mainly in the form of destruction at beam–column connections at roof level and column failure at the base or brick infill wall level. These frequently encountered problems emphasize the need for extensive investigation of the seismic performance of precast concrete structures (Ozden and Meydanlı 2003).

The loss of human lives and economic damages due to past major earthquakes indicated that the damage control should be taken into consideration in design of earthquake-resistant structures. Accordingly, the recent analytical investigations are mainly concentrated on performance-based design and evaluation procedures instead of the traditional force-based design method (Poland and Hom 1997). Nonlinear static pushover analysis is one of the frequently used method that can be employed to simulate the inelastic behavior of the structures under the seismic effects. It is commonly recommended in most of the performance-based design codes (ATC-40 1996; FEMA 2000; Eurocode 8 2004; TEC 2007). In performance-based design approach, the strength and deformation demands of a structural system is obtained under the effect of design earthquakes by using pushover analysis and these demands are compared with available capacities to determine the seismic performance of structural systems at cross-section level (Krawinkler and Seneviratna 1998).

Several analytical studies were conducted on the seismic performance assessment of precast structures in the literature. Posada and Wood (2002); Ozturk and Demiralan (2007) investigated seismic behavior of prefabricated structures for different ground motion records. Magliulo et al. (2008) carried out nonlinear static and dynamic analyses on precast buildings erected in Italy. Bozdag and Duzgun (2010) conducted pushover analyses and time history analyses on a prefabricated structure. Senel and Palanci (2013), Belleri et al. (2015) investigated the seismic performance of reinforced concrete precast structures. Ercolino et al. (2016) performed nonlinear dynamic analyses on a reinforced concrete precast industrial building damaged during the seismic event.

The main objective of this study is to compare the actual damage state of two closely located precast concrete industrial buildings after 1999 Marmara

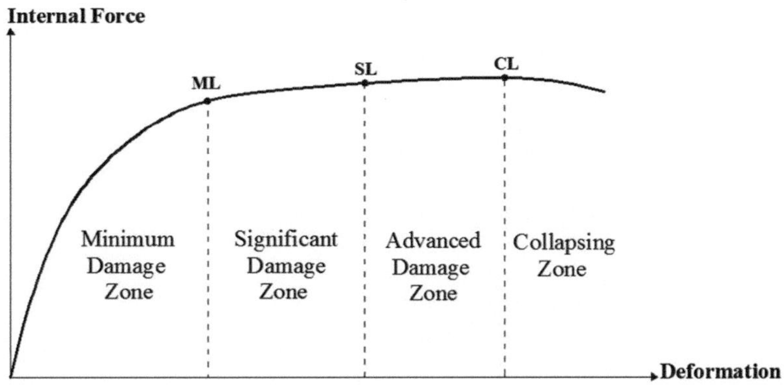

Fig. 17.1 Damage zones

Earthquake with the analytical findings from the seismic performance evaluation procedure prescribed in Turkish Earthquake Code (TEC 2007). In TEC 2007 three different performance levels were defined as Immediate Occupancy (IO), Life Safety (LS), and Collapse Prevention (CP). Code regulations require different minimum performance limits for different types of buildings. The building damage level is directly depending on the amount of damaged structural members. The damage on the structural members is classified namely as minimum damage (MN), significant damage (SD), advanced damage (AD) and state of collapse (SC) due to the compression strains for concrete and tensile strains for reinforcing steel (Fig. 17.1). Three limit conditions have been defined for ductile elements on the cross section. These are defined as minimum damage limit (ML), safety limit (SL) and collapsing limit (CL).

17.2 Structural Properties of Industrial Buildings

Several precast industrial buildings were significantly damaged after the 1999 Marmara earthquake with a moment magnitude (M_w) of 7.4. Two precast frame structures, namely precast concrete structure-I (PCS-I) and precast concrete structure-II (PCS-II) were selected for further investigation in the scope of this study (Fig. 17.2). PCS-II collapsed during the Marmara earthquake, while moderate structural damage was observed in PCS-I. Both buildings were located close to Gölcük District and the distance between those structures was only about 8 km. The buildings also located very close to the epicenter of the earthquake. Both structures were designed and constructed according to the two previous version of Turkish Earthquake Code established in 1975. The older version of TEC (1975) recommends to design such structures as a single 2D frame under the effect of the equivalent static seismic loads.

(a) (b)

Fig. 17.2 Single storey industrial buildings **a** PCS-I, **b** PCS-II

PCS-I and PCS-II have rectangular shape in plan with two bays in transverse direction and ten and nine bays in the longitudinal direction, respectively. Three-dimensional models of the structures have been demonstrated in (Fig. 17.3). Transverse bay widths of the structures are 15 and 20 m and longitudinal bay widths are 6.5 and 7.5 m respectively. The maximum storey height is 8.5 m for PCS-I and 10 m for PCS-II. Precast columns were grouted into single cast in place or precast footings and the footings were cross-tied with grave beams. The individual reinforced concrete precast frames were connected longitudinally with gutter beams and purlins that were pinned at both ends. In addition, the longitudinal facades of PCS-II were covered with brick infill walls up to 5 m height.

The concrete strength was detected to be 30 MPa, characteristic longitudinal reinforcement yield strength was equal to 420 MPa and characteristic transverse reinforcement yield strength was 220 MPa in both buildings. The column and beam cross-sections were identical in both structures. The cross-section details are shown in Fig. 17.4

(a) (b)

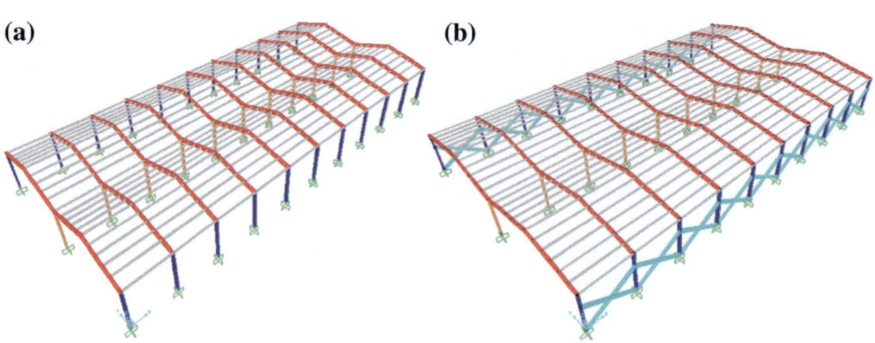

Fig. 17.3 3D model of the buildings **a** PCS-I, **b** PCS-II

Fig. 17.4 Section details of columns and beams

17.3 Structural Analysis of the Buildings

The industrial precast structures were modeled considering the dimensions, material properties and reinforcement details provided from site inspections and blueprints of the structures. Seismic performance evaluations were performed by following the incremental equivalent seismic load procedure as described in TEC-2007. During the implementation of the method, moment-curvature relationships of main frame system column and beam sections (Fig. 17.4) were obtained by using the Mander model for concrete behavior and a bilinear elasto-plastic behavior for steel was considered (Mander et al. 1988). Sectional analysis software XTRACT (2007) was employed for computations. The secondary structural elements such as gutter beams were assumed to behave linearly elastic.

While deriving the stress–strain behavior of the materials, effects of lateral confinement for concrete and strain hardening for reinforcing steel have been neglected. The effect of infill wall to the lateral stiffness of the PCS-II was taken into account by considering the equivalent strut model developed by Mainstone (1971) and Mainstone and Weeks (1972). The brick infill walls were modeled by using equivalent diagonal compression struts connected to main frame structures by pin connections.

Structural analysis program SAP2000-ver14 was employed for nonlinear static analysis. The plastic hinges were defined at maximum moment regions. Pushover analyses were performed in two horizontal directions based on the fundamental mode shape related lateral loading in the relevant direction in order to obtain capacity curves. Total of two pushover analysis performed for each building due to the symmetry of the structural systems. In other words, buildings were pushed only in one direction for each orthogonal axes, namely transverse and longitudinal. The pushover curves of the structure were converted into an equivalent single degree of freedom capacity spectrum using the modal mass coefficient and participation factor equations described in TEC 2007. The performance points in each direction were obtained according to equivalent displacement rule by using the capacity curves and the elastic response spectrum curve definition in TEC-2007 for an earthquake with a

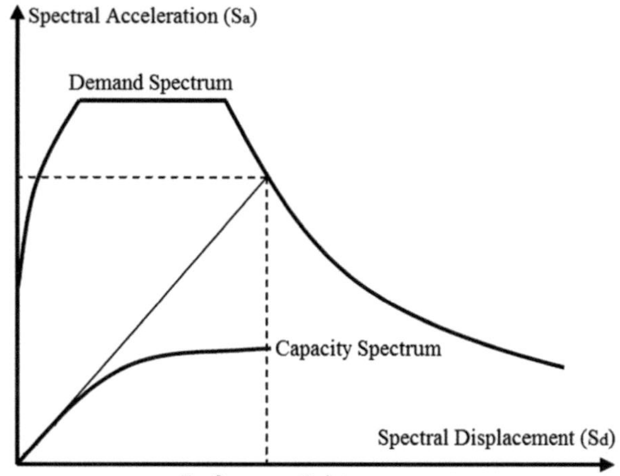

Fig. 17.5 Determination of performance point according to TEC-2007

probability of exceedance of 10% in 50 years (Fig. 17.5). The effects of building type, local site class and seismicity of the region were taken into consideration in constructing the elastic demand spectrum curve.

17.4 Seismic Performance Evaluation of the Buildings According to TEC 2007

The peak roof displacement demands at performance point are equal to 107 and 270 mm for transverse and longitudinal directions of PCS-I, respectively. Apparently, the peak roof displacement demand in longitudinal direction is greater considering the framing of the structure in transverse direction.

For PCS-II, the peak roof displacement demand is equal to 203 mm considering the loading in transverse direction. In longitudinal direction, there is significant difference between the peak roof displacement demands of interior and exterior columns due to the contribution of brick infill walls to the lateral rigidity of structure. The target displacement is 268 mm for interior columns whereas the exterior column displacement demand is 106 mm. The plastic hinge locations at performance points were presented in Figs. 17.6 and 17.7.

According to the analytical results of PCS-I, it was observed that majority of the columns experienced significant damage under loading in two orthogonal directions (Figs. 17.8, 17.9). In addition, all beams of the PCS-I were slightly damaged without exceeding the minimum damage limits. On the other hand, the all columns of PCS-II were in significant damage level for the loading in longitudinal direction (Fig. 17.10), whereas six columns were beyond the safety limits considering the one-way loading (due to symmetry) in the transverse direction (Fig. 17.11). Twelve out of 20 beams of PCS-II collapsed and four of them were within the advanced damage zone (Table 17.1).

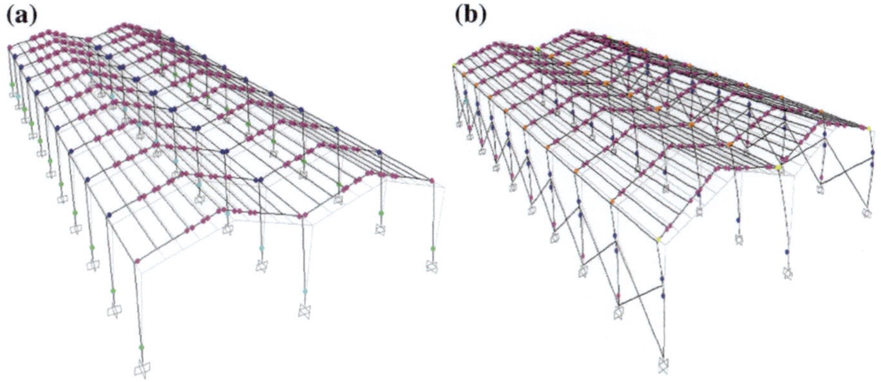

Fig. 17.6 Plastic hinge locations for loading in longitudinal direction **a** PCS-I, **b** PCS-II

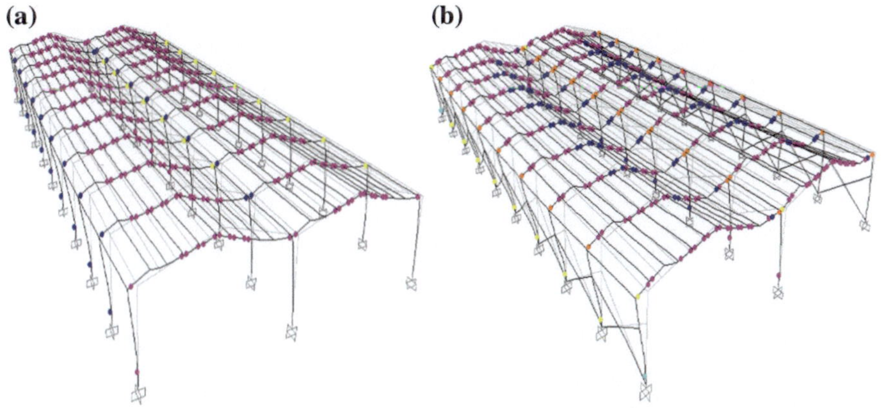

Fig. 17.7 Plastic hinge locations for loading in transverse direction **a** PCS-I, **b** PCS-II

In the field investigations, plastic hinges were detected at the base of a few columns and minimum damage were detected on the beam–column connections at the roof level for PCS-I which is consistent with the analytical results. PCS-II collapsed due to the falling of roof beams possibly because of insufficient beam–column pin connections that were not meeting the lateral deformation demands. The difference in lateral stiffness of the interior and exterior axes of the building due to existence of brick infill walls in longitudinal direction might also lead to fall of beams. In addition, some columns collapsed by hinge formations at the top level of half collapsed brick infill wall due to short column effect as presented in (Fig. 17.12). Since the damages occurred in the brick infill walls were not considered in the analytical model, this behavior cannot be captured in the seismic performance analysis.

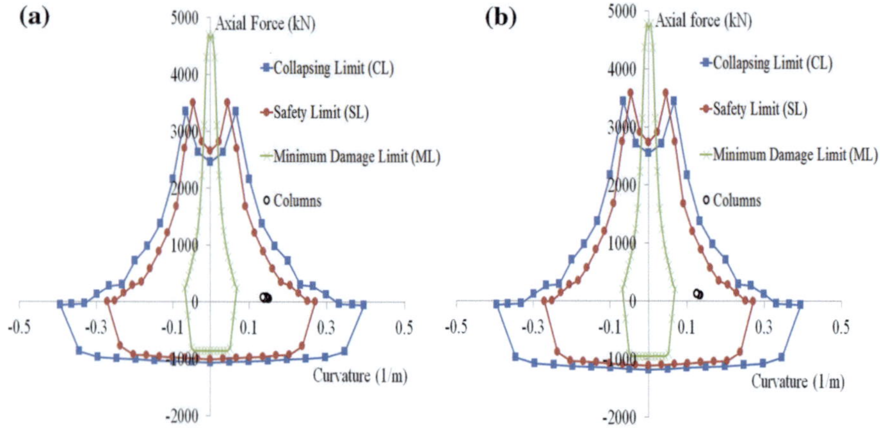

Fig. 17.8 Column damage classifications in longitudinal direction for PCS-I **a** exterior columns, **b** interior columns

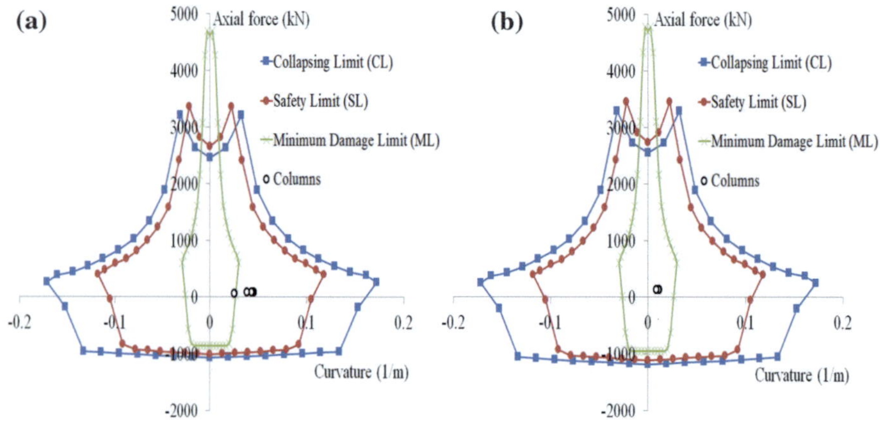

Fig. 17.9 Column damage classifications in transverse direction for PCS-I **a** exterior columns, **b** interior columns

According to TEC 2007, industrial precast buildings should satisfy the life safety performance level under a design earthquake with a probability of exceedance of 10% in 50 years. In order to satisfy the life safety performance criteria, at most 30% of the all beams excluding the secondary ones allowed to be in the advanced damage zone. Additionally, the amount shear force carried by the columns that experienced advanced damage in one storey can not exceed 20% of the total shear force acting that storey and all other beam and columns should not exceed significant damage limit.

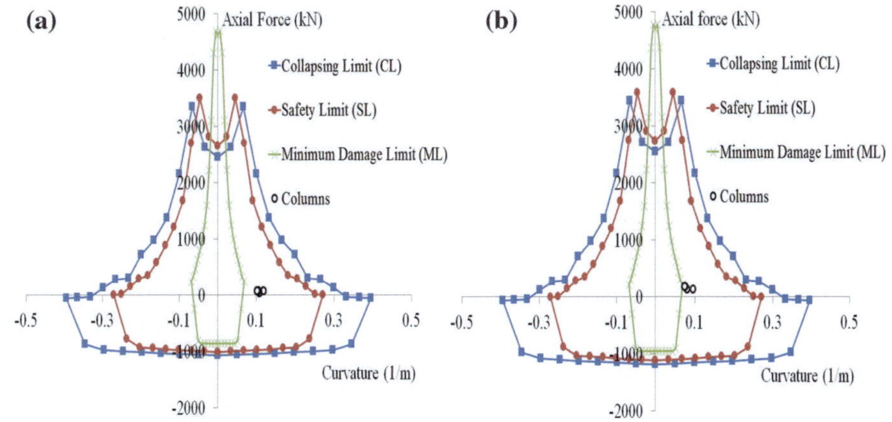

Fig. 17.10 Column damage classifications in longitudinal direction for PCS-II **a** exterior columns, **b** interior columns

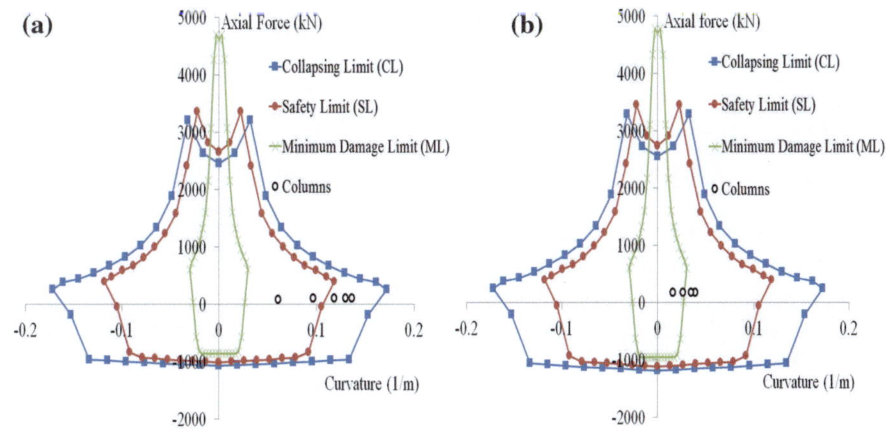

Fig. 17.11 Column damage classifications in transverse direction for PCS-II **a** exterior columns, **b** interior columns

In this context, the seismic performance assessment indicated that PCS-I satisfied the LS performance criteria. On the other hand, 60% of the beams were within the collapsing zone for PCS-II. Therefore, PCS-II did not satisfy the LS performance level considering the TEC 2007 provisions.

Table 17.1 Beam damage classification under transverse loading for PCS-II

Frame no.	Strain demands		Strain limits		Performance level
	ε_c	ε_s	ε_c	ε_s	
396 & 426	0.0191	0.0803	ML = 0.0035 SL = 0.0135 CL = 0.018	ML = 0.010 SL = 0.040 CL = 0.060	Collapsed
414 & 444	0.0182	0.0762			Collapsed
366 & 456	0.0185	0.0776			Collapsed
384 & 474	0.0170	0.0719			Collapsed
336 & 486	0.0163	0.0691			Collapsed
354 & 504	0.0152	0.0648			Collapsed
306 & 516	0.0119	0.0539			Advanced damage
324 & 534	0.0103	0.0483			Advanced damage
6 & 546	0.0019	0.0120			Significant damage
24 & 564	0.0018	0.0104			Significant damage

ε_c and ε_s denote unit deformations for concrete and steel respectively

Fig. 17.12 Observed damages in buildings **a** PCS-I, **b** PCS-II

17.5 Conclusions

In this study, the reliability and accuracy of the seismic performance evaluation method described in Turkish Earthquake Code TEC 2007 was examined for two closely located single storey precast concrete industrial buildings that were damaged during 1999 Marmara earthquake. The observed damages in the field investigations compared with the seismic performance level estimations of TEC 2007. Both buildings were having identical cross-sectional details and material properties for structural members. One of those two buildings has greater storey height and bay widths compared to the other one. It should also be noted that the building with the greater dimensions also has brick infill walls between the columns in longitudinal direction.

Nonlinear static pushover analyses were performed using structural analysis software SAP2000-ver14 in order to obtain capacity curves of the structures. The strength and deformation demands of the buildings were obtained under the effect of design earthquakes by using pushover analysis and these demands are compared with available capacities to determine the seismic performance of structural systems at cross-section level. In the analysis stage, the effect of brick infill wall is also taken into consideration.

The analytical results revealed that the overall performance levels obtained using performance evaluation procedure prescribed in TEC 2007 was in good conformity with the actual performance of the buildings. This study also indicated that the height and the bay width of industrial precast concrete structures are critical parameters that are directly affecting the seismic performance of the structures considering that both buildings have identical cross sections and material properties, located in neighboring regions and experienced the same major earthquake.

References

Arslan MH, Korkmaz HH, Gulay FG (2006) Damage and failure pattern of prefabricated structures after major earthquakes in Turkey and shortfalls of the Turkish Earthquake code. Eng Fail Anal 13(4):537–557

ATC-40 (1996) Seismic evaluation and retrofit of concrete buildings. Applied Technology Council, Redwood City, California

Belleri A, Torquati M, Riva P, Nascimbene R (2015) Vulnerability assessment and retrofit solutions of precast industrial structures. Earthquakes and structures 8(3):801–820

Bozdag O, Duzgun M (2010) Mevcut Prefabrike Bir Binanın Deprem Performansının Artımsal İtme Analizi Yöntemleri ile Belirlenmesi. Beton Prefabrikasyon 93:5–13 (in Turkish)

Computers and Structures Inc. (CSI) SAP2000: Integrated finite element analysis and design of structures. Berkeley, California

Ercolino M, Magliulo G, Manfredi G (2016) Failure of a precast RC building due to Emilia-Romagna earthquakes. Eng Struct 118:262–273

Eurocode 8 (2004) Design of structures for earthquake resistance. European Committee for Standardization, Brussels

FEMA (2000) Prestandard and commentary for the seismic rehabilitation of buildings. Federal Emergency Management Agency, Washington, DC

Ghosh SK, Cleland N (2012) Observations from the February 27, 2010, earthquake in Chile. PCI J 57(1):52–75

Iverson JK, Hawkins NM (1994) Performance of precast/prestressed concrete building structures during Northridge Earthquake. PCI J 39:38–56

Khare RK, Maniyar MM, Uma SR, Bidwai VB (2011) Seismic performance and design of precast concrete building structures: an overview. J Struct Eng 38(3):272–284

Krawinkler H, Seneviratna GDPK (1998) Pros and cons of a pushover analysis of seismic performance evaluation. Eng Struct 20:452–464

Magliulo G, Fabbrocino G, Manfredi G (2008) Seismic assessment of existing precast industrial buildings using static and dynamic nonlinear analyses. Eng Struct 30(9):2580–2588

Mainstone RJ (1971) On the stiffness and strength of infilled frames. In: Proceedings of the institute of civil engineers, Supplemental vol IV, pp 57–90

Mainstone RJ, Weeks GA (1972) The influence of bounding frame on the racking stiffness and strength of brick walls. In: Proceedings of the 2nd international conference on brick masonry, UK, 1972

Mander JB, Priestley MJN, Park R (1988) Theoretical stress-strain model for confined concrete. ASCE J Struct Div 114(8):1804–182

Muguruma H, Nishiyama M, Watanabe F (1995) Lessons learned from the Kobe earthquake—a Japanese perspective. PCI J 40:28–42

Ozden S, Meydanli HM (2003) Seismic response of pre-cast industrial buildings during 1999 Kocaeli earthquake. SE-40EEE, Skopje earthquake: 40 years of European Earthquake Engineering, Skopje, Macedonia, 2003

Ozden S, Akpinar E, Erdogan H, Atalay HM (2014) Performance of precast concrete structures in October 2011 Van earthquake Turkey. Mag Concr Res 66(11):543–552

Ozturk, B, Demiralan F (2007) Investigation of seismic behavior of a prefabricated building located at a seismically active region using various ground motion records. In: Sixth national conference on earthquake engineering, Istanbul, Turkey

Poland CD, Hom DB (1997) Opportunities and pitfalls of performance based seismic engineering. In: Proceedings of the international workshop on seismic design methodologies for the next generation of codes, Slovenia

Posada M, Wood SL (2002) Seismic performance of precast industrial buildings in Turkey. In: 7th U.S. national conference on earthquake engineering, Boston

Senel SM, Palanci M (2013) Structural aspects and seismic performance of 1-story precast buildings in Turkey. J Performance of Constructed Facilities 27(4):437–449

TEC (1975) Specification for structures to be built in disaster areas (Turkish Earthquake Code 1975). Ministry of public works and settlement, government of republic of Turkey, Ankara

TEC (2007) Specifications for buildings to be built in seismic areas (Turkish Earthquake Code 2007). Ministry of Public works and settlement, government of republic of Turkey, Ankara

XTRACT (2007) Cross section analysis program for structural engineers. Imbsen and Associates Inc., USA

Chapter 18
Influence of Polypropylene Fibers on the Shear Strength of RC Beams Without Stirrups

Mehdi Öztürk and Güray Arslan

Abstract Reinforced concrete (RC) beams with light transverse reinforcement are vulnerable to shear failure during seismic response. The codes require the use of a certain amount of transverse reinforcement to resist the expected total shear demand to prevent brittle shear failure at plastic hinge regions. Some codes ignore the contribution of concrete to shear strength when the shear demand due to seismic effects is above a certain level. This research studied the contribution of concrete to shear strength of beams reinforced with longitudinal bars and polypropylene fibers (PPF). The beams including two reference and two macro-synthetic polypropylene fibers reinforced concrete (PPFRC) beams tested under concentrated loads at mid-span to determine the shear strength. The variable parameters are volume fraction of polypropylene fibers (V_f) and shear span-to-depth ratio (a/d). Deflection of the beam and the cracking pattern were monitored during the test at different stages of the monotonic loading until failure. When the beams with a/d = 2.5 and 3.5 are compared, it is concluded that the contributions of PPF to the shear strength at ultimate state are 0.39 MPa and 0.34 MPa, respectively, in case of the beams with volume fractions of PPF equal to 1.0%. It is observed that the contribution of PPF to shear strength decreases with the increasing a/d. It can also be stated that the dissipated energies under the area of load-deflection curve by the PPFRC beams are 341 and 226 times the energy dissipated by a/d = 2.5 and a/d = 3.5 reference beams, respectively.

Keywords Reinforced concrete · Beam · Polypropylene fibers
Shear strength · Deflection

M. Öztürk
Institute of Natural and Applied Sciences, Yildiz Technical University,
Istanbul, Turkey
e-mail: mehdioztrk@gmail.com

G. Arslan (✉)
Department of Civil Engineering, Yildiz Technical University,
Istanbul, Turkey
e-mail: gurayarslan@gmail.com

A. A. Kasimzade et al. (eds.), *Seismic Isolation, Structural Health Monitoring, and Performance Based Seismic Design in Earthquake Engineering*,
https://doi.org/10.1007/978-3-319-93157-9_18

18.1 Introduction

Many experiments available in the literature show that steel fibers (SF) are significantly effective as shear reinforcement (Sharma 1986; Narayanan and Darwish 1987; Ashour et al. 1992; Imam et al. 1994; Kwak et al. 2002; Gandomi et al. 2011; Arslan 2014; Birincioglu et al. 2015). However, very few present researches on structures produced with polypropylene fibres (PPF), which can present a series of advantages (durability and corrosion avoidance above all). Today PPF might be able to impart significant toughness and ductility to concrete (Noghabai 2000; Song et al. 2005; Majdzadeh et al. 2006; Greenough and Nehdi 2008, Altoubat et al. 2009, Conforti et al. 2015).

Song et al. (2005) compared with the strength of nylon-fiber-reinforced concrete and the polypropylene fibers reinforced concrete (PPFRC). Carnovale and Vecchio (2014) investigated the shear behavior of PPFRC and steel fiber reinforced concrete (SFRC) under reversed cyclic in-plane shear loading and observed that the material response of 2.0% by volume of PPF is nominally similar to that of 1.0% SF. Altoubat et al. (2009) tested twenty-seven beams under concentrated loads at mid-span to determine the effect of macro-synthetic PPF on the shear strength and failure behavior of longitudinally reinforced concrete (RC) beams without stirrups. The test results showed that the addition of macro-synthetic PPF significantly improved the shear strength and ductility of the RC beams.

The main purpose of this study is to study the contribution of concrete to shear strength of beams reinforced with longitudinal bars and macro-synthetic PPFRC beams without stirrups. The variable parameters included shear span-to-depth ratio (a/d) and volume fraction of PPF (V_f) (0%, and 1.0%). Four RC beams including two reference and two PPFRC beams have been tested under concentrated loads at mid-span to determine shear strength.

18.2 Shear Strength of Beams

According to ASCE–ACI426 (1973), the guidelines for determining the shear strength of RC members are as follows. The governing equation given by ACI318 (2014) requires that the shear strength must exceed the shear demand as shown in Eq. 18.1.

$$\phi v_n \geq v_u \tag{18.1}$$

Most codes (ACI318 2014; TS500 2000) suggest a superposition of contributions of transverse reinforcement and concrete to determine the shear strength. Accordingly, the design shear strength is calculated as

$$v_n = v_c + v_s = \frac{\sqrt{f_c}}{6} + \rho_w f_y, \quad (18.2)$$

where v_c is the contribution of concrete to shear strength, v_s is the contribution of reinforcement to shear strength based on yield, f_c is the compressive strength of concrete in MPa. The contribution of vertical transverse reinforcement to shear strength according to ACI318 (2014) is derived from basic equilibrium considerations on a 45 degree truss model with a constant transverse reinforcement spacing and an effective depth (ASCE–ACI445 1998).

The contribution of concrete to shear strength according to TS500 (2000) is based on the compressive strength of concrete. Equation (18.3) represents the shear strength of slender beams

$$v_n = v_c + v_s = 0.2275\sqrt{f_c} + \rho_w f_{yw}, \quad (18.3)$$

where ρ_w is transverse reinforcement ratio $(\rho_w = A_v/(b_w s))$, b_w is beam width, A_v is the area of shear reinforcement within a distance s and f_y is transverse reinforcement yield strength and f_c is in MPa. In seismic design, TEC (2007) requires that the shear force taken into account for determining transverse reinforcement of beams, V_e, is determined by Eq. (18.4)

$$V_e = V_{dy} \mp (M_{pi} + M_{pj})/l_n, \quad (18.4)$$

where M_{pi} and M_{pj} are positive or negative ultimate moment capacities considering strain hardening of steel, V_{dy} is simple beam-shear developed at the column face due to vertical loads and l_n is clear span of beam. Unless a more rigorous analysis is performed, ultimate moment capacities at the beam ends may be taken as $M_{pi} \cong 1.4M_{ri}$ and $M_{pj} \cong 1.4M_{rj}$, where M_{ri} is positive or negative ultimate moment resistance calculated on left end i of a beam and M_{rj} is negative or positive ultimate moment resistance calculated on right end j. While using Eq. (18.4), the worst case among earthquake acting from left to right and from right to left is considered.

ACI318 (2014) requires that the beams shall be designed to resist shear force corresponding to the ultimate moment capacity considering strain hardening of steel M_{pr} by using conventional ACI procedures with transverse reinforcement yield stress taken equal to 1.25 times the nominal yield stress, at the both ends of the member. ACI318 (2014) and TEC (2007) ignore the contribution of concrete to shear strength when the shear demand due to seismic effects is above a certain level.

Figure 18.1 illustrates the models for predicting shear capacity degraded by inelastic deformation (Arslan and Polat 2013). The change of concrete shear strength, V_c, and shear capacity, V_n, with displacement ductility is given according to the models of Priestley et al. (1994) and Sezen and Moehle (2004). The relationship between shear capacity and drift ratio, defined as a function of transverse reinforcement ratio, plastic shear capacity and axial force, is given by the model of Elwood and Moehle (2005). According to the model of Lee and Watanabe (2003),

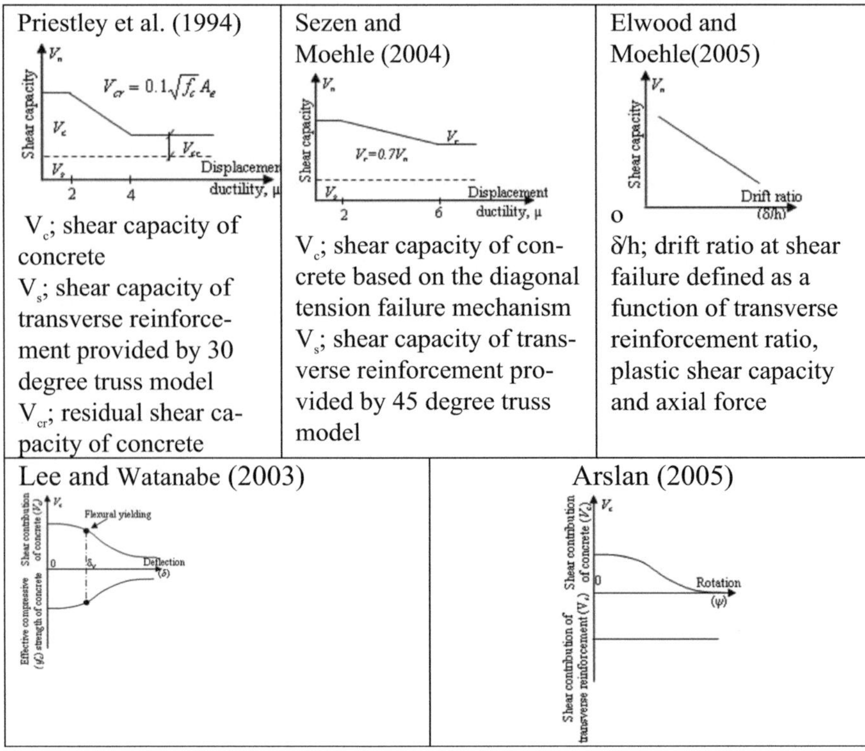

Fig. 18.1 Shear degradation models for predicting shear capacity

the shear contribution of concrete and the effective compressive strength of concrete νf_c as a function of deflection are depicted. The model of Arslan (2005) assumes that the contribution of transverse reinforcement is equal to the ultimate strength of beam. It can be seen from Fig. 18.1 that these models suggest a reduction in the contribution of concrete to shear strength for increasing displacement ductility, drift ratio, deflection and rotation, with a small residual strength at large ductility levels.

18.3 Experimental Program, Results and Discussion

18.3.1 Experimental Program

Four beams without stirrups including two reference RC beam (B2.5R and B3.5R), and two macro-synthetic PPFRC beam (B2.5P1.0 and B3.5P1.0) with 1.0% volume fraction of PPF were tested under three-point loading to determine their ultimate load capacities using a displacement-controlled loading machine.

Test beams were divided into two groups as B2.5 and B3.5 series based on a/d. The a/d ratios of the beams were chosen as 2.5 and 3.5. The geometrical properties of test specimens are shown in Fig. 18.2 The tests were carried out to evaluate the contribution of PPF to the shear strength of RC beams. Within the series under the same tensile reinforcement ratio (ρ), and section dimensions 150 mm/240 mm (b/h) have been held constant, the volume fraction of PPF varied from 0 to 1.0%.

A longitudinal reinforcement ratio (ρ) of 1.28% was chosen to avoid any premature flexural failure. Two deformed steel bars of 16 mm diameter were used as longitudinal reinforcement for all beams. The reinforcement steel bars have average yield strength of 492 MPa. Stress–strain curves for reinforcement steel bars are given in Fig. 18.3.

Tall crimped PPF were used as the only shear reinforcement. The ultimate strength of PPF is approximately 470 MPa in traction and elasticity modulus of approximately 3.6 GPa (as indicated by the manufacturer) were used in PPRC beams.

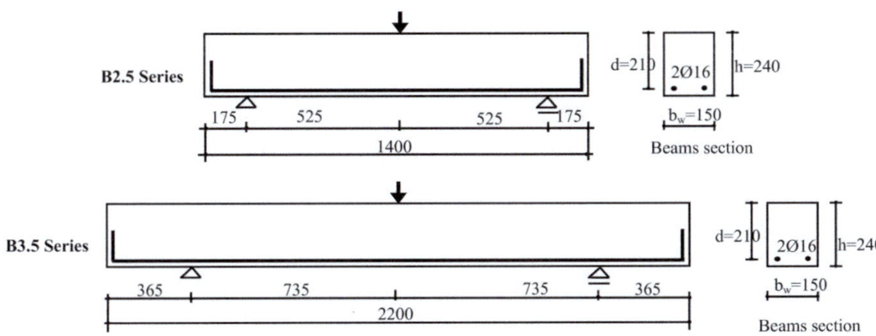

Fig. 18.2 Geometry of beams and flexural reinforcement arrangement (unit: mm)

Fig. 18.3 Stress–strain curves for reinforcement steel

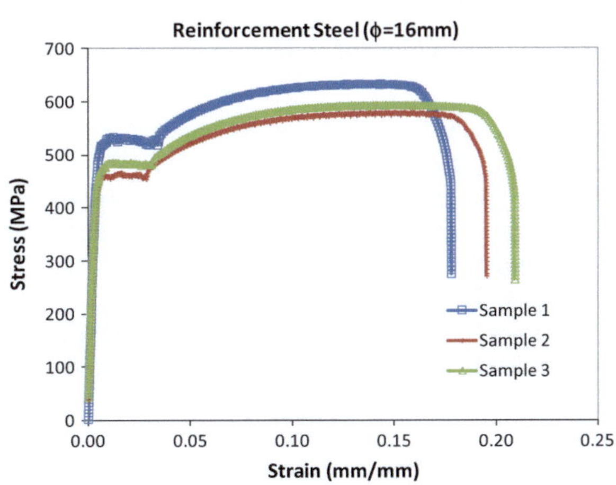

Table 18.1 Properties of test specimens

Test specimen	f_c (MPa)	V_f (%)	ρ (%)	a/d	a (mm)	l (mm)
B2.5R	26.50	0.0	1.28	2.5	525	1400
B2.5P1.0	27.00	1.0	1.28	2.5	525	1400
B3.5R	26.50	0.0	1.28	3.5	735	2200
B3.5P1.0	27.00	1.0	1.28	3.5	735	2200

Table 18.2 Mix proportions of concrete

Materials	Mixture proportions (kg/m^3)
0–1 mm natural sand	350
0–3 mm crushed sand	530
5–12 mm crushed stone	1010
Cement CEMI 42.5R	300
Water	164
Superplasticizer	3

A specimen label consists of a combination of letters and numbers. Each label starts with an "B" followed by 2.5 and 3.5 to designate a/d and continues with an "P" to indicate the volume fraction of PPF followed by 1.0. For example, a beam having a/d of 3.5 with a volume fraction of PPF (V_f) equal to 1.0% is labeled as B3.5P1.0. The specimens labeled as B2.5R and B3.5R are the reference beams that do not contain any **PPF.**

The properties of test specimens are summarized in Table 18.1, where f_c is the concrete compressive cylinder strength. Concrete composition is the same for all beams. A concrete mix consisting of Portland cement (CEM I 42.5) and a maximum aggregate size of 12 mm in diameter was used. The concrete mix proportions for 1 m^3 of concrete are given in Table 18.2.

18.3.2 Experimental Results and Discussion

All beams failed in shear and the crack patterns are shown in Fig. 18.4. During early stages of loading, fine vertical flexural cracks appeared around the mid-span of all beams, as expected. With the increase in load, new flexural cracks were formed away from the mid-span area. With further increase in load, those vertical flexural cracks appeared around the mid-span started to extend towards the loading point (Fig. 18.4).

Test results of the RC beams are given in Table 18.3, where $P_{flex,ACI}$ is nominal flexural load-carrying capacity calculated from the flexural moment capacity ($M_{flex,ACI}$) according to ACI318 (2014), P_{co} is the peak load of beam obtained experimentally, P_u is the ultimate load that is assumed to be equal to 80% of P_{co}, δ_{co} and δ_u are the deflection at the peak load and the ultimate deflection of each beam,

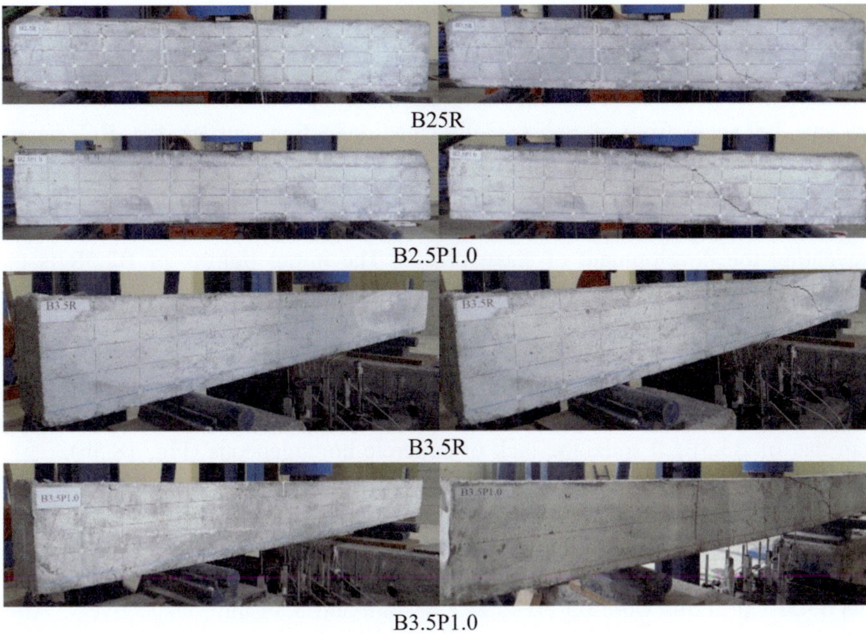

Fig. 18.4 The cracks patterns of beams at the end of the testing

Table 18.3 Theoretical/experimental load-carrying and deflection capacities of beams

Beams	$M_{\text{flex. ACI}}$ (kNm)	$P_{\text{flex. ACI}}$ (kN)	P_{co} (kN)	δ_{co} (mm)	P_u (kN)	δ_u (mm)	$v_{u,\text{Exp.}}$ (MPa)	δ_u/δ_{co}	Dissipated energy (kNm)
B2.5R	35.75	136.19	70.91	22.16	56.73	22.61	1.13	1.02	0.1155
B2.5P1.0	35.86	136.61	95.83	3.76	76.67	5.76	1.52	1.53	0.3940
B3.5R	35.75	97.28	72.05	4.82	67.84	4.89	1.14	1.01	0.2031
B3.5P1.0	35.86	97.58	93.50	5.15	74.80	7.43	1.48	1.44	0.4600

respectively, and the dissipated energy is the area under the load-deflection curve. It is deducted from Table 18.3 that the energies dissipated by the PPRC beams (B2.5P1.0 and B3.5P1.0) are 341 and 226 times the energy dissipated by the B2.5R and B3.5R beams, respectively.

It can be stated that the contributions of PPF to the shear strength at ultimate state are 0.39 MPa ($v_{u,\text{B2.5P1.0}}$- $v_{u,\text{B2.5RC}}$) and 0.34 MPa ($v_{u,\text{B3.5P1.0}}$–$v_{u,\text{B3.5RC}}$) in case of the beams with volume fractions of PPF equal to 1.0% (Table 18.3). It is observed that the contribution of PPF to shear strength decreases with the increasing **a/d.**

18.3.3 Comparison of Load–Deflection Relationships of Beams

As can be seen from experimentally load-deflection curves for RC beams (Fig. 18.5), the beam B2.5P1.0 reached a peak load of 95.83 kN whereas B2.5R beam reached a peak load of 70.91 kN. There is a 35% increase in the strength. The use of tall crimped PPF increased both the load-carrying capacity and ductility by preventing the development of shear cracks. The initial rigidity of B2.5R is the same as and B2.5P1.0 beam (Fig. 18.5a).

The beam B3.5P1.0 reached a peak load of 93.50 kN whereas B3.5R beam reached a peak load of 72.05 kN. There is a 30% increase in the strength. The use of tall crimped PPF increased both the load-carrying capacity and ductility by preventing the development of shear cracks. The initial rigidity of B3.5R is smaller than that of B3.5P1.0 since the lack of concrete homogeneous (Fig. 18.5b).

As shown in Fig. 18.5c, the initial rigidity of B3.5R is smaller than that of B2.5R but B3.5R beam exhibited even a more pronounced enhancement in ductility because of the slenderness effect. This behavior is also seen in P1.0 series beams, but it is interesting to note that drops in load-carrying capacities of the B2.5P1.0 and B3.5P1.0 beams were observed at 2.68 mm and 5.84 mm, respectively. After these initial drops, the PPFRC beams with V_f of 1.0% can sustain some increases in load-carrying capacities.

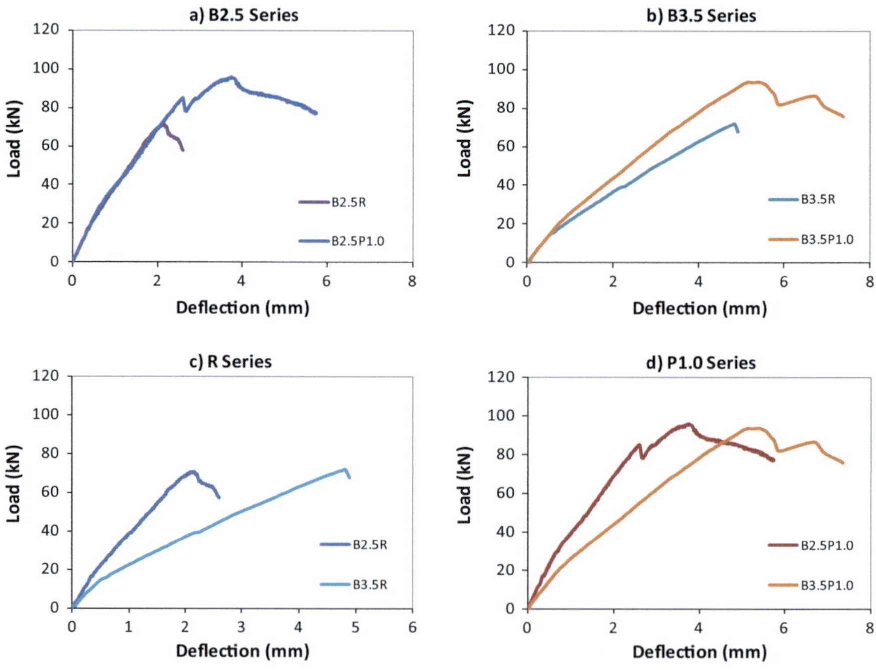

Fig. 18.5 Load-deflection curves of beams

The PPFRC beams with V_f of 1.0% exhibited a significantly enhanced behavior particularly in terms of ductility (Fig. 18.5d and Table 18.3). Peak loads of 95.83 and 93.50 kN were reached at the same drift ratio of 0.7% for B2.5P1.0 and B3.5P1.0 beams, respectively. B2.5P1.0 and B3.5P1.0 beams failed after 1.1 and 1.0% drift ratio, respectively, due to shear failure. B3.5P1.0 beam exhibited even a more pronounced enhancement in ductility.

18.4 Conclusions

Based on the results presented in this paper, the following conclusions can be drawn:

1. The energies dissipated by the PPFRC beams with V_f of 1.0% (B2.5P1.0 and B3.5P1.0) 341 and 226 times the energy dissipated by the B2.5R and B3.5R beams, respectively.
2. For the approximately same concrete compressive strength, the contributions of PPF to the shear strength at ultimate state are 0.39 MPa and 0.34 MPa in case of the beams with volume fractions of PPF equal to 1.0%. It is observed that the contribution of PPF to shear strength decreases with the increasing a/d.
3. From load-deflection curve, the initial rigidity of B3.5R is smaller than that of B2.5R but B3.5R beam exhibited even a more pronounced enhancement in ductility because of the slenderness effect. This behavior is also seen in P1.0 series beams, but it is interesting to note that drops in load-carrying capacities of the B2.5P1.0 and B3.5P1.0 beams were observed at 2.68 and 5.84 mm, respectively. After these initial drops, the PPFRC beams with V_f of 1.0% can sustain some increases in load-carrying capacities.
4. The PPFRC beams with V_f of 1.0% exhibited a significantly enhanced behavior particularly in terms of ductility. Peak loads of 95.83 and 93.50 kN were reached at the same drift ratio of 0.7% for B2.5P1.0 and B3.5P1.0 beams, respectively. B2.5P1.0 and B3.5P1.0 beams failed after 1.1 and 1.0% drift ratio, respectively, due to shear failure. B3.5P1.0 beam exhibited even a more pronounced enhancement in ductility.

Acknowledgements This research has been supported by Yıldız Technical University Scientific Research Projects Coordination Department. Project Number: 2016-05-01-YL02. In this research some tests has been made using fatigue loading machine supported by ISTKA (İstanbul Development Agency) with the Project Number: TR10/15/YNK/0034.

References

ACI Committee 318 (2014) Building code requirements for structural concrete (ACI 318–14) and commentary. *ACI*, Farmington Hills, MI

Altoubat S, Yazdanbakhsh A, Rieder K-A (2009) Shear behavior of macro-synthetic fiber-reinforced concrete beams without stirrups. ACI Mater J 106(4):381–389

Arslan G (2005) Shear strength of reinforced concrete frame members under cyclic loads. PhD thesis. Turkey: Yildiz Technical University

Arslan G, Polat Z (2013) Contribution of concrete to shear strength of RC beams failing in shear. J Civ Eng Manag 19(3):400–408

Arslan G (2014) Shear strength of steel fiber reinforced concrete (SFRC) slender beams. KSCE J Civ Eng 18(2):587–594

ASCE–ACI426 (1973) The shear strength of reinforced concrete members. Proc Am Soc Civil Eng 99(ST6):1091–1187

ASCE–ACI445 (1998) Recent approaches to shear design of structural concrete, state-of-the-art-report by ASCE–ACI committee 445 on shear and torsion. J Struct Eng ASCE 124(12):1375–1417

Ashour SA, Hasanain GS, Wafa FF (1992) Shear behavior of high-strength fiber reinforced concrete beams. ACI Struct J 89(2):176–184

Birincioglu MI, Ulusoy S, Arslan G (2015) Influence of steel fibers on the shear strength of RC beams without stirrups. *ICOCEE–Cappadocıa* 2015

Carnovale D, Vecchio FJ (2014) Effect of fiber material and loading history on shear behavior of fiber-reinforced concrete. ACI Struct J 111(5):1235–1244

Conforti A, Minelli F, Tinini A, Plizzari GA (2015) Influence of polypropylene fibre reinforcement and width-to-effective depth ratio in wide-shallow beams. Eng Struct 88:12–21

Elwood KJ, Moehle JP (2005) Axial capacity model for shear- damaged columns. ACI Struct J 102(4):578–587

Gandomi AH, Alavi AH, Yun GJ (2011) Nonlinear modeling of shear strength of SFRC beams using linear genetic programming. Struct Eng Mech 38(1):1–25

Greenough T, Nehdi M (2008) Shear behavior of fiber-reinforced self-consolidating concrete slender beams. ACI Mater J 105(5):468–477

Imam M, Vandewalle L, Mortelmans F (1994) Shear capacity of steel fiber high-strength concrete beams. In: Proceedings of the ACI international conference on high performance concrete (SP-149), Singapore

Kwak YK, Eberhard MO, Kim WS, Kim J (2002) Shear strength of steel fiber-reinforced concrete beams without stirrups. ACI Struct J 99(4):530–538

Lee JY, Watanabe F (2003) Shear deterioration of reinforced concrete beams subjected to reversed cyclic loading. ACI Struct J 100(4):480–489

Majdzadeh F, Soleimani M, Banthia N (2006) Shear strength of reinforced concrete beams with a fiber concrete matrix. Can J Civ Eng 33:726–734

Narayanan R, Darwish IYS (1987) Use of steel fibers as shear reinforcement. ACI Struct J 84 (3):216–227

Noghabai K (2000) Beams of fibrous concrete in shear and bending: experiment and mode. J Struct Eng 126(2):243–251

Priestley MJN, Verma R, Xiao Y (1994) Seismic shear strength of reinforced concrete columns. J Struct Eng ASCE 120(8):2310–2329

Sezen H, Moehle JP (2004) Shear strength model for lightly reinforced concrete columns. J Struct Eng ASCE 130(11):1692–1703

Sharma AK (1986) Influence of steel fibers on the shear resistance of lightweight concrete I-beams. ACI J Proc 83(4):624–628

Song PS, Hwang S, Sheu BC (2005) Strength properties of nylon- and polypropylene-fiber-reinforced concretes. Cem Concr Res 35:1546–1550

TEC (2007) Turkish earthquake code. Specification for structures to be built in disaster areas. Ministry of public works and settlement government of republic of Turkey, Turkey. 161 p

TS500 (2000) TS500 Requirements for design and construction of reinforced concrete structures. Ankara. Turkish Standards Institute, Turkey 67 p. (in Turkish)

MIX
Papier aus verantwortungsvollen Quellen
Paper from responsible sources
FSC® C105338

If you have any concerns about our products,
you can contact us on
ProductSafety@springernature.com

In case Publisher is established outside the EU,
the EU authorized representative is:
Springer Nature Customer Service Center GmbH
Europaplatz 3, 69115 Heidelberg, Germany

Printed by Libri Plureos GmbH
in Hamburg, Germany